Springer
New York
Berlin
Heidelberg
Barcelona
Budapest
Hong Kong
London
Milan
Paris
Tokyo

Series in Psychopathology

Series Editor
Lauren B. Alloy

Gregory A. Miller

Editor

The Behavioral High-Risk Paradigm in Psychopathology

 Springer

Gregory A. Miller, Ph.D.
Professor of Psychology and Psychiatry
Departments of Psychology and Psychiatry and Beckman Institute
University of Illinois at Urbana-Champaign
603 East Daniel Street
Champaign, IL 61820 USA

With 9 figures.

Library of Congress Cataloging-in-Publication Data
The behavioral high-risk paradigm in psychopathology / Gregory A.
 Miller, editor.
   p. cm.—(Springer-Verlag series in psychopathology)
  ISBN 0-387-94504-0.
  1. Schizophrenia—Risk factors. 2. Affective disorders—Risk
factors. 3. Schizophrenia—Pathophysiology. 4. Affective disorders—
Pathophysiology. I. Miller, Gregory A. II. Series: Series in psychopathology.
  [DNLM: 1. Schizophrenia—physiopathology. 2. Depressive Disorder—
physiopathology. WM 203 B419 1995]
RC455.4.R56B44 1995
616.89—dc20
DNLM/DLC
for Library of Congress

Printed on acid-free paper.

© 1995 Springer-Verlag New York, Inc.
All rights reserved. This work may not be translated or copied in whole or in part without the written permission of the publisher (Springer-Verlag New York, Inc., 175 Fifth Avenue, New York, NY 10010, USA), except for brief excerpts in connection with reviews or scholarly analysis. Use in connection with any form of information storage and retrieval, electronic adaptation, computer software, or by similar or dissimilar methodology now known or hereafter developed is forbidden.
The use of general descriptive names, trade names, trademarks, etc., in this publication, even if the former are not especially identified, is not to be taken as a sign that such names, as understood by the Trade Marks and Merchandise Marks Act, may accordingly be used freely by anyone.
While the advice and information in this book are believed to be true and accurate at the date of going to press, neither the authors nor the editors nor the publisher can accept any legal responsibility for any errors or omissions that may be made. The publisher makes no warranty, express or implied, with respect to the material contained herein.

Production managed by Hal Henglein; manufacturing supervised by Jeffrey Taub.
Typeset by Best-set Typesetter Ltd., Hong Kong.
Printed and bound by Braun-Brumfield, Inc., Ann Arbor, MI.
Printed in the United States of America.

9 8 7 6 5 4 3 2 1

ISBN 0-387-94504-0 Springer-Verlag New York Berlin Heidelberg

Preface

As editor of the Springer-Verlag Series in Psychopathology, Lauren Alloy knew of my work in cognitive psychophysiology to study processing anomalies in nonpatients at risk for psychopathology and invited me to edit a book for the series. This evolved into an opportunity to address an aspect of the unfortunate nature–nurture battle in the field, which too often emphasizes genes and macrolevel environment. Extreme positions are often taken (sometimes unwittingly), even though a great deal of the actual research is between the extremes, including laboratory psychological and psychophysiological studies. There is more to biology than genes and even more to it than things like brain imaging, enlarged ventricles, glucose metabolism rate, and receptor density, which have received a great deal of attention in recent years.

One goal of this book is to provide demonstrations of studies at the intersection between psychology and biology via psychophysiology. In parallel, another goal is to showcase solid psychological research that may bear directly on what are often considered biological issues. For example, Chapter 4, by Walker and colleagues, can be considered classically psychological, because the authors focus on overt behavior. Yet some of the importance of their work is its implication of a particular biological process involved in the gross motor behavior anomalies they have identified in the etiology of schizophrenia. Similarly, whereas in Chapter 7, Klein and Anderson articulate the behavioral high-risk paradigm quite well, in Chapter 10, Yee relies on their approach in pursuing psychophysiological research on risk for depression.

Another motivation for the book is to juxtapose research with patients and research with nonpatients potentially at risk. The field often views these as entirely separate research domains, but I have found in my own work with both types of samples that they bear directly on each other.

Combining these goals, the intention of this book is to explore what nongenetic (but still biologically compatible) high-risk strategies have to offer psychopathology at this time. In a sense, this is a very probiology book on psychopathology, and it takes for granted that biology will figure importantly in complete accounts of psychopathology. Yet readers should know already that biology is not simply genetics and that psychological phenomena affect biological phenomena, including gene expression. There is no assumption here that genes are more "fundamental" than or "underlie" biology or psychology or that biology "underlies" psychology. To assert any such specific causal relationship is to begin to provide a theory—an important thing to do—rather than to assert a logically necessary presumption. This point warrants some expansion.

As a cognitive neuroscientist, I felt that the onset of the "decade of the brain" several years ago held a great deal of promise for its potential to pull together disparate lines of work. This was a bandwagon that deserved much of the attention it quickly acquired. Yet, as a practicing clinician, I am finding as that decade unfolds that many portrayals of access to and study of the brain are disappointingly narrow. The breadth and integration I had anticipated are lacking, and I fear that that will cost the decade a potentially powerful punchline. The decade will end, and we will have less progress than promised.

Much modern writing in psychopathology and in neuroscience demonstrates a set of very common yet remarkably indefensible assumptions:

1. *Biology is more "fundamental" than psychology.* For example, one often runs across phrases like "the underlying brain mechanisms" or "neural substrates" or "biological basis." It may very well be that our best theorizing will involve biological mechanisms intimately associated with psychological phenomena, but we forget too quickly that this may not be the case, or that the biological mechanisms will not be able to account for the whole of the psychological phenomena. Although still rare, the converse is becoming more common—understanding "lower" level mechanisms in terms of "higher" level functions. Could we not push this just a little and assert that psychological mechanisms "underlie" biological phenomena? If one sort of "underlying" is possible, the converse seems logically just as possible.

But we may not wish to push either too far. We would not say that the gears in a clock "underlie" the keeping of time; the gears merely implement a mechanism that keeps time. That mechanism could be described in terms of gears, but it could also (and more importantly) be described in terms of the fact that it keeps time (not the fact that gears turn). Neither the gears nor the gear-level model of a clock are more "fundamental" than the notion of keeping time. Similarly, molecular genetics does not "explain" evolution. Molecular genetics tells us what happens biochemically in the course of evolution, but because biochemistry is not all there is to evolution, it cannot provide a complete account of it.

In the excitement of discovery in one area of science, we can forget that a mechanism is just that and not the concept it implements. Neural activity does not "underlie" cognition or emotion or delusions. We may be able to determine how neural activity implements these psychological phenomena. Unquestionably, we gain in our understanding of the psychological from our study of the biological, but for many of us, it is the psychological we wish to understand primarily. The biological is a convenient and sometimes valuable tool, not the criterion. An extensive treatment of aspects of this issue by Fodor (1968) is very instructive (see Kozak & Miller, 1982, and Miller & Kozak, 1993, for developments of Fodor's view in the context of emotion and psychopathology). More directly within the psychopathology literature, Lenzenweger (1993) felt compelled to explain why Meehl's famous theory of schizotaxia, schizotypy, and schizophrenia is not entirely genetic and why schizotypy as conceived by Meehl cannot be heritable. In essence, reductionism fails. But it nevertheless persists in much of our thinking.

2. *Once we work out enough of the biology, we will not need psychological methods, constructs, or data.* Much of the physics that "underlies" (sic) chemistry has been worked out, but the field of chemistry is not in jeopardy of obsolescence. Chemists deal with phenomena of a certain sort, using certain sorts of methods and relying on certain sorts of concepts. Developments in neighboring areas such as physics and biology can certainly alter the customary terrain of chemistry, but they do not render chemistry redundant. Advances in molecular genetics are not likely to do away with concepts in evolutionary theory. Physics does not replace chemistry, because it is simply not useful to try to manage the phenomena chemists are interested in by relying solely on the concepts, methods, and phenomena that physicists study and because it is not possible to capture what some chemistry concepts mean with concepts in physics. Nor will chemistry absorb biology, nor biology psychology.

In psychopathology, a classic example of this misrepresentation of the relationship between areas of science is probably the misnamed "dopamine theory" of schizophrenia. The term *schizophrenia* historically refers to phenomena that are inherently psychological (thought disorder, anhedonia, etc.). This is simply a matter of semantics, whether one refers to Kraepelin, Webster, or the DSM. No biological discovery about people with schizophrenia could alter the nature of the phenomena the word presently refers to. That is, what "thought disorder" presently refers to is psychological; even if we change what the term means, its current referent, which will surely always be of interest, remains psychological. We have already identified a number of reliable biological differences between schizophrenics and other individuals (many of them discussed in the chapters of this book), which is terrific progress. But those differences do not turn schizophrenia into a biological phenomenon. Even if it were to come to pass that we were no longer interested in (psychological) thought disorder, we should not mistake that shift in our scientific focus to mean that the phenomena that the term *thought disorder* presently refers to are no longer psychological.

I am optimistic that we will continue to identify very important biological phenomena meaningfully associated with schizophrenia, but it is apparently all too tempting to misinterpret what such progress can mean. A "dopamine theory" can never be a theory of schizophrenia. It can only be a theory about dopamine in schizophrenia. There is much more about schizophrenia that deserves theorizing than its biochemistry, and other theories will be needed to account for other aspects of schizophrenia. There is clearly a biochemical story to be told about schizophrenia, but we should not mistake it for an adequate story about schizophrenia. This again is like the relationship between molecular genetics and evolution. The former is not a biochemical theory of the latter but rather a theory of some of the biochemistry involved in the latter. No one suggests that the biochemical story is all there is to evolution.

I hasten to say that the common semantic error need not be laid at the feet of those studying dopamine in schizophrenia. By all means, it should be studied. The point is that that study need not entail a belief that all of the phenomena of schizophrenia can be reduced to biochemical phenomena. It is that belief that is disappointingly common.

To take an extreme example of why it must be fallacious: As Hooley and colleagues discuss in Chapter 3, it is quite well established that measures of a phenomenon known as "expressed emotion" (specifically, certain sorts of expressions of certain sorts of emotions in families of schizophrenics) correlate with relapse rates in schizophrenia. It is difficult to imagine that it would ever be useful to try to account for that relationship adequately in terms of dopamine. That is not to say that dopamine phenomena are unimportant to schizophrenia or even to expressed emotion or its impact on schizophrenics. It may turn out that expressed emotion, or some aspect of its impact, correlates substantially with dopamine levels. Dopamine receptor sensitivity may prove to be an important mediator in the biological story that accompanies expressed emotion phenomena. Drugs that alter dopamine metabolism may be an important means of altering the relationship between expressed emotion and relapse. But such findings would not show that dopamine "underlies" the relationship and would not make psychological accounts of schizophrenia and expressed emotion less important. Dopamine would be simply a venue in which the relationship is played out, an instance of a mechanism.

I also hasten to add that the issue is not whether the meanings of words can evolve, or the boundaries between disciplines shift. Obviously that happens. To follow the example, the present observation is not that the term *schizophrenia* can never alter its meaning but that such things as thought disorder are definitionally psychological and will always be of interest psychologically. As such, they cannot be explained away by "underlying" (sic) neurochemical phenomena. Wood slabs and iron springs do not "explain" or "underlie" the concept of a mousetrap. They provide a mechanism by which a physical entity can instantiate (the concept of) a mousetrap. The present argument is not particularly novel (see, again, Fodor, 1968, and others before and since, including Chapter 6, by Berenbaum), but it is not commonly considered.

3. *Dysfunctions conceived biologically warrant interventions conceived biologically*—and similarly for dysfunctions and interventions conceived psychologically. But it simply does not follow, for example, that having good reasons to suppose an excess of dopamine receptors in schizophrenia means that one should treat schizophrenia with appropriate receptor antagonists. It may be that the best way to manipulate the behavior of those receptors is to modulate the individual's exposure to the expression of hostility from significant others. Similarly, in Chapter 11, Hollon discusses how negative life events (generally understood psychologically) may alter biological factors in risk for depression.

The point is that one need not choose (and may even have trouble distinguishing) between exclusively biological and psychological accounts of schizophrenia. History makes the point nicely. Schizophrenia and its classic symptoms are inherently psychological constructs. Yet drugs that alter dopamine in the brain have a significant impact on those symptoms. We should study that. One payoff from the decade of the brain should be greatly increased understanding of the biochemistry of schizophrenia. But a cost from the decade of the brain should not be an overestimation of what biological stories can do, leading eventually to disaffection with biological research.

A paradox that psychophysiologists constantly confront is how to categorize types of data. Is a reaction-time button-press behavior or physiology? What about the electrophysiological activity produced by the muscle that moves the thumb—activity easily recordable by the psychophysiologist? This book aims not to champion the psychological against the biological but to foster their integration. It is not enough to give lip service to all comers and to proclaim, in effect, that everything interacts. Someone has to do the hard work of deciding which things to include in an account and of specifying how they interact.

Accordingly, a diverse group of psychopathologists was invited to write chapters discussing research on risk factors in psychopathology. Two themes were chosen to structure the book. There are what might be called "clinical" chapters, relying on traditional psychological observation and assessment. These are complemented by chapters emphasizing psychophysiological methods, inherently biological but also inherently and explicitly involving psychological variables. To make the length of the book manageable, the focus is confined to sections on mood disorders and schizophrenia. Each section ends with an integrative chapter written by a prominent researcher in the area, who was asked to work from the individual chapters in providing a commentary on the issues and findings they present.

I decided that biochemistry and genetics would not be well represented here, because they are so well represented elsewhere in treatments of psychopathology. It is already widely accepted that they have a prominent role to play. Collectively, the present chapters make a strong case that psychosocial and psychophysiological phenomena also belong in that story and that a single, rich story including all such phenomena is the eventual goal.

Thanks are due to the authors who contributed to this book as well as to Lauren Alloy, who invited me to put it together, and to Laura Gillan of Springer-Verlag, who provided valuable advice. Special thanks are also due to Howard Berenbaum, Loren Chapman, Jean Chapman, Daniel Klein, Michael Kozak, Peter Lang, and Robert Simons, who have had considerable impact on my thinking about psychopathology. Finally, preparation of this preface and editing of this book were supported in part by National Institute of Mental Health grant R01 MH39628 and by grants from the Research Board and the Department of Psychology of the University of Illinois.

Gregory A. Miller

References

Fodor, J.A. (1968). *Psychological explanation.* New York: Random House.

Kozak, M.J., & Miller, G.A. (1982). Hypothetical constructs versus intervening variables: A re-appraisal of the three-systems model of anxiety assessment. *Behavioral Assessment, 14,* 347–358.

Lenzenweger, M.F. (1993). Explorations in schizotypy and the psychometric high-risk paradigm. In L.J. Chapman, J.P. Chapman, & D.C. Fowles (Eds.), *Progress in personality and psychopathology research* (Vol. 16, pp. 66–116). New York: Springer-Verlag.

Miller, G.A., & Kozak, M.J. (1993). A philosophy for the study of emotion: Three-systems theory. In A. Öhman & N. Birbaumer (Eds.), *The structure of emotion: Physiological, cognitive and clinical aspects* (pp. 31–47). Seattle: Hogrefe & Huber.

Contents

Preface		v
Contributors		xiii
I. Schizophrenia		1
Spectrum Disorders		
Clinical	William S. Edell	3
	Chapter 1. *The Psychometric Measurement of Schizotypy Using the Wisconsin Scales of Psychosis Proneness*	
Psychophysiology	Leyan O.L. Fernandes and Gregory A. Miller	47
	Chapter 2. *Compromised Performance and Abnormal Psychophysiology Associated With the Wisconsin Scales of Psychosis Proneness*	
Family/Developmental		
Clinical	Jill M. Hooley, Laura R. Rosen, and John E. Richters	88
	Chapter 3. *Expressed Emotion: Toward Clarification of a Critical Construct*	
Clinical	Elaine Walker, Dana Davis, Jay Weinstein, Tammy Savoie, Kathleen Grimes, and Kym Baum	121
	Chapter 4. *Modal Developmental Aspects of Schizophrenia Across the Life Span*	

	Psychophysiology	Stuart R. Steinhauer and David Friedman	158
		Chapter 5. *Cognitive Psychophysiological Indicators of Vulnerability in Relatives of Schizophrenic Patients*	
	Integration	Howard Berenbaum	181
		Chapter 6. *Toward a Definition of Schizophrenia*	

II. Mood Disorders 197

Chronic			
	Clinical	Daniel N. Klein and Rochelle L. Anderson	199
		Chapter 7. *The Behavioral High-Risk Paradigm in the Mood Disorders*	
	Psychophysiology	Scott R. Sponheim, John J. Allen, and William G. Iacono	222
		Chapter 8. *Selected Psychophysiological Measures in Depression: The Significance of Electrodermal Activity, Electroencephalographic Asymmetries, and Contingent Negative Variation to Behavioral and Neurobiological Aspects of Depression*	
Cognitive			
	Clinical	Michael W. O'Hara	205
		Chapter 9. *The Cognitive Diathesis for Depression*	
	Psychophysiology	Cindy M. Yee	271
		Chapter 10. *Implications of the Resource Allocation Model for Mood Disorders*	
	Integration	Stephen D. Hollon	289
		Chapter 11. *Depression and the Behavioral High-Risk Paradigm*	

Index 303

Contributors

John J. Allen, Ph.D., Department of Psychology, University of Arizona, Tucson, AZ 85721 USA

Rochelle L. Anderson, Ph.D., North Philadelphia Health System, Girard Medical Center, 8th Street and Girard Avenue, Philadelphia, PA 19122 USA

Kym Baum, M.A., Department of Psychology, Emory University, Atlanta, GA 30322 USA

Howard Berenbaum, Ph.D., Department of Psychology, University of Illinois at Urbana-Champaign, 603 E. Daniel Street, Champaign, IL 61820 USA

Dana Davis, M.A., Department of Psychology, Emory University, Atlanta, GA 30322 USA

William S. Edell, Ph.D., Senior Vice President, Quality Improvement, Horizon Mental Health Management, Inc., 2220 San Jacinto Boulevard, Suite 320, Denton, TX 76205 USA

Leyan O.L. Fernandes, M.A., Department of Psychology, University of Illinois at Urbana-Champaign, 603 E. Daniel Street, Champaign, IL 61820 USA and Veterans Administration Medical Center, Boston, MA 02130 USA

David Friedman, Ph.D., New York State Psychiatric Institute, Box 58, 722 West 168th Street, New York, NY 10032 USA

Kathleen Grimes, Ph.D., Department of Psychology, Emory University, Atlanta, GA 30322 USA

Steven D. Hollon, Ph.D., Department of Psychology, Vanderbilt University, Nashville, TN 37240 USA

Jill M. Hooley, Ph.D., Department of Psychology, Harvard University, 33 Kirkland Street, Cambridge, MA 02138 USA

William G. Iacono, Ph.D., Department of Psychology, University of Minnesota, Minneapolis, MN 55455 USA

Daniel N. Klein, Ph.D., Department of Psychology, State University of New York at Stony Brook, Stony Brook, NY 11794-2500 USA

Gregory A. Miller, Ph.D., Departments of Psychology and Psychiatry and Beckman Institute, University of Illinois at Urbana-Champaign, 603 E. Daniel Street, Champaign, IL 61820 USA

Michael W. O'Hara, Ph.D., Department of Psychology, University of Iowa, Iowa City, IA 52242 USA

John E. Richters, Ph.D., National Institute of Mental Health, 5600 Fishers Lane, 18C-17, Rockville, MD 20857 USA

Laura R. Rosen, Ph.D., Department of Psychology, Harvard University, 33 Kirkland Street, Cambridge, MA 02138 USA

Tammy Savoie, Ph.D., Department of Psychology, Emory University, Atlanta, GA 30322 USA

Scott R. Sponheim, Ph.D., Veterans Administration Medical Center, Minneapolis, MN 55455 USA

Stuart R. Steinhauer, Ph.D., Biometrics Research, Veterans Administration Medical Center 151R, Highland Drive, Pittsburgh, PA 15206 USA

Elaine Walker, Ph.D., Department of Psychology, Emory University, Atlanta, GA 30322 USA

Jay Weinstein, M.A., Department of Psychology, Emory University, Atlanta, GA 30322 USA

Cindy M. Yee, Ph.D., Department of Psychology, University of California-Los Angeles, 405 Hilgard Avenue, Los Angeles, CA 90024-1563 USA

Part I
Schizophrenia

1
The Psychometric Measurement of Schizotypy Using the Wisconsin Scales of Psychosis Proneness

WILLIAM S. EDELL

In his presidential address to the American Psychological Association three decades ago, Paul Meehl (1962) proposed a heuristic model for the pathogenesis of schizophrenia that anticipated the widely endorsed diathesis-stress or vulnerability models of the disorder (Hartmann et al., 1984; Zubin, Magaziner, & Steinhauer, 1983; Zubin & Spring, 1977). In typical Meehlian fashion, his provocative and now classic speech suggested that the basic pathophysiology of schizophrenia was an inherited neural integrative defect, termed *schizotaxia*, which interacted with virtually any known type of social learning history to produce a personality organization called, following Rado (1956), the *schizotype*. A small minority of schizotypes, perhaps 10% (Meehl, 1989, 1990), were theorized to decompensate into clinical schizophrenia as a result of unfavorable polygenic potentiators (e.g., high anxiety readiness, low stress resistance) and adverse life experiences (e.g., schizophrenogenic mother who is highly ambivalent and hostile). To facilitate the identification of compensated or "nondisintegrated" schizotypes, Meehl (1964) developed a richly descriptive checklist of schizotypic signs, symptoms, and traits that could be used for purposes of clinical assessment and research.

Agreeing with Meehl that the "most important research need here is development of high-validity indicators for compensated schizotypy" (1962, p. 830), and recognizing the persuasive advantages of the high-risk group method for ultimately understanding the etiology and developmental course of schizophrenia (Garmezy & Streitman, 1974; Mednick & McNeil, 1968), Loren and Jean Chapman and their students at the University of Wisconsin began a program of research in the mid-1970s aimed at the development of highly reliable and valid psychometric scales that operationalized several of Meehl's schizotypic signs to identify persons at elevated risk for schizophrenia (L.J. Chapman, Chapman, Raulin, & Edell, 1978). Following the method of rational scale development as described by Jackson (1970, 1971), with an effort to maximizing scale homogeneity as reflected by high internal consistency reliability and minimizing response style variance attributable to acquiescence bias and social desirability, each scale was designed to tap long-term, characterological traits rather than the current state of the individual. Items were eliminated if their frequency of endorsement was significantly discrepant from the theoretical base rate of that trait in the population tested. Random or careless responders were identified and dropped from the sample when their score on an infrequency scale exceeded threshold. A total of seven true–false, self-report instruments aimed at measuring schizotypic signs have been developed over the past 15 years by the Chapman group, including scales of Physical (PAn) and Social (SAn) Anhedonia (Chapman & Chapman, 1978; Chapman, Chapman, & Raulin, 1976; Eckblad, Chapman, Chapman, & Mishlove, 1982), Perceptual Aberration (PER) (Chapman, Chapman, & Raulin, 1978), Somatic Symptoms (SOM) (Raulin, Chapman, & Chapman, 1978), Magical Ideation (MAG) (Eckblad & Chapman, 1983), Impulsive Nonconformity (IMP) (L.J. Chapman et al., 1984), and Intense Ambivalence (AMB) (Raulin, 1984). The choice of these particular candidate symptoms of schizophrenia-proneness was based largely on a review of the writings on schizotypy by Meehl (1964), on pseudoneurotic schizophrenia by Hoch and Cattell (1959), and on early schizophrenia by Fenichel (1945). An eighth scale, the Hypomanic Personality (HYP) scale, intended to identify a subset of persons at elevated risk for both hypomanic episodes and bipolar disorder, has also been developed (Eckblad & Chapman, 1986). With the narrowing of the definition of schizophrenia since the publication of the third edition of the *Diagnostic and Statistical Manual of Mental Disorders* (DSM-III), (American Psychiatric Association, 1980), resulting in many patients who were formerly diagnosed as schizophrenic being placed in other categories of psychosis, the schizotypy scales have been reconceptualized as measures of a more general proneness to psychosis, with the hope of reducing heterogeneity by identifying diverse pathways to the different psychotic disorders (L.J. Chapman & Chapman, 1985). Implicit in much of this work is the view of the schizotypy construct as multidimensional and multidetermined, with varying degrees of genetic and environmental contributions expected across scales (cf. Bentall, Claridge, & Slade, 1989; Kendler & Hewitt, 1992; Raine & Allbutt, 1989). A basic premise of much of the validational work to date has been that psychotic-like and schizotypal experiences often precede clinical psychosis (e.g., J. Chapman, 1966; Fenichel, 1945; Gillies, 1958; Hoch & Cattell, 1959; Strauss, 1969); therefore, finding increased rates of such experiences in nonpsychotic subjects who score deviantly on one or more of the scales provides preliminary support for the validity of the scale(s) as indices of psychosis proneness. Long-term follow-up of identified subjects through their age of risk remains of highest priority in establishing the ultimate validity of these instruments.

This chapter reviews the clinical and social findings that have addressed the question of the reliability and validity of each of the Wisconsin Scales as indices of proneness to psychosis. I briefly discuss the salient issues raised by this line of research and describe potentially fruitful areas for further research. Other chapters review the growing literature on the psychophysiological, neuropsychological, neurochemical, and attention/information processing deficits found in subjects scoring deviantly on one or more of these scales (e.g., Asarnow, Nuechterlein, & Marder, 1983; Balogh & Merritt, 1985; Bernstein & Riedel, 1987; J.P. Chapman & Chapman, 1987; Josiassen, Shagass, Roemer, & Straumanis, 1985; Jutai, 1989; Lenzenweger, Cornblatt, & Putnick, 1991; Lutzenberger, Elbert, Rockstroh, Birbaumer, & Stegagno, 1981; Merritt & Balogh, 1989; G.A. Miller, 1986; G.A. Miller, Simons, & Lang, 1984; Mo & Chavez, 1986; Overby, 1992; Silverstein, Raulin, Pristach, & Pomerantz, 1992; Simons, 1981, 1982; Simons & Katkin, 1985; Simons, MacMillan, & Ireland, 1982a, 1982b; Ward, Catts, Armstrong, & McConaghy, 1984; Yee, Deldin, & Miller, 1992; Yehuda, Edell, & Meyer, 1987).

For the reader unfamiliar with the scales, note that scores on all of the scales are positively skewed, with many more low scores than high scores in virtually all tested samples. This skew is by design, because the developers wished to approximate the expected population distribution of each trait assessed. For most of the studies reviewed, deviance on a scale is defined as scoring at least two standard deviations above the mean on that scale of the same-sex standardization sample, with nondeviant controls selected from subjects scoring no more than one-half of a standard deviation above the mean. Reliability data reported here were obtained for each scale using the original articles as well as the study by Chapman, Chapman, and Miller (1982). Of course, any investigator planning to use the scales should examine the scale's psychometric characteristics and validity in the specific populations being studied.

Physical Anhedonia Scale

Brief Description

Originally designed as a 40-item instrument, and later expanded to 61 items to increase reliability, the Physical Anhedonia (PAn) scale measures the lowered ability to experience physical or sensory pleasures, such as eating, touching, sex, temperature, movement, smell, and sound. A representative item is "The beauty of sunsets is greatly overrated" (true). Internal consistency reliability coefficients for the expanded version have ranged from .78 (women) to .83 (men), with test–retest reliability coefficients (6 weeks) of .78 (men) to .79 (women). The original scale correlated negatively with social desirability ($-.23$, men; $-.25$, women) and acquiescence bias ($-.15$, women; $-.20$, men), and positively with age (.10 men; .24, women) and social class (.13, women; .16, men). Of note, the revised PAn scale has negligible correlations with the IMP scale ($-.01$, men; .03 women), Golden and Meehl's (1979) Schizoidia scale ($r = .07$, men and women), and Eysenck and Eysenck's (1975) Psychoticism scale (.01, men; .05, women). The revised PAn scale has small, *negative* correlations with the PER scale ($-.20$ to $-.22$, men; $-.10$ to $-.15$, women) and MAG scale ($-.23$, men; $-.17$, women) in large samples of normal control subjects. It is unusual for two measures of severe psychopathology to be negatively correlated, particularly because both potentially share common method variance that is due to social desirability and acquiescence bias, suggesting that the measures are identifying different types of disorder or different manifestations of the same disorder. Indeed, there appears to be a *lower than chance* occurrence of male college students who are high on both scales. Among male schizophrenics, the PAn and PER scales have a small positive correlation ($r = .11$).

Validity Findings

COLLEGE STUDENTS

Fifteen published studies have examined characteristics of college students elevated on PAn. In comparison to same-sex nonelevated controls, male PAn subjects have demonstrated lower overall social competence using behavioral role-play instruments (Beckfield, 1985; Haberman, Chapman, Numbers, & McFall, 1979), more often giving terse responses, and have tended to be less able to recognize competent responses on a multiple-choice version of the task (Beckfield, 1985). Similarly, although female PAn subjects were comparable to control subjects on overall skill level on a role-playing measure of social skills, they were more avoidant and odd in their responses (Numbers & Chapman, 1982). Raulin and Henderson (1987) suggest that the social deficits may be attributable to differences in the anhedonics' implicit semantic trait structure (i.e., perception of trait relationships), distorting how they view the social environment.

Male PAn subjects were indistinguishable from controls on both a structured word association task (present each stimulus word 15 times and ask for one word associate to that word each time) and on an unstructured word association task (present stimulus word once and ask for continuous word associations to that word for 60 s) (E.N. Miller & Chapman, 1983). Anhedonic and control groups have also reported equivalently low rates of psychotic and psychotic-like experiences, major depression, and hypomanic episodes, with equal proportions having seen mental health professionals in their lifetime (L.J. Chapman, Edell, & Chapman, 1980). PAn subjects in that study, however, were more often socially withdrawn, reported fewer dates, were less likely to have gone steady, showed a smaller increase in spending time with opposite sex since high school, and had a higher composite score for schizotypal features (e.g., odd communication; social isolation; depersonalization or derealization experiences). Surprisingly, PAn subjects *less* often met criteria for Research Diagnostic Criteria (RDC) minor depression, perhaps as a result of the chronic characterological nature of their pleasure deficits contributing to their inability to distinguish feelings of depression from their usual anhedonic state.

On a childhood experiences scale, PAn subjects recalled less support and approval and greater disinterest from their mothers during childhood, compared with controls (Edell & Kaslow, 1991). These authors hypothesized that mothers of PAn subjects may have manifested disinterest and lack of support through minimal physical contact and holding, such that the child did not learn to feel pleasure from sensory experiences. If true, one would expect significant familial resemblance between siblings on PAn, which has recently been reported (Clementz, Grove, Katsanis, & Iacono, 1991).

The relationship of PAn scores to other scales provides further evidence for construct validity. Peterson and Knudson (1983), using what appears to be the original version of the scale, found PAn scores correlated negatively with sentience, or a sensual-hedonistic orientation to life, and with affiliation, and positively with social anhedonia and a Minnesota Multiphasic Personality Inventory (MMPI) anhedonia scale developed by Watson, Klett, and Lorei (1970). PAn scores were uncorrelated with self- and peer-ratings of anhedonia using a six-point global item. Peterson and Knudson argue that the PAn scale is the "most satisfactory of the three scales for the measurement of anhedonia" (1983, p. 548). Schuldberg (1988, 1990) found negative correlations between PAn and measures of creativity, suggesting that the lack of pleasurable affective involvement with the environment tapped by the scale may be associated with subclinical negative or deficit symptoms. In a sample of college students from Barcelona (Muntaner, Garcia-Sevilla, Fernandez, & Torrubia, 1988), PAn scores were positively correlated with the Psychoticism scale and negatively correlated with the Extraversion scale on the Eysenck Personality Questionnaire (EPQ; Eysenck & Eysenck, 1975) and with the Schizotypal Personality scale developed by Claridge and Broks (1984). A principal-components analysis found PAn to load with social anhedonia (positively) and extraversion (negatively). Peltier and Walsh (1990) found that PAn scores did not correlate significantly with independent measures of response set predictors, such as social desirability and acquiescence bias.

A study by Berenbaum, Snowhite, and Oltmanns (1987) examined subjective report and overt experience of emotion (facial expressions) in subjects viewing short excerpts from two film clips intended to elicit disgust or happiness (amusement). Contrary to expectation, Berenbaum et al. found no group differences in emotional responsiveness to the affect-evoking stimuli. The authors suggest that future studies monitor physiological responses as well.

Levin and Raulin (1991), using a shortened version of the PAn scale, reported that PAn scores were unexpectedly predictive of *low* nightmare frequency. This finding was attributable largely to the differences in mean ratings for men. Interestingly, female anhedonics had higher percentages of subjects in both the highest and lowest frequency categories of nightmares (Raulin, personal communication, May 14, 1992). It would be of interest to know if anhedonics have fewer dreams in general, given their diminished creativity and social withdrawal, although this appears not to be the case (Raulin, personal communication, May 14, 1992).

Finally, and perhaps most significantly, L.J. Chapman and Chapman (1985, 1987) interviewed PAn subjects ($n = 74$) on two occasions approximately 25 months apart. During this short-term follow-up period, PAn subjects were no more likely to have sought professional help for their problems than were normal controls. Although a few of the PAn subjects showed poor social adjustment, no evidence of increased rates of psychotic or psychotic-like symptoms was seen at follow-up. More recently, in a 10-year follow-up study (J.P. Chapman & Chapman, 1992), college students elevated on PAn ($n = 70$) continued to demonstrate no increased liability for psychosis as compared with control subjects ($n = 153$).

PSYCHIATRIC PATIENTS AND THEIR BIOLOGICAL RELATIVES

Eleven published studies have examined the performance of psychiatric patients and/or their biological relatives on PAn. In the original scale development paper (L.J. Chapman et al., 1976), a sample of hospitalized male schizophrenics was found to score more deviantly than a group of normal male subjects. Of note, the distribution of PAn scores in the schizophrenic group appeared bimodal, as Meehl (1964, p. 24) predicted, suggesting the presence of two underlying distributions, with approximately one third of the schizophrenics scoring in the anhedonic cluster. In the schizophrenic sample, PAn scores were uncorrelated with age, history of delusions or hallucinations, current depression, days in hospital over the prior 5 years, age at first hospitalization, and months since first hospitalization. Anhedonic and hedonic schizophrenics, however, did differ on premorbid adjustment in the expected direction (i.e., hedonics were much more likely to have good premorbid adjustment, whereas the anhedonics were somewhat more likely to have poor premorbid adjustment). Carpenter (1983) found that although premorbid status was related to PAn scores and to performance on Gorhams' Proverbs Test in schizophrenic patients, PAn scores were unrelated to performance on the Proverbs Test.

Given the heterogeneity of symptoms among schizophrenics, some investigators have looked at the relationship of PAn scores to clinical manifestations of the disorder. Kirkpatrick and Buchanan (1990), in a sample of clinically stable outpatient chronic schizophrenics, found much higher PAn scores among schizophrenics with the deficit syndrome (presence of prominent, enduring negative symptoms such as restricted affect, poverty of speech, and diminished social drive that are not attributable to secondary factors such as depression, anxiety, or drug effects) as compared with nondeficit schizophrenics. The proportion of schizophrenics with the deficit syndrome (27.6%) was comparable to the proportion of schizophrenics with high anhedonia scores in L.J. Chapman et al. (1976) and is consistent with the earlier reported bimodal distribution of PAn scores. In contrast, in a sample of British schizophrenics, Cook and Simukonda (1981) found no relationship between PAn scores and type of schizophrenic symptom (paranoia; apathy–withdrawal; hallucinations–delusions; aggression–violence; thought disorder–impaired concentration). Note that symptom ratings obtained in the former study used a reliable diagnostic research instrument to measure the deficit syndrome, while the latter study used standard clinical case histories not written originally for research purposes to measure symptoms. The unreliability of such case histories in documenting specific symptoms is well-known and may account for the failure to find any significant relationships between symptoms and PAn scores. In addition, although schizophrenics exceeded normal controls on mean PAn score (males only), Cook and Simukonda (1981) found extensive overlap in the distribution of scores for the two groups, with none of the 52 schizophrenic patients scoring more than 2 SD above the control mean. They suggest that high anhedonia scores may be an effect, not cause, of schizophrenic illness.

Similarly, Schuck, Leventhal, Rothstein, and Irizarry (1984) have argued that PAn is the consequence of such liabilities as poor premorbid status, psychiatric disturbance, low educational level, and low intellectual ability, rather than being an indicator of liability to schizophrenia or to psychosis. Their failure to find differences in PAn scores among outpatient paranoid schizophrenics, nonparanoid schizophrenics, unipolar affectives, and other psychiatric control subjects (bipolar and personality disordered) raises the question of whether the scale is merely tapping generalized psychopathology rather than being specific to psychosis. Consistent with college student studies, PAn was negatively correlated with premorbid social status and with years of education and vocabulary. PAn scores were unrelated to number or length of hospitalizations or to age. The specificity of anhedonia to schizophrenia was further called into question by the study of Fawcett, Clark, Scheftner, and Gibbons (1983), who reported a strong negative correlation ($r = -.42$) between the "Chapman Anhedonia Scale" (unspecified) and their Pleasure Scale for depressed inpatients. The highest scorers on what appears to be a sum score derived from the physical and social anhedonia scales were found in the patients with major depression, not in the schizophrenic, bipolar, or unspecified functional psychoses groups.

Recent evidence examining PAn scores in schizophrenic probands and their relatives argues against the interpretation that PAn scores have no etiologic significance. For example, Grove et al. (1991) found that schizophrenic probands exceeded normal controls on PAn, as did their first-degree relatives compared with relatives of controls. The estimated heritability (twice the sibling intraclass correlation) for PAn scores was quite high (h squared = .79), substantially higher than that found for smooth-pursuit eye movements performance and comparable to d' deficits on the Continuous Performance Test (CPT). Similarly, Clementz et al. (1991), collecting data across three sites (New York, Minneapolis, & Vancouver), found that schizophrenics had higher PAn scores than their first-degree relatives, who had higher scores than the normal controls. Surprisingly, affected relatives did not differ from unaffected relatives on PAn scores. PAn scores also demonstrated significant familial resemblance between siblings of probands (intraclass r = .50), although not between parents and offspring, suggesting that shared environment rather than genetic factors account for these findings. These findings strongly suggest that PAn may be a useful indicator of liability for schizophrenia. However, Katsanis, Iacono, and Beiser (1990) raise important questions about the specifity of the PAn scale as an indicator of liability for schizophrenia. Although psychotic patients scored higher than their first-degree relatives, who in turn scored higher than nonpsychiatric controls, PAn scores did not distinguish among the four groups of first-episode psychotics examined (schizophrenics, schizophreniforms, major depressives, bipolars) or between their first-degree relatives. This study provides strong evidence for the construct validity of the PAn scale as an index of psychosis proneness, rather than schizophrenia proneness per se. Interestingly, the reason for elevated scores among first-episode psychotic patients appears to be quite different across schizophrenic and mood-disordered patients. Among schizophrenic patients, anhedonia scores appear related to premorbid functioning and enduring personality traits, while such scores in affective disordered patients appear more related to current clinical state and level of adjustment (Katsanis, Iacono, Beiser, & Lacey, 1992). What appears clear from these studies is that PAn runs in families, although it may have little specificity across severe forms of psychopathology.

Several studies have examined correlates of PAn scores in nonpsychotic populations. Garnet, Glick, and Edell (1993) found that PAn scores predicted poor social competence in nonpsychotic psychiatric inpatients. Penk, Carpenter, and Rylee (1979) examined the relationship of MMPI scores to PAn scores in a sample of male inpatient drug abusers. PAn scores correlated positively with MMPI validity scale F; clinical scales measuring hypochondriasis, depression, paranoia, psychasthenia, schizophrenia, and social introversion; content scales measuring social maladjustment, depression, psychoticism, organic symptoms, and poor health;

and pure scales 3, 5, and 7, which are associated with MMPI profile types 2-7, 8-6, and 2-7-8, respectively, commonly found in borderline schizophrenics. PAn was correlated negatively with pure scale 6, associated with profile type 9 (hypomania). The authors conclude that PAn is associated with social maladjustment and intrapsychic aspects of disordered thinking. Edell and Joy (1989), in a sample of young, nonpsychotic psychiatric inpatients, found PAn scores were uncorrelated with a variety of Symptom Checklist (SCL-90) and Brief Psychiatric Rating Scale (BPRS) factors and negatively correlated with histrionic and narcissistic personality disorder criteria in males. In women, PAn scores were correlated positively with obsessive–compulsive, depression, anxiety, phobic anxiety, and psychoticism factor scores on the SCL-90 and with schizoid, antisocial, and avoidant personality disorder criteria. PAn scores were uncorrelated with PER ($r = .06$) and IMP ($r = .19$) in men but positively correlated with PER ($r = .39$) and IMP ($r = .37$) in women. These findings illustrate dramatically the importance of analyzing the data separately for males and females, because the scale appears to measure different things in men and women.

Indeed, although men tend to score higher on PAn than women, the impact of high PAn in women may be more strongly predictive of future psychopathology. Recent evidence from the New York High Risk Project (Erlenmeyer-Kimling et al., 1993) suggests that high anhedonia in the offspring of a schizophrenic parent is predictive of future psychotic episodes in women but not men. In addition, anhedonia was not associated with being an offspring of an affective disordered parent or with the later development of a major affective illness. Further evidence for the specificity of PAn to schizophrenia was their finding that attentional disturbance in childhood was related to anhedonia in adolescence and social isolation in adulthood only in subjects at risk for schizophrenia.

Conclusion

Reviewing early validational evidence, Grove argued that the PAn scale is "not powerful as a schizotypy indicator" (1982, p. 35). More recent evidence, particularly examining familial patterns in schizophrenic probands and their first-degree relatives, is more promising, although the issue of specificity requires further inquiry. The weight of the evidence to date suggests that the PAn scale is identifying a cohort of individuals who manifest a variety of social deficits, poor premorbid adjustment, negative symptoms, introversion, and somewhat lowered verbal abilities. In the absence of a genetic relationship with a schizophrenic proband, the predictive power of the PAn scale for identifying individuals at increased risk for future psychosis appears weak.

Social Anhedonia Scale

Brief Description

Originally designed as a 48-item instrument to measure the lowered ability to experience interpersonal or social pleasures, the Social Anhedonia (SAn) scale was later revised because of the high negative correlations with social desirability ($-.36$, men; $-.33$, women) as well as the failure to find a relationship between scores on the original scale and psychotic-like symptoms in college students. A representative item is "Writing letters to friends is more trouble than it's worth" (true). The revised 40-item scale eliminated items tapping social anxiety, avoidant withdrawal, and social desirability variance, while maintaining and adding items correlated highly with schizoid withdrawal and asociality. The revised SAn scale has a coefficient alpha of .79 (men and women), low negative correlations with social desirability ($-.04$, men; $-.22$, women), and low positive correlations with PER, MAG, and IMP, ranging from .04 to .21. Correlations between the original SAn and PAn scales were substantially higher (.60, men; .51, women) than are correlations between the revised versions (.25, men; .24, women).

Validity Findings

COLLEGE STUDENTS

Four published papers have examined characteristics of college students elevated on SAn. Mishlove and Chapman (1985) reported a series of studies using the revised SAn scale. In study 1, SAn subjects had poorer overall social adjustment, when compared to nonelevated controls. They reported having fewer friends, being more reticent with their friends, engaging in fewer social interactions, having somewhat less interest in dating, and experiencing greater discomfort in social situations. They were indistinguishable from controls on hypersensitivity or loneliness. They acknowledged deriving less pleasure from the company of others because of a reduced need for interpersonal contact. The scale had a sensitivity of 92% and a specificity of 75% for identifying socially maladjusted persons. In study 2, female SAn subjects (but not male) reported more schizotypal features and greater deviancy of the highest rated psychotic-like experience when compared with controls. Groups did not differ on history of depression. Sensitivity and specificity for the SAn scale in identifying women with psychotic-like symptoms were 100% and 59%, respectively. In study 3, PER-MAG subjects were divided into high and low SAn groups using 1 SD above the same-sex mean as the cutting score. PER-MAG subjects high on SAn had poorer overall social adjustment, fewer friends, greater reticence with their friends and (in men) with their families, tended to engage in fewer social interactions, reported a diminished interest in dating or interpersonal contact, reported enjoying the company of other people less than most people do, and had poorer leisure and familial adjustment, but not poorer global academic adjustment, when compared with PER-MAG subjects who scored low on SAn. Male PER-MAG subjects who scored high on SAn reported more schizotypal, features and more deviant psychotic-like experiences than male PER-MAG subjects who score low on SAn. The sensitivity and specificity for identifying PER-MAG men who had moderately deviant psychoticlike symptoms were .80 and .67, respectively. Nonanhedonic PER-MAG subjects did report more schizotypal features and, in females only, more psychotic and psychoticlike experiences, than nonelevated control subjects. They concluded that the revised SAn scale alone may be useful in identifying females who are psychosis-prone and may be combined with the PER-MAG scale in identifying males who are at elevated risk.

Muntaner et al. (1988) found that SAn correlated positively with the Psychoticism and Neuroticism scales and correlated negatively with the Extraversion scale on the EPQ (Eysenck & Eysenck, 1975) and with the Schizotypal Personality and Borderline Personality trait scales (Claridge & Broks, 1984). A principal-components analysis found SAn to load with PAn (positively) and Extraversion (negatively). Peterson and Knudson (1983) were more critical of the scale, arguing that it appears to provide little information beyond that which can be gleaned from existing measures of affiliation and introversion, correlating most strongly with neurotic maladjustment. The scale correlated negatively with affiliation, surgency (i.e., behaviors that are cheerful, lively, social, and happy-go-lucky), and extraversion, and correlated positively with repression-sensitization, depression, and neuroticism.

More recently, Leak (1991) reported good convergent validity data and mixed discriminant validity data for the SAn scale. As predicted, scores on SAn were positively correlated with measures of anomia (alienation from others), drivenness (self-defeating interpersonal behaviors), loneliness, avoidance, oppositional tendencies (tendencies to resist authority through antagonistic behavior and to be critical of others) and power (aggressiveness and authoritarianism in an effort to control people). Scores on SAn were negatively correlated with measures of affiliative tendency (measuring social skills that facilitate pleasurable social interaction and a need for friendly and cooperative relations with others), social interest, faith in people, and empathy. Finally, SAn scores were uncorrelated with involvement, hopelessness, sensitivity to rejection, self-esteem, and social desirability. Contrary to expectation, SAn score were positively correlated with anxiety, rigidity, competent (defined as desire to achieve high standards of performance through independent effort) and competitive life-styles, and a task/security factor (measuring one's attempt to gain security while working toward a task, characterized by a lack of cooperation in working with people). SAn scores were uncorrelated with humanistic-helpful attitudes (positive philosophy of human nature coupled with desire to help and teach others). Leak concluded that the SAn scale is not simply a measure of pervasive pathology, global undifferentiated dissatisfaction, or interpersonal psychic distress. Persons scoring high on the SAn scale would appear to be lacking social interest, social motivation and satisfactions, social skills, and other positive orientations toward others; to possess traits reflecting social alienation, anxiety, and rigidity; and to have negative orientations to both people and tasks.

Finally, J.P. Chapman and Chapman (1992), in a 10-year follow-up study, reported data suggesting that the presence of elevated SAn scores in college students defined initially by high PER-MAG scores may be strongly predictive of future psychosis and psychotic-like symptoms. Of the 10 PER-MAG subjects who had become clinically psychotic during the follow-up period, 7 (70%) had elevated SAn scores (>7).

Psychiatric Patients and Their Biological Relatives

Five published studies have examined the performance of psychiatric patients and/or their biological relatives on SAn. In a sample of hospitalized male schizophrenics (L.J. Chapman et al., 1976), the original SAn scale was uncorrelated with age, history of delusions or hallucinations, current depression, days in hospital over the prior 5 years, age at first hospitalization, and months since first hospitalization. The schizophrenics scored more deviantly when compared with normal male controls. Anhedonic and hedonic schizophrenics also differed on premorbid adjustment in the expected direction (i.e., hedonics were much more likely to have good premorbid adjustment, anhedonics were somewhat more likely to have poor premorbid adjustment). Similarly, Cook and Simukonda (1981) found higher SAn scores in their sample of British schizophrenics when compared with normal controls for both men and women and found no correlation of SAn scores to type of schizophrenic symptom. In contrast, Kirkpatrick and Buchanan (1990) found a relationship between SAn scores and type of symptoms in a sample of clinically stable outpatient schizophrenics. Those with the deficit syndrome (defined previously) had higher scores on SAn than did nondeficit schizophrenics. The methodologic advantages of this study when contrasted with Cook and Simukonda (1981) were addressed earlier. The specificity of SAn scores to diagnosis was examined by Katsanis et al. (1990), who found that scores on SAn were indistinguishable among four groups of first-episode psychotics (schizophrenics, schizophreniforms, major depressives, bipolars) or their firsts degree relatives. Psychotics scored higher than their relatives and the normal control subjects, and relatives scored higher than controls, supporting the construct validity of SAn as an index of psychosis proneness.

Finally, Penk et al. (1979) examined MMPI correlates of SAn scores in male inpatient drug abusers. Using the original SAn scale, they found positive correlations with MMPI validity scales F; with clinical scales measuring hypochondriasis, depression, paranoia, psychasthenia, schizophrenia, and social introversion; with content scales measuring social maladjustment, depression, poor morale, psychoticism, organic symptoms, and poor health; and with pure scales 3, 5, and 7, associated with MMPI profile types 2-7, 8-6, and 2-7-8, often found in borderline schizophrenia, and negatively with pure scales 4 and 6, associated with profile types 4 and 9, respectively. Penk et al. concluded that SAn is associated with social maladjustment and interpersonal aspects of disordered thinking and, indeed, may be a more promising measure than PAn for identifying proneness to schizophrenia, given its stronger associations with MMPI scales typically associated with schizophrenia.

Conclusion

Although the SAn scale appears to identify a cohort similar in many respects to that described for the PAn scale, some emerging data suggest that SAn may be associated with increased risk for psychosis in individuals concurrently elevated on the PER-MAG scale.

Perceptual Aberration Scale

Brief Description

The Perceptual Aberration (PER) scale was originally designed as a 28-item instrument to measure body-image aberrations commonly described by schizophrenics and borderline schizophrenics, such as unclear body boundaries and feelings of unreality, estrangement, deterioration, or changes in the appearance of parts of one's body. A representative item is "I have sometimes had the feeling that my body is decaying inside" (true). Coefficient alpha reliability figures have ranged from .88 to .94 across schizophrenics, nonpsychotic clinic clients, and noncollege and college normal controls. The PER scale correlates negatively with social desirability ($-.08$ in male schizophrenics to $-.34$ in matched male normal controls) and positively with acquiescence bias (.17 in male clinic clients to .27 in male normal control subjects). It is uncorrelated with age, education, and social class. A separate 7-item scale of changes in visual and auditory experiences (e.g., "My hearing is sometimes so sensitive that ordinary sounds become uncomfortable" [true]) was found to correlate .76 with the 28-item body-image aberration scale, despite its modest internal consistency reliability (.60), suggesting that it may be tapping a similar construct, called *perceptual aberration*. The coefficient alpha figures for the 35-item PER scale are quite comparable to the 28-item scale, as are the correlations with demographic and method variance factors. Virtually all published studies have used the 35-item version, because of its inclusion of a broader range of perceptual distortions, although researchers interested primarily in body-image distortions have used the shorter version (e.g., Bornstein, O'Neill, Galley, Leone, & Castrianno, 1988). Test–retest reliability coefficients (6 weeks) in college students are .75 (men) to .76 (women). The PER scale correlates positively with MAG (.68 men; .70 women), IMP (.41, men; .44 women), Golden and Meehl's (1979) Schizoidia scale (.20 men; .18 women), and Eysenck and Eysenck's (1975) Psychoticism scale (.33, men; .34, women) and *negatively* with PAn (−.20 to −.22, men; −.10 to −.15, women) in large samples of normal controls.

Validity Findings

COLLEGE STUDENTS

Fourteen published studies have examined characteristics of college students elevated on the PER scale (studies using a combination of PER and MAG scale scores are reviewed in a subsequent section). In general, subjects scoring high on the PER have not displayed deficits in social competence, when compared to same-sex nonelevated controls. Subjects with high PER scores and controls were equally socially competent on an interpersonal behavior role-playing test in men (Haberman et al., 1979) and in women (Numbers & Chapman, 1982), although female PER responses were more odd and hostile. No differences have been noted on reported frequency of dates, likelihood to have gone steady, or overall heterosexual adjustment on structured interview (L.J. Chapman et al., 1980). Nor were the PER subjects more variable in their implicit semantic trait structures, which Raulin and Henderson (1987) suggested may account for social deficits in schizotypes.

Evidence for mild cognitive slippage among PER subjects has been more compelling. PER subjects give more schizophrenia-like thought-disordered responses on the Rorschach (Edell & Chapman, 1979) and have shown evidence of deviance similar to that found in schizophrenics on a task of referential communication believed to reflect a failure to self-edit or pretest utterances to facilitate communication (Allen, Chapman, Chapman, Vuchetich, & Frost, 1987; Martin & Chapman, 1982). Particularly poor performance on this task was found in those PER subjects with concurrent elevations of IMP in one study (Martin & Chapman, 1982) but not another (Allen, Chapman, & Chapman, 1987).

On structured psychiatric interview, L.J. Chapman et al. (1980) found that male and female PER subjects report many more psychotic and psychotic-like experiences, than controls, including thought transmission, passivity experiences, voice experiences and other auditory hallucinations, aberrant beliefs, visual experiences, and other schizotypal symptoms (e.g., depersonalization, derealization, ideas of reference, extreme suspiciousness and paranoid ideation, out-of-body experiences, difficulty concentrating, complaints of speech being mixed up, deviant vocalizations, odd communication as observed by examiner, and social withdrawal). They more often met criteria for major depressive syndrome and hypomanic episodes. They tended more often to have seen a mental health professional. Allen, Chapman, Chapman, Vuchetich, et al. (1987) found also that PER exceeded controls on psychotic and psychotic-like symptoms. Of note, subjects scoring above the cutoff point on at least two of three scales (PER, General Behavior Inventory [GBI], IMP) yielded an especially deviant group of subjects, of whom 85% evidenced psychotic or psychotic-like symptoms. Subjects reporting symptoms of psychotic severity were generally elevated on all three scales. Selecting subjects solely on the basis of the GBI, without considering their PER scores, did not identify subjects with increased psychotic-like symptoms.

On a childhood experiences scale, PER subjects recalled more criticism from their mothers and fathers and tended to have felt less support, greater disinterest, and greater distance from their mothers during childhood, compared with controls (Edell & Kaslow, 1991). Interestingly, PER subjects were criticized by their mothers for dependent (attachment) behaviors and by their fathers for independent (initiating) behaviors, leaving them in a "double bind" (Bateson, Jackson, Haley, & Weakland, 1956) where criticism was inevitable no matter how they behaved.

Further support for construct validity comes from studies that have examined the relationship of PER scores to other scales. Muntaner et al. (1988) found PER was positively correlated with the Psychoticism and Neuroticism scales and negatively correlated with the Lie scale on the EPQ (Eysenck & Eysenck, 1975); it was positively correlated with the Schizotypal Personality and Borderline Personality trait scales (Claridge & Broks, 1984). A principal-components analysis with Varimax rotation found PER to load with MAG, Neuroticism, and the Schizotypal Personality and Borderline Personality trait scales. Peltier and Walsh (1990), in a study highly critical of the PER scale, found that variance attributable to response style, such as social desirability and acquiescence bias, was significantly related to scores on PER, and that normal subjects could easily fake their scores in either a pathological or normal direction. They concluded that PER largely measures the tendency to give socially desirable responses in self-description. However, marked shrinkage in the amount of variance accounted for by response set predictor variables was noted in a cross-validation study (from 17.6% to 2.8%), suggesting that their initial results were highly unstable and require further examination. One study suggests that high scores on PER are not necessarily associated with socially undesirable responding. Schuldberg (1988) found that PER scores correlate *positively* with two measures of creativity among male students. Most recently, Lenzenweger and Korfine (1992b), in a randomly ascertained sample of college students, found a strong relationship between elevated scores on PER and what they refer to as schizophrenia-related personality disorders (SRPD) (i.e., schizotypal and paranoid personality disorders), as defined by personality disorder scales derived from the MMPI by Morey, Waugh, and Blashfield (1985). It should be noted, however, that subjects scoring high on the PER scale also displayed elevations on several non-SRPD scales measuring borderline, antisocial, passive-aggressive, compulsive, histrionic, and narcissistic traits, and were not elevated on one SRPD (schizoid). This finding suggests that subjects who endorse large numbers of PER items are also likely to endorse a broad spectrum of deviant experiences across several psychopathological domains.

Two studies using an object relations, psychodynamic perspective have employed shortened versions of the PER scale. Bornstein et al. (1988) examined the relationship of orality, which is theoretically associated with development of a weak, unstable body ego, and scores on PER (using only the 28 items pertaining to body-image aberrations). As predicted, level of orality (measured by group Rorschach) was positively correlated to PER scores in men ($r = .52$), although not women ($r = .06$). Levin and Raulin (1991), using a 10-item version of the scale, found that PER subjects were more likely to report high nightmare frequency compared to controls, although here the effects were due largely to female subjects. Although both studies support the notion that ego boundary disturbances in high-PER subjects may be comparable to that found among psychotics and prepsychotics, the sex differences obtained are puzzling and require further investigation.

PSYCHIATRIC PATIENTS AND THEIR BIOLOGICAL RELATIVES

Nine published studies have examined the performance of psychiatric patients and/or their biological relatives on PER. L.J. Chapman, Chapman, & Realin (1978) found that schizophrenics had a higher mean score on the original 28-item body-image aberration scale than did matched normal controls, with about 15% of the schizophrenics and 5% of the controls scoring at least 2 SD above the mean of the normal control group. The distribution of scores for the schizophrenics was slightly skewed but not obviously bimodal. Most of the patients were receiving antipsychotic medication, which may have reduced the self-reports of such symptoms. Only a tendency was noted for body-image aberration to be higher in the poor premorbid than in good premorbid schizophrenics. Body-image aberration was negatively correlated with months since first hospitalization ($r = -.30$), suggesting that reports of body-image aberration are greater early in the illness. Of note, body-image aberration scores correlated $-.10$, $-.20$, and $.11$ with PAn scores in female college students, male college students, and male schizophrenics, respectively, despite their mutually high internal consistency reliability figures, and were positively correlated ($r = .30$) with scores on the Beck (1967) Depression Inventory. Male nonpsychotic clinic clients and college students were indistinguishable on body-image aberration scores.

Several studies have reported higher scores on the full PER scale in schizophrenics when compared with normal control subjects (Clementz et al., 1991; Cook & Simukonda, 1981; Grove et al., 1991; Katsanis et al., 1990). No relationship between scores on PER and type of schizophrenic symptoms displayed has been found (Cook & Simukonda, 1981; Kirkpatrick & Buchanan, 1990). More puzzling has been the performance of first-degree relatives of schizophrenic probands on PER. For example, Grove et al. (1991) found that the first-degree relatives of schizophrenic probands did not differ from normal relatives, which the authors attributed to the relatives' wish not to acknowledge having the obvious pathologies tapped by the PER items. Clementz et al. (1991), examining data collected across three sites (New York, Minneapolis, and Vancouver), unexpectedly found that first-degree relatives of the schizophrenics had *lower* scores than did the normal controls. Relatives with higher PER scores, however, were more frequently diagnosed with a schizophrenia-related disorder, and their scores were indistinguishable from schizophrenic probands. Nor did PER scores show familial resemblance (correlations between parental–parental, sibling–parental, and sibling–sibling correlations were nonsignificant). They concluded that PER may contribute little beyond affection status to identify relatives at risk for schizophrenia (but see Lenzenweger & Loranger, 1989a, 1989b). Instead, they believe that PER may help to identify persons among nonschizophrenic proband and relative groups who are at risk for disorders with prominent cognitive-perceptual type symptoms rather than being an indicator of liability for schizophrenia among families ascertained through a schizophrenic proband.

The specificity of PER was examined by Katsanis et al. (1990), who found that scores were indistinguishable among four groups of first-episode psychotics (schizophrenics, schizophreniforms, major depressives, bipolars) or their first-degree relatives. The psychotic patients scored higher than their relatives and the normal control subjects, but the relatives of patients once again scored *lower* than controls. Katsanis et al. believe that this may reflect defensiveness on the part of relatives, who knew they were selected because they were related to a psychotic patient.

Two studies providing the strongest evidence to date of the construct validity of the PER scale as a useful measure of schizotypy and proneness to schizophrenia were conducted by Lenzenweger and Loranger (1989a, 1989b). In one study, in a sample of mostly white, single, middle- and upper-class inpatients, they used the group median on PER as a cutoff score to define schizotypy-positive and schizotypy-negative groups. They found significantly more treated schizophrenia, but not more treated unipolar depression or bipolar disorder, among first-degree relatives of schizotypy positive probands when compared with schizotypy-negative probands. Indeed, all cases of treated schizophrenia found ($n = 5$; 4 siblings, 1 parent) were in the first-degree relatives of the schizotypy-positive group, coming from four different families. The morbid risk for schizophrenia among siblings was greater among family members of the schizotypy-positive group compared with the schizotypy-negative group and was greater than the morbid risk among parents of the probands in the schizotypy-positive group. This finding is similar to the known greater morbid risk for schizophrenia among siblings than among parents of actual schizophrenics. In a second study, Lenzenweger and Loranger examined associations between scores on PER and several measures of clinical psychopathology in a carefully diagnosed, nonpsychotic, predominantly white, female, single, inpatient sample. Scores on PER were positively correlated with anxiety, depression, global impairment ratings, number of prior psychiatric hospitalizations, and number of diagnostic criteria met for schizotypal, schizoid, avoidant, and obsessive–compulsive personality disorders. PER scores were not significantly correlated with age at onset of psychiatric difficulties (index of illness chronicity) or impairment in current adult social competence. Similarly, Garnet et al. (1993) found that nonpsychotic psychiatric inpatients, divided into high- and low-risk groups using the median score on PER, were indistinguishable on premorbid social competence. Hierarchical regression analyses suggested that schizotypal symptoms and clinical anxiety are the two underlying psychopathological processes most useful in explaining variance in PER scores (Lenzenweger & Loranger, 1989a).

One possible limitation of this latter study was the failure to analyze data separately by sex or hospitalizations status, the inclusion of patients up to age 55 who are well past the age of risk for schizophrenia, and the failure to examine the relationship of PER scores to a broader range of psychopathology than anxiety and depression. Edell and Joy (1989) examined the relationship of PER scores to a broad spectrum of clinically significant psychopathology in a sample of young, nonpsychotic, mostly middle-class psychiatric inpatients. To their surprise, PER scores were positively correlated with virtually every factor score on the SCL-90 in men and women, with virtually every BPRS factor score in men but not in women, with paranoid, obsessive–compulsive, avoidant, and borderline personality disorder criteria in men, with narcissistic personality disorder criteria in women, and with schizotypal personality disorder criteria in both sexes. PER scores were positively correlated with MAG for men and women, and with PAn in women only. Despite the large number of significant intercorrelations obtained, the PER scale performed better than the PAn, IMP, and MAG scales in correlating positively with those characteristics hypothesized to reflect a proneness to psychosis (i.e., psychoticism and schizotypal personality disorder criteria) in both men and women. Of interest, and consistent with the findings of lower scores among the relatives of psychotic probands than nonpsychiatric controls, mean scores on PER among these nonpsychotic inpatients were lower than those for a large college student normative sample in men (but not women), perhaps reflecting their defensiveness about revealing symptoms of perceptual distortions, believing that it might extend their hospital stay.

Finally, one study (Carpenter, 1983) found no relationship between PER scores and performance on Gorhams' Proverbs Test by schizophrenics (but see Allen & Schuldberg's, 1989, study on PER-MAG college students). Another study (Bornstein et al., 1988), consistent with the college student findings described previously using the 28-item body-image aberration scale, reported a significant positive correlation ($r = .44$) between scores on PER and level of orality among male voluntary psychiatric inpatients (with affective, schizophrenic, and substance use disorders) who were administered the Rorschach individually, but not among women ($r = .03$).

Conclusion

In an earlier review of psychometric methods for detecting schizotypy, Grove suggested that the PER scale "seems quite capable of isolating a group of subjects at high risk for schizophrenia" (1982, p. 35). Overall, the data are far more encouraging for the PER scale as an index of psychosis proneness than for either of the anhedonia scales, with some intriguing data provided by Lenzenweger and Loranger (1989a, 1989b) that the scale has predictive specificity for schizotypy and schizophrenia.

Magical Ideation Scale

Brief Description

The Magical Ideation (MAG) scale was designed as a 30-item instrument to measure the belief in forms of causation that by conventional standards are invalid. A representative item is "Some people can make me aware of them just by thinking about me" (true). Internal consistency reliability coefficients for the scale have ranged from .82 (men) to .85 (women). Test–retest reliability coefficients (6 weeks) are .80 (men) and .82 (women). The scale correlates −.18 with social desirability and .26 with acquiescence bias. No correlations with age or social class are reported in the original papers. The MAG scale correlates negatively with the PAn scale (−.23 to −.29, men; −.15 to −.17, women) and correlates positively with the PER scale (.68, men; .70 to .71, women), the IMP scale (.45, men and women), Golden and Meehl's (1979) Schizoidia scale (.25, men; .26, women), and Eysenck and Eysenck's (1975) Psychoticism scale (.32, men; .32, women) in large samples of normal control subjects. As noted earlier, it is unusual for two measures of severe psychopathology such as MAG and PAn to be negatively correlated, suggesting that the measures are identifying different types of disorder or different manifestations of the same disorder.

Validity Findings

COLLEGE STUDENTS

Four published studies have examined characteristics of college students elevated on the MAG scale (studies using a combination of PER and MAG scale scores are reviewed in a subsequent section). In the initial study describing the scale's development, Eckblad and Chapman (1983) contrasted MAG subjects who were not concurrently deviant on PER, which is about half of the deviant MAG subjects, with nonelevated controls using structured diagnostic interview. The MAG subjects exceeded controls on evidence of magical thoughts; psychotic and psychotic-like experiences (e.g., thought-broadcasting experiences, voice and other auditory experiences, aberrant beliefs); schizotypal experiences; episodes of depression, mania, and hypomania; trends toward affective personality disorders; and difficulties concentrating. Subjects showing affective symptoms did not tend, above chance, to be the same subjects who showed psychotic-like symptoms, suggesting that some individuals elevated on the MAG scale may be prone to affective disorder and others toward schizophrenia.

Muntaner et al. (1988) found the MAG scale to be positively correlated with the Psychoticism, Extraversion (women only), and Neuroticism scales and negatively correlated with the Lie scale on the EPQ (Eysenck & Eysenck, 1975) and positively correlated with the Schizotypal Personality and Borderline Personality scales (Claridge & Broke, 1984). A principal-components analysis with Varimax rotation found MAG to load with the PER, Neuroticism, and Schizotypal Personality and Borderline Personality trait scales. Schuldberg (1988) found no relationship between MAG scores, which purportedly measure the cognitive aspect of schizotaxia, and measures of creativity in male undergraduates.

Peltier and Walsh (1990), in a study highly critical of the MAG scale, found that variance attributable to response style, such as social desirability and acquiescence bias, was significantly related to scores on the MAG scale, and that normal subjects could easily fake their scores in either a pathological or normal direction. They concluded that the scale largely measures the tendency to give socially desirable responses in self-description. However, marked shrinkage in the amount of variance accounted for by response set predictor variables was noted in a cross-validation study (from 25.1% to 12.2%), suggesting that their initial results were highly unstable and require further examination.

PSYCHIATRIC PATIENTS AND THEIR BIOLOGICAL RELATIVES

Two published studies have examined the performance of psychiatric patients on the MAG scale, and none have examined performance by their relatives. George and Neufeld (1987) found that hospitalized nonparanoid and paranoid schizophrenics scored higher on the MAG scale than did a mixed psychiatric control inpatient group (predominantly affective disorders) and a normal control group. Of interest, the mean scores obtained by the mixed psychiatric control group was somewhat *lower* than those obtained by the normal control group, which raises the question again as to why normals may score higher than some psychiatric control groups. Kirkpatrick and Buchanan (1990) found no differences on MAG scores between clinically stable outpatient schizophrenics with and without the deficit syndrome (defined earlier). Finally, in a sample of young, nonpsychotic psychiatric inpatients, Edell and Joy (1989) found MAG scores to be positively correlated with SCL-90 factor scores measuring anxiety in men and women, hostility in men, and somatization, obsessive–compulsive, depression, phobic anxiety, and psychoticism in women. MAG scores were also positively correlated with BPRS factors measuring hostility and activation and with paranoid, schizotypal, histrionic, and borderline personality disorder criteria in men but not in women. MAG scores were positively correlated with PER for men and women and with IMP in men only. Of interest, mean scores on the MAG scale were lower than those for the college student normative sample in men, perhaps reflecting their defensiveness about revealing symptoms of magical thinking, believing that it might extend their hospital stay.

Conclusion

The MAG scale appears to identify a cohort of subjects comparable to that identified by the PER scale, although each scale may identify a slightly different group of individuals at risk for psychosis. One possible advantage of the MAG scale over PER is that the MAG items are less obviously pathological, although lowered mean scores by nonpsychotic patients compared with college students as well as the findings of Peltier and Walsh (1990) for both scales suggest that subjects are aware of the pathology tapped by each set of items and can alter their responses to suit their purposes.

Perceptual Aberration–Magical Ideation Combined

Brief Description

Some investigators have combined the PER and MAG scales into a single scale because they share about half of their variance, select many of the same subjects, and probably identify the same syndrome (Eckblad & Chapman, 1983). Most studies using the two scales together have defined deviance as a score at least 2.0 SD above the same-sex mean on either of its constituent scales or 3.0 SD on the sum of the two Z-scores.

Validity Findings

COLLEGE STUDENTS

A total of 13 published studies have examined characteristics of college students elevated on the PER-MAG scales. Although one study found no differences between males with elevated PER-MAG score and controls in level of social adequacy and interpersonal competence (Beckfield, 1985), several studies have demonstrated increased schizophrenia-like cognitive slippage in PER-MAG subjects. E.N. Miller and Chapman (1983), for example, found that whereas male PER-MAG subjects were indistinguishable from controls on a structured word association task, they gave fewer popular responses, proportionally more deviant and nondeviant idiosyncratic responses, proportionally fewer common responses, and lower response communality scores (proportion of normative sample that produced that same associate) on an unstructured word association task. PER-MAG subjects scoring above a median Z-score of 1.0 on IMP were most likely to show the cognitive slippage. Similarly, Allen, Chapman, and Chapman (1987) found that male and female PER-MAG subjects gave increased numbers of unusual responses and decreased response communality scores on an unstructured continued word association task and gave less popular clues on a word communication task believed to reflect a failure to self-edit or pretest utterances to facilitate communication. PER-MAG subjects scoring above the mean on the GBI depression subscale (Depue et al., 1981) were most deviant on both cognitive slippage measures. Unlike earlier studies (Martin & Chapman, 1982; E.N. Miller & Chapman, 1983), PER-MAG subjects divided into high- and low-IMP groups performed similarly, which the authors attribute to administering the IMP scale separately rather than embedded with other scales, resulting in lower mean IMP scores. In another study, PER-MAG subjects gave more bizarre-idiosyncratic scores on unfamiliar (different culture) proverbs but not on familiar (same culture) proverbs. As proverbs became less familiar or more ambiguous, PER-MAG subjects showed increased subclinical positive thought disorder (Allen & Schuldberg, 1989). It is worth noting that lack of structure increased the probability of finding cognitive slippage in these subjects, not unlike findings obtained by borderline syndrome patients (Edell, 1987).

Comparing two methods of identifying psychosis-prone college students (PER-MAG vs. 2-7-8 MMPI profile [highest elevations on scales measuring Depression, Psychasthenia, and Schizophrenia]), Fujioka and Chapman (1984) found different subjects were identified by each. Both groups exceeded controls on number and proportion of subjects reporting psychotic and psychotic-like experiences, presence of either major or minor depressive disorder, mania, or personality disorders on structured interview. PER-MAG subjects exceeded 2-7-8 subjects and controls on mean number of schizotypal experiences reported (oddness of speech, out-of-body experiences, depersonalization, frequent déjà vu experiences), with male PER-MAG subjects more likely to receive a probable or definite RDC diagnosis of schizotypal features and more likely to have a history of hypomania. Both groups were equally likely to have seen a mental health professional for psychiatric reasons. Only 3 of 80 subjects met criteria for *both* PER-MAG and 2-7-8 and were especially deviant. All had RDC-probable or definite schizotypal features. One woman had a chronic major depressive disorder and had been hospitalized twice following suicide attempts after hearing an inner voice urging her to kill herself; another woman had a history of recurrent depressive episodes, reported three different psychotic-like experiences, including believing others could hear her thoughts and hearing her mother as an outer voice calling to her; one man met RDC criteria for hypomania and minor depressive disorder and believed incorrectly that strangers were laughing at him.

As noted earlier, Mishlove and Chapman (1985) found that PER-MAG subjects high on SAn had poorer overall social adjustment and reported more schizotypal features and more deviant psychotic-like experiences than did low SAn PER-MAG subjects. Nonanhedonic PER-MAG subjects reported more schizotypal features and, in women only, more psychotic and psychotic-like experiences, than nonelevated control subjects.

Frost and Chapman (1987) found that male PER-MAG subjects exceeded controls on a composite index of polymorphous sexuality consistent with clinical reports of the preschizophrenic personality. They acknowledged sexual arousal to a broader range of stimuli on interview and reported a wider range of sexual behaviors, arousal, and fantasies on questionnaire.

On the Personality Research Form (PRF, Jackson, 1984), Stone and Schuldberg (1990) found that PER-MAG subjects exceeded controls on aggression and change, suggesting that they could be described as adaptable, flexible, irritable, and unpredictable, and were lower on desirability. PER-MAG females scored higher than controls on understanding and lower on succorance and nurturance, while these differences were *reversed* for men. Surprisingly, the tendency of PER-MAG subjects to receive higher scores on Achievement and Autonomy suggests that they may also be described as capable, productive, ambitious, and unconstrained. These findings are consistent with those of Schuldberg, French, Stone, and Heberle (1988), who found that PER-MAG subjects (particularly women) scored higher than controls on creativity tests measuring preference for novel and complex figures as well as biographical and personality factors relevant to creative behavior. PER-MAG subjects who scored above the median on IMP tended to receive the highest creativity scores. Schuldberg (1990) also found that PER-MAG was *positively* correlated with a test of creativity but negatively correlated with ego strength, arguing that PER-MAG taps the perceptual/cognitive aspects of positive symptoms. Finally, Schuldberg and London (1989) found no evidence of a distinct perceptual style between PER-MAG and control groups on the Group Embedded Figures Test, even when IMP was elevated.

The strongest evidence supporting the validity of the PER-MAG scales as indices of psychosis proneness comes from L.J. Chapman and Chapman (1985, 1987), who interviewed college students scoring deviantly on PER-MAG ($n = 162$) on two occasions approximately 25 months apart. During this short-term follow-up period, PER-MAG subjects were more likely than controls to have sought professional help for their problems, which included depression, anxiety, interpersonal difficulties, and adjustment problems. They continued to have higher rates of psychotic and psychotic-like symptoms than controls, reported a higher number of schizotypal symptoms, and were more likely to use street drugs. Indeed, during the short follow-up period, 10% of the PER-MAG group and none of the control group reported having had symptoms that were severe enought to be rated psychotic (L.J. Chapman & Chapman, 1980). PER-MAG women were more likely to report depression, while PER-MAG men had lower overall global adjustment, when compared with same-sex controls. PER-MAG subjects also exceeded controls on rate of nonpsychiatric hospitalization that was due to illness, although not for accidents and injuries. Most noteworthy was that three subjects, all from the PER-MAG group, received their first clinical attention for psychosis during the follow-up period. One man met DSM-III criteria for schizophrenia and had been hospitalized twice in the year following the initial interview, presenting with a variety of bizarre psychotic symptoms including religious hallucinations, delusions, passivity experiences, grandiosity, social withdrawal, affective blunting, and thought disorder. One woman developed a clear-cut bipolar disorder and was hospitalized twice during the follow-up period, once for a manic episode with psychotic features (delusions, hallucinations) and once for a major depressive episode with psychotic features (delusions, hallucinations). Another woman was receiving outpatient treatment for a paranoid delusion that she found aberrant and distressing, believing that two scientists had put her in her local community to test her daily reactions. Of note, the specific symptoms reported at follow-up had been reported by each of the three subjects in attenuated forms at the time of their initial interview, suggesting a continuity of psychotic and premorbid symptoms over time. A fourth (female) PER-MAG subject was hospitalized for depression without psychotic features (Mishlove & Chapman, 1985). Nine other PER-MAG subjects also reported isolated psychotic and psychotic-like symptoms at follow-up that were more deviant than similar symptoms reported at the earlier interview, and they may have been heading toward a full-blown psychosis.

More recently, J.P. Chapman and Chapman (1992) reported on a 10-year follow-up study of college students elevated on the PER-MAG scale. Of note, of the 182 PER-MAG subjects followed, 10 (5.5%) had developed a clinically psychotic disorder and an additional 3 (1.6%) subjects were "possibly psychotic." Of 153 control subjects followed, only 2 (1.3%) had developed a psychotic disorder. Psychotic-like symptoms were exhibited by approximately 43% of PER-MAG subjects (compared with 14% of controls). In addition, 19% of the relative of PER-MAG subjects had psychotic disorders, compared with 8% of the relatives of controls. Finally, the MAG scale appeared to be a somewhat better predictor than the PER scale in predicting psychotic and psychotic-like symptoms at follow-up.

Psychiatric Patients and Their Biological Relatives

Only one study has examined psychiatric patients who scored deviantly on PER-MAG. Satel and Edell (1991) found that PER-MAG scores were able to predict with very high sensitivity and specificity which subjects in a sample of cocaine-dependent patients will develop clinical paranoia during prolonged cocaine use.

Conclusion

The combined PER-MAG scale appears to identify an extremely deviant cohort of individuals among college students, some of whom have already displayed evidence of psychotic decompensation. Of all the scales reviewed, PER-MAG appears to hold the greatest promise as a powerful tool in identifying individuals at elevated risk for psychosis. Interestingly, remarkably little there is known about the performance of psychiatric patients and their relatives on the PER-MAG scale or MAG scale alone. Whether the combined PER-MAG scale would provide sufficient incremental validity over the PER scale alone to warrant its use in these groups awaits further empirical investigation.

Impulsive Nonconformity Scale

Brief Description

The Impulsive Nonconformity (IMP) scale was designed as a 51-item instrument, of which 30 items measure nonconformity (i.e., a lack of concern for prevailing social and ethical standards of society and of conventional mores, hostility and a lack of remorse for injuries to others, a lack of empathy for the pain of others, and an unrestrained pursuit of self-gratification) and 21 items measure impulsivity (i.e., habitual acting on impulse, difficulty in delaying any sort of gratification, lack of consideration of the consequences of behavior, and episodes of explosive, uncontrolled rage). Representative items are "I usually laugh out loud at clumsy people" (true) and "When I start out in the evening, I seldom know what I'll end up doing" (true). Internal consistency reliability coefficients for the scale have ranged from .83 (women) to .84 (men). Test–retest reliability coefficients (6 weeks) are .84 (men and women). The scale correlates −.30 with social desirability and .004 with acquiescence bias. No correlations with age or social class are reported in the original papers. Of note, the IMP scale has negligible correlations with the PAn scale (−.01 to −.05, men; .03 to .06, women), and positive correlations with the PER scale (.40 to .41, men; .43 to .44, women), the MAG scale (.45, men and women), Golden and Meehl's (1979) Schizoidia scale (.20, men; .24, women), Eysenck and Eysenck's (1975) Psychoticism scale (.65 to .68, men; .63 to .67, women), the MMPI scales F (.46, men; .57, women) and Pd (K-corrected; .26, men; .41, women), and several scales from Zuckerman's (1971) Sensation Seeking Scale (measuring disinhibition, experience seeking, and boredom susceptibility) and Jackson's (1974) PRF (measuring aggression and impulsivity) in large samples of college students.

Validity Findings

College Students

Eight published studies have examined characteristics of college students elevated on the IMP scale. In the original paper describing the scale's development (L.J. Chapman et al., 1984), high-scoring IMP subjects, as compared with nonelevated controls, reported on structured interview more antisocial behaviors, more drug and alcohol abuse, more failures to keep scheduled appointments (men only), more psychotic and psychotic-like experiences (i.e., passivity experiences and aberrant beliefts among men; auditory and visual experiences and aberrant beliefs among women), more schizotypal experiences (i.e., feeling odd or different, impulsivity, anger, depersonalization, loss of feelings, confusing dreams with reality (women), out-of-body experiences (women), and feelings of confusion (women), more major depressive disorder (men), more mania or hypomania, more other disorders of affect (cyclothymic personality, labile personality, intermittent depressive disorder), and poorer overall social adjustment (including highly conflictual relationships with family, friends [men], and schoolwork [men]). Persons scoring high on the IMP scale may be prone to both affective and schizophrenic disorders, because affective symptoms tended to co-occur with the psychotic-like and schizotypal symptoms reported. The authors suggested using IMP in conjunction with PER-MAG to identify psychosis-prone subjects to reduce false-positives, because subjects elevated on IMP alone tend not to demonstrate evidence of cognitive slippage, although they are quite deviant when their scores are combined with the PER and/or MAG scales. For example, Martin and Chapman (1982) found that IMP subjects were indistinguishable from controls on a word-communication task, although they were quite deviant when the IMP score was combined with elevated PER scores. E.N. Miller and Chapman (1983) found no differences between male IMP subjects and controls on both a structured and unstructured word-association task but found that they were deviant when combined with elevated PER-MAG scores.

Beckfield (1985) also found lower social competence of male IMP subjects compared to controls, as evidenced by more often giving hostile responses and displaying more oddness to the problematic situations. Although IMP subjects were more depressed than controls, this did not explain their lower competence. They were also less able to recognize competent responses and more likely to select hostile responses on a multiple-choice version of a problem-solving skills inventory, which was not due to group differences in verbal ability. Numbers and Chapman (1982), while finding equivalent overall skill levels on a role-playing task of social skills among women, did find female IMP subjects to be more hostile than controls.

In an article critical of the IMP scale, Peltier and Walsh (1990) found that variance attributable to response style, such as social desirability and acquiescence bias, was significantly related to scores on IMP, although the marked shrinkage in cross-validation suggested that their initial results were highly unstable (from 22.4% to 9.3% of the variance attributable to the response set predictor variables). That IMP is not merely tapping social desirability is suggested by Schuldberg's (1990) finding of a positive association between high IMP and creativity. The IMP scale was also found to be negatively correlated with ego strength, arguing that IMP reflects behavioral aspects of positive symptoms.

The strongest evidence addressing the validity of IMP as an index of psychosis proneness comes from L.J. Chapman and Chapman (1985, 1987), who interviewed college students scoring deviantly on IMP ($n = 74$) on two occasions approximately 25 months apart. During this period, IMP subjects were no more likely to have sought professional help for their problems than normal controls. Although a few IMP subjects showed poor social adjustment, no evidence of increased rates of psychotic or psychotic-like symptoms at follow-up was observed; indeed, they had declined over time. At follow-up, IMP subjects did exceed controls on drug use and (in men) alcohol use.

PSYCHIATRIC PATIENTS AND THEIR BIOLOGICAL RELATIVES

Only two published studies have examined IMP scores in psychiatric patients, and none have examined scores in their relatives. L.J. Chapman et al. (1984) found that scores on IMP in schizophrenic and schizoaffective patients exceeded normal controls for men and women. Kirkpatrick and Buchanan (1990) found no differences in IMP scores in clinically stable outpatient schizophrenics with or without the deficit syndrome (defined earlier). Finally, in an unpublished study of young, nonpsychotic psychiatric inpatients, Edell and Joy (1989) found that IMP scores were positively correlated with SCL-90 factor scores measuring hostility, paranoid ideation, and psychoticism in men and women, with somatization in men, and with obsessive–compulsive, interpersonal sensitivity, depression, and anxiety in women. IMP scores were uncorrelated with BPRS factor scores. IMP scores were also positively correlated with antisocial, borderline, and passive–aggressive personality disorder criteria for both sexes. Of the four scales examined (PER, PAn, MAG, and IMP), the IMP scale showed the least evidence of differential performance by men and women.

Conclusion

These data provide slim support for the predictive validity of IMP alone as an index of psychosis proneness, although the scale does appear to serve as a potentiator of pathology in PER and PER-MAG subjects.

Somatic Symptoms Scale

Brief Description

No published papers have described the development of the Somatic Symptoms (SOM) scale. One study using a shortened version of the scale reported a coefficient alpha coefficient of .84 (Levin & Raulin, 1991).

Validity Findings

COLLEGE STUDENTS

Two studies have examined characteristics of college students elevated on the SOM scale. Raulin and Henderson (1987) found that SOM subjects were more variable than controls in their ratings of trait relationships, failing to share semantic trait structure accepted by most normal subjects. Levin and Raulin (1991), using a 10-item version of the test, found that high SOM in females was predictive of high nightmare frequency compared to controls, suggesting a possible ego boundary disturbance.

PSYCHIATRIC PATIENTS AND THEIR BIOLOGICAL RELATIVES

No published studies have used the SOM scale in this population.

Conclusion

Insufficient data are available to evaluate the SOM scale.

Intense Ambivalence Scale

Brief Description:

The Intense Ambivalence (AMB) scale was designed as a 45-item instrument to measure the existence of simultaneous or rapidly interchangeable strong positive and negative feelings toward the same object or activity. Internal consistency reliability coefficients have ranged from .86 (men) to .88 (women). Test–retest reliability coefficient (10 to 12 weeks) is .81. The scale correlates negatively with social desirability (−.26 to −.29, men; −.24 to −.31, women) and positively with acquiescence bias (.22 to .23, men; .21 to .26, women) in large college student samples. AMB scores correlate negatively with age (range, from −.09 in psychology outpatient clinic clients to −.29 in hospitalized depressed patients) and education (range, from −.06 in psychology clinic clients to −.21 in schizophrenics) and positively with social class (range, from .14 in normal controls to .37 in hospitalized depressed patients). The AMB scale correlates positively with the PAn scale (range, from .14 in depressed inpatients to .46 in schizophrenics) and with the PER scale (range, from .38 in male college students to .63 in schizophrenics).

Validity Findings

COLLEGE STUDENTS

Three studies have examined characteristics of college students elevated on the AMB scale. In the original paper describing the scale's development, Raulin (1984) reported that college students with elevated AMB scores, when compared with nonelevated controls, on structured interview more often contradicted themselves about their feelings, more often had troubles with roommates and parents, described themselves as ambivalent, were less likely to have close friends, and had more difficulty dating.

Raulin and Henderson (1987) reported that AMB subjects were more variable than controls in their ratings of trait relationships, failing to share semantic trait structure accepted by most normal subjects. Finally, Levin and Raulin (1991), using a shortened version of the test (10 items), found that high AMB is predictive of high nightmare frequency in women.

PSYCHIATRIC PATIENTS AND THEIR BIOLOGICAL RELATIVES

Two studies have examined the AMB scale in psychiatric patients, none with their relatives. Raulin (1984) found that hospitalized depressed patients scored most deviantly on AMB, followed by schizophrenics and psychology clinic clients (who were indistinguishable), and then by normal controls. In schizophrenics, AMB scores were strongly correlated with Beck depression scores ($r = .52$), but uncorrelated with Phillips premorbid adjustment scores ($r = .04$). Acute schizophrenics were more likely to score deviantly on AMB than were chronic schizophrenics. Raulin concluded that AMB may be a transient feature of acute psychosis, particularly when depressive components are present. AMB was unrelated to performance by schizophrenics on Gorham's Proverbs Test (Carpenter, 1983).

Conclusion

Insufficient data are available to evaluate the AMB scale.

Hypomanic Personality Scale

Brief Description

The Hypomanic Personality (HYP) scale was designed as a 48-item instrument to identify upbeat, gregarious, confident, and energetic people who sometimes become maladaptively euphoric, hypersociable, grandiose, and overactive, with occasional episodic hypomanic symptoms (Eckblad & Chapman, 1986). A representative item is "I can usually slow myself down when I want to" (false). Internal consistency reliability coefficient is .87 (men and women). Test–retest reliability coefficient (15 weeks) is .81. The scale correlates. .05 with social desirability and .08 with acquiescence bias in large college student samples. The HYP scale correlates negatively with the PAn scale ($r = -.18$) and positively with the PER scale ($r = .43$), the MAG scale ($r = .49$), the IMP scale ($r = .44$), and the GBI scale ($r = .47$), which was designed to identify persons at risk for bipolar disorder. In a principal components analysis with Varimax rotation, the first two factors (measuring episodic hypomanic features and attention-seeking behavior in groups) accounted for 20.6% of the variance.

Validity Findings

COLLEGE STUDENTS

Two published studies have examined characteristics of college students with elevated HYP scores. In an interview study, Eckblad and Chapman (1986) found that high scorers on HYP, when compared with nonelevated controls, showed evidence of greater gregariousness and social activity, demonstrated greater social poise and less anxiety during the interview, and showed evidence of higher self-esteem and ambitiousness, although their actual academic achievement was not better. In addition, HYP subjects were less well-adjusted socially (e.g., had more fights with friends, were more often hurt or offended by friends, were more often lonely around people, and were more likely to call attention to themselves by acting oddly in public or dressing provocatively). HYP subjects were rated as having more elevated or speeded mood and being more restless, consistent with greater self-report of being higher in their energy and activity levels (although not higher in self-confidence or optimism). High-HYP subjects were more likely to have had poor physical health, perhaps because of their overactive life-styles. They tended to see themselves as unique and creative individuals, with special aptitudes or traits, such as strong artistic interests. Their rated overall social adjustment was poorer compared with controls, with a greater number of HYP subjects having seen a psychotherapist for evaluation or treatment of emotional problems. More than three fourths of the high-HYP subjects met criteria for hypomanic episodes (compared with none of the controls); none met criteria for a full manic syndrome. HYP subjects were also more likely to report week-long depressive episodes and to meet criteria for affective personality disorder (cyclothymic personality, intermittent personality disorder, or labile personality). Drug and alcohol use was higher in the HYP group. They were more likely to report psychotic and psychotic-like experiences and schizotypal signs. The authors argue that HYP is a valid measure of hypomanic personality and may be a useful risk indicator for bipolar disorder. Schuldberg (1990) found HYP was *positively* correlated with several tests of creativity (Jamison, 1990), but negatively correlated with ego strength, arging that HYP taps affective aspects of positive symptoms.

PSYCHIATRIC PATIENTS AND THEIR BIOLOGICAL RELATIVES

No published studies have used the HYP scale in this population.

Conclusion

Insufficient data are available to evaluate the HYP scale, although the early work with college students looks quite promising.

Salient Issues and Future Directions

Absence of Guiding Theory

Despite the impressive array of findings linking high scores on the scales reviewed to a variety of interpersonal, behavioral, cognitive, and emotional deficits, strikingly absent is any coherent or unifying theory that links the particular trait(s) studied to the later development of psychosis. For example, what are the underlying processes that account for deviant scores on each scale? And how do such processes become transformed over time in a subgroup of individuals to become severe psychopathology? Why should individuals deficient in their capacity to experience physical pleasures be at greater risk for developing psychosis than individuals who are overly and exquisitely sensitive to sensory stimulation? How does magical thinking develop into psychotic thought disorder? What are the developmental processes that protect most individuals from developing serious psychopathology despite the presence of these maladaptive traits? What neural mechanisms might account for aberrations in perceptual experience in otherwise well-functioning individuals? Meehl (1990) has recently provided an updated and highly elaborated version of his grand theory of schizotypy and schizophrenia that only begins to address some of these questions. Without such theory, it is unclear how far the current work correlating trait scores to other variables will take us. The formulation of testable developmental theories of psychosis that incorporate our current knowledge in such areas as cognition, personality, genetics, neurochemistry, and psychophysiology should be of the highest priority. More sophisticated statistical methods, such as taxometric analytic procedures and admixture analyses, may be needed to examine the underlying latent structure of these instruments (Lenzenweger & Korfine, 1992a; Lenzenweger & Moldin, 1990).

The Use of Self-Report Measures to Identify High-Risk Subjects

Many experimental psychopathologists would probably assert that the identification of individuals at risk for psychosis using self-report scales is an exercise in futility. Perhaps they are right. Because genes are generally viewed as accounting for the largest proportion of variance in the development of psychosis, any strategy that defines risk using nongenetic markers is considered inferior or inadequate. Yet it is an empirical fact that the vast majority of psychotic individuals do not have psychotic relatives, and thus, defining risk by the presence of familial psychosis may seriously jeopardize the generalizability of the findings obtained. The use of biological markers such as platelet monoamine oxidase (MAO) activity level or smooth-pursuit eye-tracking dysfunction to define risk (endophenotypic strategy) may be the next best thing, except that it is highly impractical and costly to screen large representative samples using such methods in the hopes of finding the true-risk subjects. Pragmatically speaking, as a first-stage case identification procedure, nothing beats the exophenotypic paradigm of identifying risk subjects using some behavioral index, such as the self-report "schizotypy" scales, to mass screen large groups of individuals in their late adolescence or early adulthood who may be close to the average age of onset of the disorder of interest. It offers enormous advantages of great economy and efficiency and potentially yields a more representative sample of future psychotics when compared with traditional genetic paradigms of selecting for study the biological relatives of affected probands.

Of course, a major disadvantage of this strategy is the absence of solid empirical evidence for determining which of the many possible overt behavioral dispositions to select for study (L.J. Chapman & Chapman, 1985), combined with the potential for choosing highly fallible markers that are far removed from the action of the putative etiologic gene(s) for the disorders (Depue et al., 1981). The exophenotypic strategy therefore must be viewed only as a first step in the identification of at-risk subjects. No single strategy is likely to identify a large proportion of at-risk subjects as well as an integrative strategy the defines risk by multiple criteria. For example, an investigator might choose for intensive study individuals who are high on PER-MAG, come from families displaying high expressed emotions (Leff & Vaughn, 1985), and who show concomitant impairments in sustained attention (Nuechterlein & Dawson, 1984) or in smooth-pursuit eye movements (Holzman & Levy, 1977). Another fruitful strategy is to identify subjects with a known biological relationship to a psychotic proband, and examine whether other risk factors such as attentional dysfunction and elevated schizotypy scores add predictive power in identifying individuals who ultimately break down (cf. Erlenmeyer-Kimling et al., 1993). Clearly, more work must be done examining the interrelationships of risk factors (e.g., Lenzenweger et al., 1991). Studies that use a multistep procedure for defining risk, perhaps moving from the molar to the molecular at each succeeding step, can function synergistically rather than competitively in helping to isolate the powerful variables accounting for most of the variance in the understanding and prediction of psychosis.

The Problem of False-Positives, False-Negatives, and Cutting Scores

Where one sets the cutoff for defining risk will have a major impact on the obtained rates of false positives (i.e., individuals falsely identified as being at high risk) and false-negatives (i.e., individuals falsely identified as being at low risk). Generally speaking, an increase in the rates of one is associated with a decrease in the rates of the other. A more stringent cutoff score will typically result in an increase in false-negatives and decrease in false-positives, whereas a more liberal cutoff score will do the reverse. Thus, a cost–benefit analysis must be performed to determine whether one would do less harm by erring on one side or the other. The "cost" of identifying an individual as being at "high risk" must be weighed against the "benefit" of studying that individual with the ultimate hope of reducing risk. To the extent that one has an accurate method for defining risk, the benefits are likely to outweigh the costs, both to the individual and to the society at large. With more experimental methods for which the validational data are only beginning to accrue, the situation is far more complicated. Investigators using the Wisconsin scales should exercise extreme caution in what they communicate to subjects, because any mention of psychosis proneness at this point is premature and likely to arouse unnecessary anxiety and possibly anger in individuals so identified. On the other hand, intentionally misleading subjects is unethical and inappropriate. Informing subjects that one is examining the relationship of "attitudes and experiences" to other aspects of functioning is adequate.

Most investigators using the Wisconsin scales in nonclinical populations have adopted the Chapman standard of defining deviance on a scale as any score at least 2 SD above the mean for the same-sex standardization sample and nondeviance as scores no more than .5 SD above the mean. While quite reasonable, these cutoff criteria are arbitrary and raise multiple questions. First, because of the skewness of the distributions, approximately 4% to 5% of the subjects are defined as deviant on each scale (L.J. Chapman & Chapman, 1985). Obviously, the total number of subjects identified as deviant by these criteria is greater than the total number who will become psychotic, resulting in an unknown but probably substantial number of false-positives. The separation of true-positives from false-positives in these identified groups using additional measures of risk remains of highest priority (Edell & Struckus, 1986). Second, the criterion for selecting control subjects is intended to reduce the possibility of selecting false-negatives (i.e., individuals who are high on the trait of interest but not as high as the cutoff score used to select experimental subjects) (E.N. Miller & Chapman, 1983). However, little is known about individuals who score in the "gray area" of the scales (i.e., between .5 and 2 SD above the mean). It would be fascinating to examine how individuals across the full spectrum of scores perform on the dependent variables of interest, which might be a more powerful strategy than is the traditional method of dichotomizing the sample into high- and low-risk groups for understanding the full meaning of scale scores. Balogh and Merritt (1990) have made similar arguments for the need for descriptive research using clinical and nonclinical samples aimed at identifying the type and degree of clinical symptomatology associated with specific score ranges. Evidence exists that splitting groups at the median score on the scale can yield intriguing results, at least in clinical populations where there are no previously established cutoff scores for use in classifying subjects as deviantly elevated on one or more of the scales (e.g., see Lenzenweger & Loranger, 1989a; Satel & Edell, 1991).

Third, Merritt and Balogh (1986) have argued that the 2-SD selection criterion for the PER-MAG scales in college student samples may increase the likelihood of selecting individuals who are already psychotic rather than being psychosis prone. As they note, the potentially confounding effects of acute and/or chronic disturbance on performance would then play a role in findings obtained and would make impossible the separation of state from trait characteristics. They suggest determining a specific range of PER-MAG scores to identify vulnerable, prepsychotic individuals and not those who are already psychotic. In my experience, the number of college students who are elevated on the PER-MAG scales who are actively psychotic is extremely small and unlikely to substantially bias the findings obtained. In addition, Merritt and Balogh's suggestion of identifying a range of scores that would separate vulnerable from already psychotic subjects may be impossible, given the large overlap of scores across these groups.

Finally, when selecting subjects on one scale, should one examine scores on the other Wisconsin scales or use the scales in isolation? No consensus has been reached across studies on this issue. One must consider scores on the other scales if they are available because failure to do so may inadvertently include a substantial number of deviant subjects (false-negatives) in the "normal" control group. Along that line, some evidence indicates that administering one of the scales in isolation may produce lower mean scores on that scale than when the scales are administered conjointly with their items interspersed, which may obviously alter the findings obtained (see Allen, Chapman, & Chapman, 1987).

The Role of Sociodemographic Variables

Virtually all of the validational work to date has focused on white, middle-class, academically advantaged, English-speaking college students under the age of 25. Obviously, psychosis afflicts individuals irrespective of their race, social class, primary language, or level of education. How subjects with other sociodemoographic characteristics perform on the scales is largely unknown. L.J. Chapman and Chapman (1985) have some data to suggest that groups of East Asian students score higher on PAn and that black students score higher on MAG, when compared with white students. Data also suggest that age, sex, and educational achievement may affect scores as well (Clementz et al., 1991; Lenzenweger & Loranger, 1989a; Schuck et al., 1984) and that the scales perform differently in men and women (Bornstein et al., 1988; Edell & Joy, 1989;

Erlenmeyer-Kimling et al., 1993; Mishlove & Chapman, 1985; Stone & Schuldberg, 1990). Unfortunately, such differences are often treated as error variance to be partialed out or explained away rather than as reflecting something important about the scales and perhaps the development of psychosis. There is a particularly strong need to examine the performance of these scales in young people from diverse socioeconomic backgrounds and experiences who are not attending college, because they may be at higher risk for psychopathology.

State Versus Trait Measurements

Virtually no published data examine the stability of scores on the Wisconsin scales for periods exceeding 4 months. Nor are there much data on how scale scores are affected by such variables as changes in clinical status, use of psychotropic medications, or response to psychotherapy in clinical samples. The scales are described as measuring traits rather than momentary states of the individual, but the data are generally lacking to assess the veracity of this claim.

The Puzzling Performance of the First-Degree Relatives of Psychotic Probands and of Nonpsychotic Patients on the PER and MAG Scales

A number of studies have reported no difference or even *lower* scores on the PER scale among the first-degree relatives of psychotic probands compared with the relatives of controls (Clementz et al., 1991; Grove et al., 1991; Katsanis et al., 1990). Edell and Joy (1989) found *lower* mean scores on the PER and MAG scales (but not on the PAn or IMP scales) among young, nonpsychotic psychiatric inpatients when compared with mean scores for college students. George and Neufeld (1987) found that the mean scores obtained by their mixed psychiatric inpatient control group (predominantly affective disorders) was somewhat *lower* than that obtained by the normal control group on the MAG scale. Because it is unlikely that the proportion of schizotypes or psychosis-prone individuals among the relatives of psychotic probands or among nonpsychotic psychiatric inpatients is lower than the proportion found in the relatives of the normal controls or in college students, such findings do suggest a possible problem with these scales. Katsanis et al. (1990) argued that their findings may reflect defensiveness by the relatives who knew they were selected because they were related to a psychotic patient. Edell and Joy (1989) suggested that their patients may have been reluctant to acknowledge having the more deviant experiences tapped by the PER and MAG scales, fearing that it might prolong hospitalization. Peltier and Walsh (1990) argue that the scales are simply measuring the tendency of subjects to give socially desirable responses in self-description, although this does not explain why this tendency would necessarily be greater among first-degree relatives of psychotic patients or among nonpsychotic psychiatric inpatients than among control subjects. Despite these puzzling findings, the PER and MAG scales have been remarkably successful in identifying persons with characteristics one would expect among the psychosis prone.

Reporting the Psychometric Efficiency of the Scales

Few of the published studies have reported measures of the psychometric efficiency of the scales or groups of indicators as diagnostic instruments. It would be useful for readers to be informed of the sensitivity (i.e., probability of a particular symptom or deficit given a deviant score on the psychosis-proneness scale), specificity (i.e., probability of not having a particular symptom or deficit given a nondeviant score on the psychosis-proneness scale), positive predictive power (i.e., probability of scoring deviantly on the psychosis proneness scale given the presence of a particular symptom or deficit), and negative predictive power (i.e., probability of not scoring deviantly on the psychosis-proneness scale given the absence of a particular symptom or deficit) of the scale examined. The degree of overlap in the distribution of scores among groups is also quite informative, as are the base rates of the groups examined, because these will obviously affect the psychometric efficiency levels obtained.

Creativity or Madness?

As noted earlier, Schuldberg and colleagues (1988) have found positive relationships between scores on some of the Wisconsin scales and measures of creativity. This raises the intriguing possibility that high scores may have widely divergent implications for different individuals. For example, subjects high on PER-MAG and IMP have been noted to show the highest levels of cognitive slippage (E.N. Miller & Chapman, 1983) and the highest levels of creativity (Schuldberg et al., 1988). The overlap between subjects high on both cognitive slippage and creativity in the PER-MAG group is unknown. These data suggest that one should consider examining both the strengths and weaknesses in subjects selected for scoring deviantly on the scales.

Concluding Comment

Over 15 years have passed since the initial publication of the first scales of psychosis proneness. The preliminary evidence to date has been encouraging for some but not all of the instruments studied. Long-term follow-up studies of individuals identified as scoring deviantly on at least some of these putative indicators of psychosis proneness are clearly needed to provide the strongest and ultimate validational evidence for their efficacy. A few suggestions have been offered regarding directions future research might pursue. The efficiency and value of the scales as first-step screening tools have been emphasized, distinct from the mixed evidence to date for their utility as stand-alone diagnostic instruments. Although the scales have not as yet yielded the bountiful fruit hoped for, they still may play an important role in helping to unravel the mysteries of the major psychoses.

References

Allen, J., & Schuldberg, D. (1989). Positive thought disorder in a hypothetically psychosis-prone population. *Journal of Abnormal Psychology, 98*, 491–494.

Allen, J.J., Chapman, L.J., & Chapman, J.P. (1987). Cognitive slippage and depression in hypothetically psychosis-prone college students. *Journal of Nervous and Mental Disease, 175*, 347–353.

Allen, J.J., Chapman, L.J., Chapman, J.P., Vuchetich, J.P., & Frost, L.A. (1987). Prediction of psychoticlike symptoms in hypothetically psychosis-prone college students. *Journal of Abnormal Psychology, 96*, 83–88.

American Psychiatric Association. (1980). *Diagnostic and statistical manual of mental disorders* (3rd ed.). Washington, DC: Author.

Asarnow, R.F., Nuechterlein, K.H., & Marder, S.R. (1983). Span of apprehension performance, neuropsychological functioning, and indices of psychosis-proneness. *Journal of Nervous and Mental Disease, 171*, 662–669.

Balogh, D.W., & Merritt, R.D. (1985). Susceptibility to Type A backward pattern masking among hypothetically psychosis-prone college students. *Journal of Abnormal Psychology, 94*, 377–383.

Balogh, D.W., & Merritt, R.D. (1990). Accounting for schizophrenics' magical ideation scores: Are college-student norms relevant? *Psychological Assessment: A Journal of Consulting and Clinical Psychology, 2*, 326–328.

Bateson, G., Jackson, D.D., Haley, J., & Weakland, J. (1956). Toward a theory of schizophrenia. *Behavioral Science, 1*, 251–264.

Beck, A.T. (1967). *Depression: Clinical, experimental and theoretical aspects*. New York: Hoeber.

Beckfield, D.F. (1985). Interpersonal competence among college men hypothesized to be at risk for schizophrenia. *Journal of Abnormal Psychology, 94*, 397–404.

Bentall, R.P., Claridge, G.S., & Slade, P.D. (1989). The multidimensional nature of schizotypal traits: A factor analytic study with normal subjects. *British Journal of Clinical Psychology, 28*, 363–375.

Berenbaum, H., Snowhite, R., & Oltmanns, T.F. (1987). Anhedonia and emotional responses to affect evoking stimuli. *Psychological Medicine, 17*, 677–684.

Bernstein, A.S., & Riedel, J.A. (1987). Psychophysiological response patterns in college students with high physical anhedonia: Scores appear to reflect schizotypy rather than depression. *Biological Psychiatry, 22*, 829–847.

Bornstein, R.F., O'Neill, R.M., Galley, D.J., Leone, D.R., & Castrianno, L.M. (1988). Body image aberration and orality. *Journal of Personality Disorders, 2,* 315–322.

Carpenter, B.N. (1983). Relationship of scales of schizophrenia proneness and premorbid adjustment to thinking deficits in schizophrenia. *Journal of Clinical Psychology, 39,* 311–315.

Chapman, J. (1966). The early symptoms of schizophrenia. *British Journal of Psychiatry, 112,* 225–251.

Chapman, J.P., & Chapman, L.J. (1987). Handedness of hypothetically psychosis-prone subjects. *Journal of Abnormal Psychology, 96,* 89–93.

Chapman, J.P., & Chapman, L.J. (1992, November). *Follow-up study of psychosis-prone subjects.* Paper presented at the annual meeting of the Society for Research in Psychopathology, Palm Springs, CA.

Chapman, L.J., & Chapman, J.P. (1978). *Revised physical anhedonia scale.* Unpublished test.

Chapman, L.J., & Chapman, J.P. (1980). Scales for rating psychotic and psychotic-like experiences as continua. *Schizophrenia Bulletin, 6,* 476–489.

Chapman, L.J., & Chapman, J.P. (1985). Psychosis proneness. In M. Alpert (Ed.), *Controversies in schizophrenia: Changes and constancies* (pp. 157–174). New York: Guilford Press.

Chapman, L.J., & Chapman, J.P. (1987). The search for symptoms predictive of schizophrenia. *Schizophrenia Bulletin, 13,* 497–503.

Chapman, L.J., Chapman, J.P., & Miller, E.N. (1982). Reliabilities and intercorrelations of eight measures of proneness to psychosis. *Journal of Consulting and Clinical Psychology, 50,* 187–195.

Chapman, L.J., Chapman, J.P., Numbers, J.S., Edell, W.S., Carpenter, B.N., & Beckfield, D. (1984). Impulsive nonconformity as a trait contributing to the prediction of psychotic-like and schizotypal symptoms. *Journal of Nervous and Mental Disease, 172,* 681–691.

Chapman, L.J., Chapman, J.P., & Raulin, M.L. (1976). Scales for physical and social anhedonia. *Journal of Abnormal Psychology, 85,* 374–382.

Chapman, L.J., Chapman, J.P., & Raulin, M.L. (1978). Body-image aberration in schizophrenia. *Journal of Abnormal Psychology, 87,* 399–407.

Chapman, L.J., Chapman, J.P., Raulin, M.L., & Edell, W.S. (1978). Schizotypy and thought disorder as a high risk approach to schizophrenia. In G. Serban (Ed.), *Cognitive deficits in the development of mental illness* (pp. 351–360). New York: Brunner/Mazel.

Chapman, L.J., Edell, W.S., & Chapman, J.P. (1980). Physical anhedonia, perceptual aberration, and psychosis proneness. *Schizophrenia Bulletin, 6,* 639–653.

Claridge, G., & Broks, P. (1984). Schizotypy and hemisphere function: I. theoretical considerations and the measurement of schizotypy. *Personality and Individual Differences, 5,* 633–649.

Clementz, B.A., Grove, W.M., Katsanis, J., & Iacono, W.G. (1991). Psychometric detection of schizotypy: Perceptual aberration and physical anhedonia in relatives of schizophrenics. *Journal of Abnormal Psychology, 100,* 607–612.

Cook, M., & Simukonda, F. (1981). Anhedonia and schizophrenia. *British Journal of Psychiatry, 139,* 523–525.

Depue, R.A., Slater, J.T., Wolfstetter-Kausch, H., Klein, D., Goplerud, E., & Farr, D. (1981). A behavioral paradigm for identifying persons at risk for bipolar depressive disorders: A conceptual framework and five validation studies. *Journal of Abnormal Psychology, 90*, 381-437.

Eckblad, M., & Chapman, L.J. (1983). Magical ideation as an indicator of schizotypy. *Journal of Consulting and Clinical Psychology, 51*, 215-225.

Eckblad, M., & Chapman, L.J. (1986). Development and validation of a scale for hypomanic personality. *Journal of Abnormal Psychology, 95*, 214-222.

Eckblad, M., Chapman, L.J., Chapman, J.P., & Mishlove, M. (1982). *The revised Social Anhedonia Scale*. Unpublished scale, University of Wisconsin-Madison.

Edell, W.S. (1987). Role of structure in disordered thinking in borderline and schizophrenic disorders. *Journal of Personality Assessment, 51*, 23-41.

Edell, W.S., & Chapman, L.J. (1979). Anhedonia, perceptual aberration, and the Rorschach. *Journal of Consulting and Clinical Psychology, 47*, 377-384.

Edell, W.S., & Joy, S.P. (1989, November). *Psychosis proneness in young hospitalized psychiatric patients*. Paper presented at the annual meeting of the Society for Research in Psychopathology, Coral Gables, FL.

Edell, W.S., & Kaslow, N.J. (1991). Parental perception and psychosis proneness in college students. *Americal Journal of Family Therapy, 19*, 195-205.

Edell, W.S., & Struckus, J.E. (1986, October). *The problem of false positives in high risk research*. Paper presented at the first annual meeting of the Society for Research in Psychopathology, Cambridge, MA.

Erlenmeyer-Kimling, L., Cornblatt, B.A., Rock, D., Roberts, S., Bell, M., & West, A. (1993). The New York High Risk Project: Anhedonia, attentional deviance, and adult psychopathology *Schizophrenia Bulletin., 19*, 141-153.

Eysenck, H.J., & Eysenck, S.B.G. (1975). *Manual of the Eysenck personality ouestionnaire*. London: Hodder and Stoughton.

Fawcett, J., Clark, D.C., Scheftner, W.A., & Gibbons, R.D. (1983). Assessing anhedonia in psychiatric patients: The Pleasure Scale. *Archives of General Psychiatry, 40*, 79-84.

Fenichel, O. (1945). *The psychoanalytic theory of neurosis*. New York: W.W. Norton.

Frost, L.A., & Chapman, L.J. (1987). Polymorphous sexuality as an indicator of psychosis proneness. *Journal of Abnormal Psychology, 96*, 299-304.

Fujioka, T.A., & Chapman, L.J. (1984). Comparison of the 2-7-8 MMPI profile and the perceptual aberration-magical ideation scale in identifying hypothetically psychosis-prone college students. *Journal of Consulting and Clinical Psychology, 52*, 458-467.

Garmezy, N., & Streitman, S. (1974). Children at risk: The search for antecedents of schizophrenia: Part I. Conceptual models and research methods. *Schizophrenia Bulletin, 8*, 14-90.

Garnet, K., Glick, M., & Edell, W.S. (1993). Anhedonia and premorbid competence in young, nonpsychotic psychiatric inpatients. *Journal of Abnormal Psychology, 102*, 580-583.

George, L., & Neufeld, R.W.J. (1987). Magical ideation and schizophrenia. *Journal of Consulting and Clinical Psychology, 55*, 778-779.

Gillies, H. (1958). The clinical diagnosis of early schizophrenia. In T.F. Rodger, R.M. Mowbray, & J.R. Roy (Eds.), *Topics in psychiatry* (pp. 47-56). London, Cassell.

Golden, R.R., & Meehl, P.E. (1979). Detection of the schizoid taxon with MMPI indicators. *Journal of Abnormal Psychology, 88*, 217–233.
Grove, W.M. (1982). Psychometric detection of schizotypy. *Psychological Bulletin, 92*, 27–38.
Grove, W.M., Lebow, B.S., Clementz, B.A., Cerri, A., Medus, C., & Iacono, W.G. (1991). Familial prevalence and coaggregation of schizotypy indicators: A multitrait family study. *Journal of Abnormal Psychology, 100*, 115–121.
Haberman, M.C., Chapman, L.J., Numbers, J.S., & McFall, R.M. (1979). Relation of social competence to scores on two scales of psychosis proneness. *Journal of Abnormal Psychology, 88*, 675–677.
Hartmann, E., Milofsky, E., Vaillant, G., Oldfield, M., Falke, R., & Ducey, C. (1984). Vulnerability to schizophrenia: Prediction of adult schizophrenia using childhood information. *Archives of General Psychiatry, 41*, 1050–1056.
Hoch, P., & Cattell, J. (1959). The diagnosis of pseudoneurotic schizophrenia. *Psychiatric Quarterly, 33*, 17–43.
Holzman, P.S., & Levy, D.L. (1977). Smooth-pursuit eye movements and functional psychoses: A review. *Schizophrenia Bulletin, 3*, 15–27.
Jackson, D.N. (1970). A sequential system for personality scale development. In C.N. Spielberger (Ed.), *Current topics in clinical and community psychology* (Vol. 2, pp. 61–69). New York: Academic Press.
Jackson, D.N. (1971). The dynamics of structured personality tests. *Psychological Review, 78*, 229–248.
Jackson, D.N. (1974). *Manual for the personality research form*. Goshen, NY: Research Psychologists Press.
Jackson, D.N. (1984). *Personality research form manual* (3rd ed.). Port Huson, MI: Research Psychologists Press.
Jamison, K.R. (1990). Manic-depressive illness and accomplishment: Creativity, leadership, and social class. In F.K. Goodwin & K.R. Jamison (Eds.), *Manic-depressive illness*. New York: Oxford University Press.
Josiassen, R.C., Shagass, C., Roemer, R.A., & Straumanis, J.J. (1985). Attention-related effects on somatosensory evoked potentials in college students at high risk for psychopathology. *Journal of Abnormal Psychology, 94*, 507–518.
Jutai, J.W. (1989). Spatial attention in hypothetically psychosis-prone college students. *Psychiatry Research, 27*, 207–215.
Katsanis, J., Iacono, W.G., & Beiser, M. (1990). Anhedonia and perceptual aberration in first-episode psychotic patients and their relatives. *Journal of Abnormal Psychology, 99*, 202–206.
Katsanis, J., Iacono, W.G., Beiser, M., & Lacey, L. (1992). Clinical correlates of anhedonia and perceptual aberration in first-episode patients with schizophrenia and affective disorder. *Journal of Abnormal Psychology, 101*, 184–191.
Kendler, K.S., & Hewitt, J. (1992). The structure of self-report schizotypy in twins. *Journal of Personality Disorders, 6*, 1–17.
Kirkpatrick, B., & Buchanan, R.W. (1990). Anhedonia and the deficit syndrome of schizophrenia. *Psychiatry Research, 31*, 25–30.
Leak, G. (1991). An examination of the construct validity of the social anhedonia scale. *Journal of Personality Assessment, 56*, 84–95.
Leff, J., & Vaughn, C. (1985). *Expressed emotion in families: Its significance for mental illness*. New York: Guilford Press.

Lenzenweger, M.F., Cornblatt, B.A., & Putnick, M. (1991). Schizotypy and sustained attention. *Journal of Abnormal Psychology, 100*, 84–89.

Lenzenweger, M.F., & Korfine, L. (1992a). Confirming the latent structure and base rate of schizotypy: A taxometric analysis. *Journal of Abnormal Psychology, 101*, 567–571.

Lenzenweger, M.F., & Korfine, L. (1992b). Identifying schizophrenia-related personality disorder features in a nonclinical population using a psychometric approach. *Journal of Personality Disorders, 6*, 256–266.

Lenzenweger, M.F., & Loranger, A.W. (1989a). Detection of familial schizophrenia using a psychometric measure of schizotypy. *Archives of General Psychiatry, 46*, 902–907.

Lenzenweger, M.F., & Loranger, A.W. (1989b). Psychosis proneness and clinical psychopathology: Examination of the correlates of schizotypy. *Journal of Abnormal Psychology, 98*, 3–8.

Lenzenweger M.F., & Moldin, S.O. (1990). Discerning the latent structure of hypothetical psychosis proneness through admixture analysis. *Psychiatry Research, 33*, 243–257.

Levin, R., & Raulin, M.L. (1991). Preliminary evidence for the proposed relationship between frequent nightmares and schizotypal symptomatology. *Journal of Personality Disorders, 5*, 8–14.

Lutzenberger, W., Elbert, T., Rockstroh, B., Birbaumer, N., & Stegagno, L. (1981). Slow cortical potentials in subjects with high or low scores on a questionnaire measuring physical anhedonia and body image distortion. *Psychophysiology, 18*, 371–380.

Martin, E.M., & Chapman, L.J. (1982). Communication effectiveness in psychosis-prone college students. *Journal of Abnormal Psychology, 91*, 420–425.

Mednick, S.A., & McNeil, T.F. (1968). Current methodology in research on the etiology of schizophrenia: Serious difficulties which suggest the use of the high-risk group method. *Psychological Bulletin, 70*, 681–693.

Meehl, P.E. (1962). Schizotaxia, schizotypy, schizophrenia. *American Psychologist, 17*, 827–838.

Meehl, P.E. (1964). *Manual for use with check-list of schizotypic signs.* Minneapolis, MN: University of Minnesota Medical School, Psychiatric Research Unit.

Meehl, P.E. (1989). Schizotaxia revisited. *Archives of General Psychiatry, 46*, 935–944.

Meehl, P.E. (1990). Toward an integrated theory of schizotaxia, schizotypy, and schizophrenia. *Journal of Personality Disorders, 4*, 1–99.

Merritt, R.D., & Balogh, D.W. (1986). Methodological and interpretive constraints resulting from the use of the per-mag classification in high-risk information-processing research. *Journal of Nervous and Mental Disease, 174*, 42–43.

Merritt, R.D., & Balogh, D.W. (1989). Backward masking spatial frequency effects among hypothetically schizotypal individuals. *Schizophrenia Bulletin, 15*, 573–583.

Miller, E.N., & Chapman, L.J. (1983). Continued word association in hypothetically psychosis-prone college students. *Journal of Abnormal Psychology, 92*, 468–478.

Miller, G.A. (1986). Information processing deficits in anhedonia and perceptual aberration: A psychophysiological analysis. *Biological Psychiatry, 21*, 100–115.

Miller. G.A., Simons, R.F., & Lang, P.J. (1984). Electrocortical measures of information processing deficit in anhedonia. *Annals of the New York Academy of Sciences, 425*, 598–602.

Mishlove, M., & Chapman, L.J. (1985). Social anhedonia in the prediction of psychosis proneness. *Journal of Abnormal Psychology, 94*, 384–396.

Mo, S.S., & Chavez, M.R. (1986). Perceptual aberration and brain hemisphere reversal of foreperiod effect on time estimation. *Journal of Clinical Psychology, 42*, 787–792.

Morey, L.C., Waugh, M.H., & Blashfield, R.K. (1985). MMPI scales for DSM-III personality disorders: Their derivation and correlates. *Journal of Personality Assessment, 49*, 245–251.

Muntaner, C., Garcia-Sevilla, L., Fernandez, A., & Torrubia, R. (1988). Personality dimensions, schizotypal and borderline personality traits and psychosis proneness. *Personality and Individual Differences, 9*, 257–268.

Nuechterlein, K.H., & Dawson, M.E. (1984). Information processing and attentional functioning in the developmental course of schizophrenic disorders. *Schizophrenia Bulletin, 10*, 160–203.

Numbers, J.S., & Chapman, L.J. (1982). Social deficits in hypothetically psychosis-prone college women. *Journal of Abnormal Psychology, 91*, 255–260.

Overby, L.A., III. (1992). Perceptual asymmetry in psychosis-prone college students: Evidence for left-hemisphere overactivation. *Journal of Abnormal Psychology, 101*, 96–103.

Peltier, B.D., & Walsh, J.A. (1990). An investigation of response bias in the Chapman scales. *Educational and Psychological Measurement, 50*, 803–815.

Penk, W.E., Carpenter, J.C., & Rylee, K.E. (1979). MMPI correlates of social and physical anhedonia. *Journal of Consulting and Clinical Psychology, 47*, 1046–1052.

Peterson, C.A., & Knudson, R.M. (1983). Anhedonia: A construct validation approach. *Journal of Personality Assessment, 47*, 539–551.

Rado, S. (1956). *Psychoanalysis of behavior.* New York: Grune & Stratton.

Raine, A., & Allbutt, J. (1989). Factors of schizoid personality. *British Journal of Clinical Psychology, 28*, 31–40.

Raulin, M.L. (1984). Development of a scale to measure intense ambivalence. *Journal of Consulting and Clinical Psychology, 52*, 63–72.

Raulin, M.L., Chapman, L.J., & Chapman, J.P. (1978). *Somatic symptoms scale.* Unpublished scale. (Available from Michael Raulin, Psychology Department, SUNYAB, Buffalo, NY 14260).

Raulin, M.L., & Henderson, C.A. (1987). Perception of implicit relationships between personality traits by schizotypic college subjects: A pilot study. *Journal of Clinical Psychology, 43*, 463–467.

Satel, S.L., & Edell, W.S. (1991). Cocaine-induced paranoia and psychosis proneness. *American Journal of Psychiatry, 148*, 1708–1711.

Schuck, J., Leventhal, D., Rothstein, H., & Irizarry, V. (1984). Physical anhedonia and schizophrenia. *Journal of Abnormal Psychology, 93*, 342–344.

Schuldberg, D. (1988). Abstract: Perceptual-cognitive and affective components of schizotaxia and creativity in a group of college males. *Journal of Creative Behavior, 22*, 73–74.

Schuldberg, D. (1990). Schizotypal and hypomanic traits, creativity, and psychological health. *Creativity Research Journal, 3*, 219-231.

Schuldberg, D., French, C., Stone, B.L., & Heberle, J. (1988). Creativity and schizotypal traits: Creativity test scores, perceptual aberration, magical ideation, and impulsive nonconformity. *Journal of Nervous and Mental Disease, 176*, 648-657.

Schuldberg, D., & London, A. (1989). Psychological differentiation and schizotypal symptoms: Negative results with the Group Embedded Figures Test. *Perceptual and Motor Skills, 68*, 1219-1226.

Silverstein, S.M., Raulin, M.L., Pristach, E.A., & Pomerantz, J.R. (1992). Perceptual organization and schizotypy. *Journal of Abnormal Psychology, 101*, 265-270.

Simons, R.F. (1981). Electrodermal and cardiac orienting in psychometrically defined high-risk subjects. *Psychiatry Research, 4*, 347-356.

Simons, R.F. (1982). Physical anhedonia and future psychopathology: An electrocortical continuity? *Psychophysiology, 19*, 433-441.

Simons, R.F., & Katkin, W. (1985). Smooth pursuit eye movements in subjects reporting physical anhedonia and perceptual aberrations. *Psychiatry Research, 14*, 275-289.

Simons, R.F., MacMillan, F.W., III, & Ireland, F.B. (1982a). Anticipatory pleasure deficit in subjects reporting physical anhedonia: Slow cortical evidence. *Biological Psychology, 14*, 297-310.

Simons, R.F., MacMillan, F.W., III, & Ireland, F.B. (1982b). Reaction-time crossover in preselected schizotypic subjects. *Journal of Abnormal Psychology, 91*, 414-419.

Stone, B.L., & Schuldberg, D. (1990). Perceptual aberration, magical ideation, and the Jackson personality research form. *Journal of Nervous and Mental Disease, 178*, 396.

Strauss, J.S. (1969). Hallucinations and delusions as points on continua function. *Archives of General Psychiatry, 21*, 581-586.

Ward, P.B., Catts, S.V., Armstrong, M.S., & McConaghy, N. (1984). P300 and psychiatric vulnerability in university students. *Annals of the New York Academy of Sciences, 425*, 645-652.

Watson, C.G., Klett, W.G., & Lorei, T.W. (1970). Toward an operational definition of anhedonia. *Psychological Reports, 26*, 371-376.

Weissman, M.M., & Paykel, E.S. (1974). *The depressed woman.* Chicago: University of Chicago Press.

Yee, C.M., Deldin, P.J., & Miller, G.A. (1992). Early stimulus processing in dysthymia and anhedonia. *Journal of Abnormal Psychology, 101*, 230-233.

Yehuda, R., Edell, W.S., & Meyer, J.S. (1987). Platelet MAO activity and psychosis proneness in college students. *Psychiatry Research, 20*, 129-142.

Zubin, J., Magaziner, J., & Steinhauer, S.R. (1983). The metamorphosis of schizophrenia: From chronicity to vulnerability. *Psychological Medicine, 13*, 551-571.

Zubin, J., & Spring, B. (1977). Vulnerability—A new view of schizophrenia. *Journal of Abnormal Psychology, 86*, 103-126.

Zuckerman, M. (1971). Dimensions of sensation seeking. *Journal of Consulting and Clinical Psychology, 36*, 45-52.

2
Compromised Performance and Abnormal Psychophysiology Associated With the Wisconsin Scales of Psychosis Proneness

LEYAN O.L. FERNANDES AND GREGORY A. MILLER

The debilitating effects and poor recovery rates associated with schizophrenia encourage the development of valid and reliable methods for early identification of and intervention with individuals at risk for it. As Edell discussed in Chapter 1, reliable assessment of vulnerability to schizophrenia is not possible at present, and additional work is needed to develop adequate etiological models of psychosis (see also Knight, 1984, 1992, 1993; Lenzenweger, 1993). Thus, multiple assessment strategies are warranted to maximize the opportunity for developing sensitive and specific measures of psychosis vulnerability. In addition to this screening and diagnostic goal, multiple measurement may prove valuable for understanding the specific mechanisms involved in psychosis vulnerability and their possible role in schizophrenia. This chapter reviews evidence for behavioral performance and psychophysiological anomalies in subjects potentially at risk for psychosis on the basis of psychometric criteria, in the service of understanding and preventing severe psychosis.

The Chapman Research Program on Psychosis Proneness

Chapman and Chapman (1985, 1987) at the University of Wisconsin developed a series of self-report scales with strong psychometric properties as a means of identifying subjects potentially vulnerable to psychosis. Validation studies have demonstrated a variety of symptom, familial, communication skills, social skills, and clinical test findings for nonpatient individuals selected with these scales (Edell, Chapter 1; Chapman & Chapman, 1985, 1987, 1992; Chapman, Chapman, Kwapil, Eckblad, & Zinser, 1994; Edell & Kaslow, 1991; Grove, 1982; Lenzenweger, 1993; Levin & Raulin, 1991; Rockstroh, Elbert, Birbaumer, & Lutzenberger, 1982). Furthermore, studies of patients and their relatives have confirmed that scores on these scales are elevated or are associated with significant symptoms in clinical samples and their affected relatives (Clementz, Grove, Katsanis, & Iacono, 1991; Edell & Joy, 1989; Grove et al., 1991; Katsanis, Iacono, & Beiser, 1990; Lenzenweger & Loranger, 1989a, 1989b; Lutzenberger, Birbaumer, Rockstroh, & Elbert, 1983). In a review of schizotypy assessment methods, Lenzenweger (1993) concluded that these scales were among the best validated.

We discuss research on two distinct populations defined by the scales as having subclinical characteristics that may predispose them to schizophrenia or other forms of psychopathology. The most thoroughly studied group, anhedonics, are individuals who report a substantial deficit in the ability to experience physical pleasure (Chapman & Chapman, 1978; Chapman, Chapman, & Raulin, 1976). Anhedonia has long been suspected as a basic defect in schizophrenia (e.g., Meehl, 1962, 1990; for review, see Lenzenweger, 1993). Bernstein (1987, p. 626) termed it a "quintessential 'negative' feature" of schizophrenia. Two other scales, aimed at other possible core features of schizophrenia, are often used in combination: perceptual aberration (Chapman, Chapman, & Raulin, 1978) and magical ideation (Eckblad & Chapman, 1983). These two scales correlate highly with each other but have small, negative correlations with anhedonia. Of interest is that no attempt was made during scale development to make them independent. Thus, the populations identified by these scales may face different etiological trajectories or may be vulnerable to different types of stressors. Comparison of these populations may provide an opportunity to address the heterogeneity of schizophrenia. Indeed, various findings suggest that anhedonics exhibit some of the negative symptoms of schizophrenia such as withdrawal and flat affect, whereas the perceptual aberration-magical ideation group (commonly referred to as PER-MAGs or simply as perceptual aberrators) is more likely to demonstrate mild to moderate versions of positive symptoms such as hallucinations (e.g., Bernstein, 1987; Chapman & Chapman, 1985, 1987; Öhman, 1981; Zahn, 1988). At least one study suggested that high scorers on perceptual aberration and magical ideation should not be treated as a single population (Kendler & Hewitt, 1992). Their recommendation was based on a large sample of monozygotic and dizygotic twins, with the two scales apparently having different relationships to genetic factors. Most studies to date, however, have not distinguished the scales.

When examining how the symptoms observed in these groups relate to schizophrenia, we should consider whether such symptoms are unique to schizophrenia or are characteristic of more than one disorder. To date, the picture is mixed for the Chapman scales. Anhedonia is often prominent in depression (Lutzenberger et al., 1983) and has been proposed as a core feature of some types of depression (Fawcett, Clark, Scheftner, & Hedeker, 1983; Klein, 1974) as well as schizophrenia. Furthermore, follow-up data have not yet established that anhedonics are at elevated risk for severe psychopathology, although on nearly all of the outcome measures in the one available long-term prospective study, anhedonics' mean scores indicated more psychopathology than controls' (Chapman et al., 1994). In contrast, high scorers on the perceptual aberration scale clearly appear to be at risk for both mood disorders and schizophrenia (Chapman et al., 1994). These initial uncertainties about the clinical sensitivity and specificity of the scales make it important to establish the sensitivity and specificity of any laboratory findings obtained for subjects selected with the Chapman scales.

Multiple Measures and Convergent Validity

Because it can often be difficult to detect milder symptoms, precursors, or subsyndromal features in a vulnerable population, it is of utmost importance to examine multiple measures of phenomena to provide convergent validity for specific findings. Although the traditional clinical interview may provide useful information about subjects who score high on the Chapman scales, examining these subjects' behavioral performance on laboratory tasks can provide convergent validity for our understanding of these groups. For example, examining specific processes "on-line" with psychophysiological measurements provides a very different window from self-report interviews. Indeed, the importance of analyzing deviant processes themselves, rather than merely the overt output of these processes, has been emphasized as a means of differentiating psychopathology subgroups (Neale, 1982). More recently, Frith asserted: "the starting point must be the symptoms of schizophrenia and not the diagnostic category of schizophrenia.... It is not the symptoms themselves that we should be examining, however, but the cognitive abnormalities that underlie them" (1991, p. 28). Psychophysiological investigation of information processing is particularly well suited for studying cognitive processes insofar as it offers unobtrusive (at least in some tasks), immediate, multidimensional measurement. Intermediate processes can be examined, thus going beyond sole reliance on final behavioral output (Coles, 1989). This is particularly valuable for the study of individuals at risk, because their overt behavior may be normal. Indeed, at-risk subjects often achieve normal behavioral performance by resorting to abnormal task strategies (Giese-Davis, Miller, & Knight, 1993; G.A. Miller, 1986; Yee & Miller, 1988, 1994).

Accordingly, this chapter provides the first comprehensive review of laboratory cognitive performance in individuals identified with the Chapman scales (see Edell, Chapter 1, for a more detailed examination of the scales themselves and a review of self-report and clinical evidence of their reliability and validity; see also Grove, 1982, and Lenzenweger, 1993, for selective reviews). In addition, we summarize and update a recent review of the available psychophysiological literature on such individuals (G.A. Miller & Yee, 1994). In combination with the more clinical review provided by Edell in Chapter 1, we hope to provide a thorough portrayal of these individuals. We hope that these chapters will encourage future studies that will establish empirically derived, reliable predictors of psychosis. Ideally, such studies will eventually contribute to theoretical models, etiology, and prevention of psychosis.

Behavioral Performance Evidence for a Processing Deficit in Anhedonia

Individuals who score high on the Chapman scales tend to perform normally on a variety of laboratory tasks, consistent with their ability to function adequately in daily life. In fact, none of the many psychophysiological studies we located observed group differences in performance. However, several laboratory studies not involving psychophysiological measurement observed processing deficits in these individuals.

Reaction Time Crossover

Simons, MacMillan, and Ireland (1982b) examined the reaction-time (RT) crossover effect in a sample of anhedonic subjects as identified by the Chapman scales. The presence of an RT crossover effect in schizophrenia and its absence in normal control subjects is well documented (see Nuechterlein, 1977, for a review). Typically, normal control subjects display faster RTs to fixed-length preparatory intervals than to variable-length intervals. Schizophrenics, however, display faster RTs to fixed-length intervals only when these preparatory intervals are relatively short. As these fixed intervals are lengthened, schizophrenics display slower RTs. Simons et al. (1982b) found that anhedonics, like schizophrenics, demonstrated this RT crossover effect. In addition, they found slower absolute RTs for anhedonics than those for either normal control subjects or perceptual aberrators, although this latter difference was not statistically significant (see also Simons, MacMillan, & Ireland, 1982a). The RT crossover effect has since been replicated by Drewer and Shean (1993) in a new sample of anhedonic subjects identified by the Chapman scales.

Visual Processing

Visual backward-masking deficits similar to those demonstrated in schizophrenia (reviewed by Balogh & Merritt, 1987; Merritt & Balogh, 1989) have also been examined in anhedonic subjects. In these studies, the presentation of a "masking" stimulus after the onset of a target is found to interfere with the processing of the target. Typically, schizophrenics identify fewer correct targets than normal control subjects. Anhedonics similarly have been found to correctly identify fewer target stimuli than control subjects. As in the preliminary work with RT crossover, this finding suggests a possible deficit in speed of information transfer for anhedonics (Balogh & Merritt, 1985). Further evidence for such a deficit comes from a recent study by Wilkins and Venables (1992). They found that RTs were longer for anhedonics than for control subjects when subjects were asked to switch their attention between auditory and visual modalities. Additional study is warranted to assess the generality of such findings across paradigms and stimulus modalities.

In addition to possible deficits in speed of information transfer among anhedonics, Jutai (1989) suggested that anhedonics exhibit evidence of a lateralized cerebral hemisphere abnormality. Jutai (1989) examined the visual search patterns of subjects who scored high on the Chapman scales using four cancellation tasks consisting of structured and unstructured arrays of verbal and nonverbal stimuli. He found that psychosis-prone subjects displayed erratic visual search patterns, similar to right-brain-damaged patients, and that anhedonic subjects relied more on erratic search strategies than did other subjects. This finding was particularly strong when the task consisted of structured verbal stimuli and random nonverbal stimuli. Jutai concluded that such erratic visual search patterns were evidence for a disturbance in right-hemisphere attentional control among anhedonic subjects.

Although these studies have found preliminary evidence for information processing deficits in anhedonics, other visual perception studies have failed to find fundamental group performance differences for anhedonics. For example, Silverstein, Raulin, Pristach, and Pomerantz (1992) examined the perceptual organization processes of anhedonics in a series of studies designed to further understand visual information processing in these subjects. Each study used a different paradigm that focused on early stimulus evaluation in figure–ground discrimination. These studies found no performance differences between anhedonics and controls on tasks that required preattentive visual information processing. Similarly, Schuldberg and London (1989) found that anhedonics performed similarly to controls on the Group Embedded Figures Test, a perceptual differentiation test.

Emotional Processing

Because anhedonia is typically associated with a diminished capacity to experience pleasure, the emotional processing of anhedonics is an area of research in which one might expect to find performance differences between anhedonics and controls. In a study of emotional stimulus processing, Berenbaum, Snowhite, and Oltmanns (1987) examined whether Chapman-selected anhedonics would differ from nonanhedonics in their emotional responsivity. Anhedonics and nonanhedonics did not differ in self-report of emotional experience or in emotional facial expressions in response to a film clip.

Communication and Social Skills

Because uncommon word usage has long been associated with thought disorder (specifically cognitive slippage) and schizophrenia (Meehl, 1962), several laboratory studies have examined the communication skills of subjects selected with the Chapman scales in an effort to support the validity of these scales. On the Rosenberg-Cohen word communication task, believed to be a measure of schizophrenic failure to self-edit utterances to facilitate communication, anhedonics performed similarly to control subjects (Martin & Chapman, 1982). In addition, no performance differences were found between anhedonics and normal controls on either the Kent-Rosanoff Word Association Test or Word Halo Test used as measures of thought disorder (E.N. Miller & Chapman, 1983; Ward, McConaghy, & Catts, 1991).

However, the communication of anhedonics has often been found to be accompanied by social skills deficits and inappropriate social behavior (Beckfield, 1985; Haberman, Chapman, Numbers, & McFall, 1979; Numbers & Chapman, 1982; reviewed by Edell, Chapter 1). Anhedonics have also been found to report greater variance than controls in their ratings of social words (Raulin & Henderson, 1987) and to describe their family of origin as having a more deviant social structure (Perosa & Simons, 1991). Such deficits in social functioning are consistent with Jutai's (1989) hypothesis of right-hemisphere dysfunction in anhedonics.

Summary

The behavioral performance data on processing deficits in anhedonia show enough inconsistency in paradigms and results that clear patterns are not apparent. Nevertheless, we regard the existence of *some* overt performance deficit as established because of the number of studies that have found anhedonics to differ from controls in the direction of patient groups. Particularly promising are findings for a deficit in later stages of visual processing. Still, the specific nature of any overt performance deficit remains to be determined.

Behavioral Performance Evidence for a Processing Deficit in Perceptual Aberration

Reaction Time Crossover and Visual Processing

Many of the studies reviewed here also examined the performance of perceptual aberrators. Like schizophrenics, perceptual aberrators were found to demonstrate the RT crossover effect and visual backward-masking deficits. Interestingly, these findings were more pronounced for perceptual aberrators than for anhedonics (Balogh & Merritt, 1987; Merritt & Balogh, 1989; Simons et al., 1982b), further evidence that these subjects represent two distinct groups.

Several investigators have reported lateralized processing deficits in perceptual aberrators. Similar to anhedonics and patients with right-brain damage, perceptual aberrators have demonstrated erratic visual search patterns (Jutai, 1989). However, unlike anhedonics, these patterns were less generalized and were typically associated with tasks using only unstructured verbal and nonverbal stimuli. Perceptual aberrators have also displayed lateralized processing on left- and right-hemisphere dichotic-listening tasks. Overby (1992) found perceptual aberrators displayed an exaggerated right-ear advantage for consonant-vowel (left hemisphere) stimuli and a reduced left-ear advantage for tonal contour (right hemisphere) stimuli. Although Overby reported that these results were consistent with left-hemisphere overactivity in perceptual aberrators, this finding is also consistent with Jutai's (1989) claim that psychosis-prone subjects demonstrate deficits in right-hemisphere processing.

In a study similar to those demonstrating RT corssover effects, Mo and Chavez (1986) used time interval estimation, instead of RT, as their dependent measure to assess whether crossover effects occur in the absence of motoric responses. In their study, subjects were asked to estimate the duration of a dot presented tachistoscopically following an auditory warning signal presented monaurally to the left or right ear. Mo and Chavez (1986) found subjects differed in their reported time interval estimates of the dot according to whether they scored high or low on the perceptual aberration-magical ideation scale. Subjects who scored high on the perceptual-aberration scale reported longer duration times for warning signals presented to the right ear, while subjects with low scores reported longer duration times for warning signals presentd to the left ear. No group effects were found for the visual field in which the dot was presented. Moreover, anhedonics did not differ from controls in estimates of dot duration, suggesting that any differential lateralized processing of the stimuli was specific to the perceptual aberration group. Determining the nature of such lateralized processes warrants additional study.

Although the term *perceptual aberration* suggests that these subjects display unusual perceptual processes, perceptual aberrators typically perform as well as controls on visual perceptual tasks of figure–ground discrimination (Silverstein et al., 1992) and field dependence (Schuldberg & London, 1989). However, evidence for a subtle deficit in visual sustained attention in perceptual aberrators, similar to that noted in schizophrenics, has been published. In a study by Lenzenweger, Cornblatt, and Putnick (1991), subjects were asked to identify and respond to two successive identical visual stimuli while watching a series of rapidly presented numbers and shapes. Perceptual aberrators exhibited significantly poorer sustained attention performance on this continuous performance test

(CPT) than did control subjects. Asarnow, Nuechterlein, and Marder (1983) examined separate groups scoring high on the perceptual aberration or magical ideation scales and found that both groups of subjects exhibited poorer span of apprehension performance on a forced-choice visual target identification task. In an unpublished study, Simons and Chapman did not observe span-of-apprehension deficits in perceptual aberrators (Simons, personal communication, August 3, 1994). A study that did not differentiate high scorers on the perceptual aberration, magical ideation, and physical anhedonia scales reported that such subjects tended to interject distractors on a selective attention task more often than controls (Spring, Lemon, Weinstein, & Haskell, 1989). The specificity of such a finding to the perceptual aberration-magical ideation group remains to be determined. Overall, the evidence for a visual sustained attention deficit in perceptual aberrators should be considered preliminary, because an additional CPT study with visual stimuli found perceptual aberrators to perform similarly to controls (Simons & Russo, 1987).

Communication and Social Skills

The pattern of communication skills and social functioning in perceptual aberrators differs from that of anhedonics, in that perceptual aberrators tend to perform worse than controls on putative measures of thought disorder but as well as controls on tasks of social functioning. In contrast to the normal performance of anhedonics on tasks of word communication and word association, perceptual aberrators performed worse than controls on such tasks (Martin & Chapman, 1982; E.N. Miller & Chapman, 1983). These findings were replicated by Allen, Chapman, and Chapman (1987), using a larger and more diverse sample of subjects. In addition, Ward, McConaghy, and Catts (1991) found similar results for perceptual aberrators on both a word communication task and a word association task. As suggested by some investigators, these findings may have been due to a select group of perceptual aberrators who scored high on an impulsive nonconformity measure (Martin & Chapman, 1982; E.N. Miller & Chapman, 1983) or who reported high levels of depression (Allen, Chapman, & Chapman, 1987; Allen, Chapman, Chapman, Vuchetich, & Frost, 1987). Further evidence for thought disorder, or cognitive slippage, in perceptual aberrators can be found in a study by Allen and Schuldberg (1989). They reported that perceptual aberrators exhibited responses reflecting bizarre and idiosyncratic thinking as they interpreted relatively unfamiliar proverbs. The deviant Rorschach responses of perceptual aberrators also provide evidence for cognitive slippage in this group (Edell & Chapman, 1979).

Unlike anhedonics, the social communication of perceptual aberrators tends to be normal (Beckfield, 1985; Haberman et al., 1979). In addition, their ratings of social words do not differ in variance from normal controls (Raulin & Henderson, 1987). However, several studies of other social factors have turned up some differences for perceptual aberrators. One study found perceptual aberrators to provide more odd and hostile responses to a role-playing task than did controls (Numbers & Chapman, 1982). These differences in communication between perceptual aberrators and anhedonics lend further support for these subjects representing two distinct psychosis-prone groups. Furthermore, Zborowski and Garske (1993) reported that interviewers rated perceptual aberrators as more odd and avoidant and as leading the interviewer to feel more anxious, more angry, and less interested, relative to controls. (Anhedonics were not studied.) Additional research is warranted to follow up on these replicated findings about oddness and hostility. Finally, perceptual aberrators have reported unusual family structures (Perosa & Simons, 1991).

Summary

Although several laboratory studies provide evidence for processing deficits in both the anhedonic and the perceptual aberration groups, more pronounced effects appear for the perceptual aberration group. This differential is in line with results from a longitudinal study indicating that individuals selected with the perceptual aberration scale, but not those selected with the anhedonia scale, eventually display an elevated rate of psychotic symptoms (Chapman & Chapman, 1992; Chapman et al., 1994). However, we lack a comprehensive theory for understanding these apparent processing differences in the Chapman groups. Future studies may wish to examine the behavioral performance of anhedonics and perceptual aberrators with more process-oriented paradigms that use continuous, "on-line" assessment. Such methodology could further enhance our understanding of the mechanisms underlying the deficits reported for these groups. Thus, it is to the process-oriented methodology of psychophysiology that we next turn.

Psychophysiological Evidence for Processing Deficits in Anhedonia

Emotional Dysregulation

Most of the psychophysiological research on psychometrically defined subjects at risk for severe psychopathology has been on anhedonia. The first psychophysiological assessment of anhedonics (G.A. Miller, Simons, & Lang, 1981) evaluated the construct validity of the questionnaire using a variant of the contingent negative variation (CNV) paradigm developed by Simons, Öhman, and Lang (1979). The CNV, a portion of the event-related brain potential (ERP), consists of two components first distinguished by Weerts and Lang (1973). The earlier component, often called the O-wave because of its possible associations with the orienting response (Loveless, 1979; Rohrbaugh & Gaillard, 1983; but see Simons, 1988, for a contrary view), and a later component, sometimes known as the E-wave, for expectancy (Walter, Cooper, McCallum, & Winter, 1964), have been differentiated in numerous studies.

Simons (1988) argued that the E-wave is a function of the emotional significance of an expected stimulus in the absence of an RT requirement, making the CNV a ready means for assessing a pleasure deficit in anhedonia. G.A. Miller et al. (1981) used a version of the paradigm in which the pitch of a 6-s warning tone accurately predicted a slide (nude or household object) to appear at tone offset. Slide duration depended on RT to slide onset, with fast responses rewarded with longer exposures. Normal performance and CNV responses for anhedonics would argue against a pleasure deficit, at least with regard to sexual stimuli. Nonspecifically reduced CNV would demonstrate similarity with a variety of psychiatric groups (reviewed by Rockstroh et al., 1982; Sponheim, Allen, & Iacono, Chapter 8).

G.A. Miller et al. (1981) found a normal O-wave but a diminished E-wave in anhedonics in anticipation of the nude stimuli and normal responses to household objects. This effect, replicated by Simons and co-workers (1982a) in an expanded sample, suggested that anhedonics anticipated both nudes and household objects in the same manner as control subjects anticipated household objects. Both anhedonics and controls responded faster on nude than on neutral trials, and RT did not differentiate anhedonics and controls. Although both groups distinguished nude and neutral stimuli in ratings of interest and in anticipatory heart rate response, these differences were smaller for anhedonics than for controls. Because anhedonics' RT was normal, it would be difficult to argue that the E-wave and heart rate effects were secondary to a failure in anhedonics' ability to distinguish the warning tones, distinguish the two classes of slides, understand the task, or engage actively in the task. The effects seemed to be confined to higher-level processing specifically associated with something about anticipation of normally emotion-evoking slides.

These early CNV studies using sexual stimuli did not establish whether anhedonics' psychophysiological hyporesponsiveness would generalize to processing of pleasant stimuli in general. Accordingly, G.A. Miller and Yee (1985) modified the CNV paradigm to include aversive slides (injury victims and surgical procedures) and a wide variety of pleasant slides (such as cute pets and attractive landscapes) instead of household and nude slides, respectively. In this case, anhedonics' CNV did not differ from that of controls. RT, ratings, and anticipatory heart rate responses also did not differentiate the groups. However, anhedonics' heart rate during slide presentation was less responsive to pleasant slides than was controls.

In another attempt to broaden the assessment of psychophysiological response deficits for pleasant stimuli, G.A. Miller, Yee, and Anhalt (submitted) attached a monetary incentive to simple visual stimuli, using a paradigm described by Brecher and Begleiter (1983) in a study of schizophrenics. No anticipation was involved in the task. Performance on a choice RT task earned or cost the subject a bonus. Anhedonics did not differ significantly from controls in either overt performance or amplitude of components of the ERP (N200, P300), although the latency of the components in anhedonics was not as responsive to the incentive manipulation as in controls. In another study of monetary incentive, Rockstroh et al. (1982) found anhedonics to be autonomically hyporesponsive during a CNV biofeedback paradigm. In both heart rate and skin conductance, controls but not anhedonics discriminated between positive and negative reinforcement trials, even though anhedonics learned CNV self-control as well as controls did.

Finding CNV and autonomic hyporesponsiveness in these studies provided some empirical evidence for pleasure processing deficits in subjects who scored deviantly on the Chapman anhedonia scale. However, as for anhedonics' normal emotional responsivity in the Berenbaum et al. (1987) study, no specific anomaly in terms of information processing could be pinpointed. Indeed, overt performance and at least some aspects of psychophysiological responding were entirely normal in the anhedonic groups, which is what might be expected from assessing individuals believed to be at risk for psychosis and not patients themselves.

Birbaumer and colleagues at the University of Tübingen found additional evidence for anomalous psychophysiological responding in anhedonia when they translated the Chapman scales into German and conducted a series of CNV studies (Lutzenberger, Elbert, Rockstroh, Birbaumer, & Stegagno, 1981; Rockstroh et al., 1982). In contrast to G.A. Miller et al. (1981) and Simons et al. (1982a), who found an E-wave *decrement* in anhedonics, the Tübingen group consistently found normal CNVs (like G.A. Miller & Yee, 1985) and *enhanced* postimperative negative variations (PINVs; not assessed by Miller or Simons). Similarly, Eikmeier et al. (1992) reported that a topographic PINV index correlated marginally ($P < .06$) with physical anhedonia score in remitted inpatient schizophrenics. Rockstroh et al. interpreted such findings as suggesting that anhedonics "may develop expectations or models about stimulus probabilities in a different way" than controls (1982, p. 218) and that they represent "a basic, probably unspecific disturbance of attentional processes and contingency processing" (p. 221), as evidenced by the reduced skin conductance responses for anhedonics in the first of the three Tübingen studies and the deviant heart rate responses in the third study. Jutai's (1989) finding of erratic visual search patterns among anhedonics supports such a disturbance of attentional processes.

Considered together, these studies across several laboratories provide considerable, though somewhat inconsistent, evidence of psychophysiological hyporesponsiveness in anhedonic samples. This was seen in CNV for the nude slides of G.A. Miller et al. (1981) and Simons et al. (1982a); in CNV in a monetary incentive study by Pierson, Ragot, Ripoche, and Lesevre (1987; see also Pierson, Loas, & Lesevre, 1990; Pierson, Ragot, Ripoche, & Lesevre, 1988); in heart rate during anticipation of sexual slides in G.A. Miller et al. (1981); in response to nonsexual pleasant slides in G.A. Miller and Yee (1985); in response to monetary incentive in Rockstroh et al. (1982); and in skin conductance in one of the Tübingen studies (Rockstroh et al., 1982). No instances of hyperresponsiveness were noted in any study for any of these measures.

Although it is tempting to conclude that psychophysiological validation of a pleasure deficit in nonpatient anhedonics is well established and not confined to sexual stimuli, such a conclusion is only preliminary for several reasons. First, anhedonics' extended PINV runs counter to this generalization about hyporesponsiveness. Second, among the other psychophysiological measures, there is considerable variability in which measures show group effects in the different paradigms. It cannot be assumed that measures are interchangeable or equally meaningful for assessment of a pleasure deficit.

Three studies described by Simons, Fitzgibbons, and Fiorito (1993; Fitzgibbons & Simons, 1992) addressed the issue of psychophysiological validation of a pleasure deficit in anhedonics in a very different manner from the studies reviewed so far. Their first study used slides from a set that has been carefully validated (Lang, Öhman, & Vaitl, 1988). Pleasant slides were primarily nonsexual. Simons's paradigm involved presentation of acoustic startle probes during slide presentation. Anhedonics produced normal reflex blinks, but they rated pleasant slides as less interesting and less pleasant (though not less arousing) than controls did. Ratings of unpleasant slides were similar to those of controls. The second study of pleasant and unpleasant slides measured heart rate, skin conductance, and facial electromyogram (EMG) during slide viewing. Anhedonics again rated pleasant slides as less pleasant than controls did, and again the groups did not differ in ratings of unpleasant slides. Anhedonics' heart rate failed to respond to the valence dimension, yet facial EMG indicated an exaggerated differentiation of pleasant and unpleasant slides. Skin conductance was similar for the two groups. The third study employed these same physiological measures during emotional imagery rather than slide presentation. Anhedonics scored more poorly on a self-report measure of imagery ability. They also rated pleasant images as less pleasant and unpleasant images as less unpleasant. No group differences emerged for skin conductance or facial EMG, but anhedonics were less responsive in heart rate than were controls. In combination with the CNV studies already discussed, these three diverse studies provide compelling evidence that anhedonics process pleasant stimuli (not just nude slides) differently from controls and that their self-described anhedonia is not merely a self-report response bias.

Nevertheless, lacking a consensus on the psychophysiology of pleasant emotions, including an understanding of the relationship among response systems, it is difficult to determine the set of empirical pattern(s) that would justify a firm conclusion that self-described anhedonics do have a pleasure deficit. It is a further challenge to integrate such empirical findings into a comprehensive story about anhedonia. Simons et al. (1993) have come closest to doing so. They proposed that the key is not simply an attenuation of psychophysiological response in the face of pleasant stimuli but a disorganization of the various response components in emotion. An implication of this view is that anhedonics are doing qualitatively different things—not merely doing normal things, albeit less of them, when positive stimuli arise. Whether those different things are outside the range of normal subjects' repertoire, or simply an abnormal selection among normal strategies and responses, is a question to which we will return after summarizing additional research on anhedonia that has not employed emotional stimuli.

Several issues warrant investigation before evidence for a pleasure deficit is considered fully understood. First, the operationalization of "pleasure" has varied considerably across studies. For example, it is not clear whether anhedonics are hyporesponsive across the full range of intensity of pleasant stimuli or even how to represent that range in the laboratory. Second, it is not clear whether anhedonics fail to appreciate the pleasantness of stimuli or simply fail to generate the self-report and physiological efference normally associated with such an experience. Third, the finding of a pleasure deficit in anhedonics appears to be better established for self-report than for psychophysiological measures. A comprehensive assessment of the psychophysiology of pleasant emotions is needed to determine whether the psychophysiological findings for anhedonics can be strengthened via appropriate elicitation and measurement. Obtaining psychophysiological evidence of anhedonia per se is somewhat constrained by the limited and controversial picture currently available of the normal psychophysiology of pleasure. In a recent review of autonomic nervous system differences among emotions in normal subjects (Levenson, 1992), cross-study consistencies for positive emotions consisted simply of smaller heart rate and skin conductance responses than for negative emotions, rather than a qualitatively distinct pattern for pleasure. In a review of electroencephalography (EEG) alpha findings related to emotion, Davidson (1992) suggests that approach/withdrawal rather than valence is crucial in emotion. That dimension has not been systematically assessed in anhedonics.

Auditory Perceptual Processing Deficits

Inferences about emotional processing in the studies already reviewed would presumably depend to varying degrees on intact perceptual processing; anhedonics would not be expected to process the emotional import of stimuli normally if they do not perceive the physical stimulus characteristics normally. A variety of studies have been undertaken to better understand how anhedonics might evaluate auditory warning stimuli. Because the CNV studies employed an auditory stimulus, Yee, Deldin, and Miller (1992) wished to rule out simple sensory deficits in tone perception. Anhedonics and controls underwent a basic auditory augmenting–reducing paradigm. Tones of five different intensities were presented in random order, with no overt task. As expected, the amplitude of exogenous ERP components was larger for louder stimuli. Importantly, anhedonics proved to have the same augmenting slope as controls. That is, peak-to-peak ERP component amplitude (P50 to N100 and N100 to P200) increased as a function of stimulus intensity. Overall, anhedonics were statistically indistinguishable from controls.

Because schizophrenics have deficient inhibition of the startle blink reflex (Braff et al., 1978), Simons and Giardina (1992) evaluated facilitation and inhibition of the startle blink reflex as a function of stimulus-onset asynchrony in anhedonics and controls. Anhedonics showed a normal modulation of blink magnitude and latency. Schell, Dawson, Hazlett, and Filion (1995) recently replicated this in a similar, no-task condition. Preliminary analyses from our laboratory (Chmielewski, 1994) of a study of suppression of the P50 component of the ERP found no group differences when tones were presented in pairs. The negative findings of Yee et al. (1992), Simons and Giardina (1992), and Chmielewski (1994) regarding warning and intensity effects argue against the possibility that heart rate, skin conductance, N200, P300, O-wave, E-wave, and PINV results for anhedonics in other studies were secondary to some basic perceptual problem for auditory stimuli.

To investigate possible pitch discrimination effects in the CNV studies, G.A. Miller (1986) added a third tone pitch (720 Hz) to the two easily discriminable tones (700 and 1,200 Hz) used in several of the CNV studies. The imperative stimulus was not included, and no overt action was required, aside from a rating, which the subject made well after the 6-s tone had terminated. The task was simply to judge which of the three tones was presented on each trial. The paradigm was based on earlier work that had developed a model of information processing stages during tone discrimination and of the associated psychophysiological outputs at each stage (Bull & Lang, 1972; Gatchel & Lang, 1973; Lang, Gatchel, & Simons, 1975; Simons & Lang, 1976; see also Simons, Russo, & Hoffman, 1988). Those studies had established that during more difficult discriminations P300 is diminished, O-wave is enhanced, and heart rate decelerates.

Anhedonics were found to differ from controls in several ways in this paradigm. During discrimination trials, effects involving Group were obtained for N200 amplitude, P300 amplitude and latency, O-wave, and heart rate, though not skin conductance. Because anhedonics' overt discrimination performance resembled controls', a cognitive model of the task (G.A. Miller, 1986; Simons & Lang, 1976) suggested that anhedonics were able to perform the task adequately by using different strategies than controls. In particular, the psychophysiological data suggested that anhedonics had found all trials to be difficult and had processed the stimuli accordingly.

The nature of this difficulty was not clear in these findings—whether a perceptual or cognitive problem. Previous studies using this tone discrimination paradigm had not quantified the N200 component of the ERP. In G.A. Miller (1986), conventional and principal components analyses indicated that the P300 attenuation for difficult trials reported previously for normal subjects in pitch discrimination paradigms actually reflects an overlapping N200 enhancement. Interestingly, anhedonics had a significantly larger N200 than controls. Given studies of N200 by Näätänen (1990, 1992) and others, N200 enhancement in this paradigm could reflect the detection of a mismatch between incoming stimuli and some template maintained by the subject. The implication would be that anhedonics treated every stimulus as novel or unfamiliar. The data from this study did not indicate whether the processing anomaly was localized at perceptual, template-maintenance, or template-comparison stages. However, given the well-established findings suggesting a deficit in the orienting response in schizophrenic and mood disorder patients (Sponheim et al., Chapter 8; Bernstein, 1987, 1992; Dawson & Nuechterlein, 1984; Dawson, Nuechterlein, & Schell, 1992; Öhman, 1981; Venables & Bernstein, 1983; Zahn, 1986) and the suggestion that N200 reflects an aspect of orienting (Näätänen, 1990), this finding of an N200 enhancement in anhedonia appeared very promising.

Orienting

Given the prominence of orienting deficits in schizophrenia and mood disorders, it would be useful to determine how much of anhedonics' apparent pleasure deficit is attributable to a more general orienting deficit. Yet a puzzle was apparent in the possibility of an orienting problem in anhedonia. Based on the orienting data from relevant patient groups, *hyporesponsiveness* would be expected for any index of orienting in anhedonic subjects. If N200 reflects orienting, then anhedonics' N200 *enhancement* is contrary to the patient data. However, the patient data generally involved autonomic rather than ERP measures. Needed are both a survey of available autonomic data for anhedonics in orienting/

habituation paradigms and a more intensive look at the N200 enhancement in anhedonia.

According to the studies reviewed so far, basic perceptual processes manifested in exogenous ERP components as well as overt performance appear to be intact in anhedonics. Most of the psychophysiological evidence for anomalies points to orienting (N200 and O-wave enhancement, heart rate deceleration, skin conductance hyporesponding) or anticipatory (E-wave) processes. The data suggesting an orienting deficit were promising but required support from studies better designed to study classic orienting phenomena.

Simons (1981) employed a traditional autonomic orienting paradigm. Mild, nonsignal tones were repeated at 30- to 60-s intervals, while heart rate and skin conductance responses were recorded. Anhedonics produced significantly smaller heart rate and skin conductance responses. Simons (1980) also tested these subjects' habituation to repetitions of tape-recorded baby cries and telephone rings. Reduced skin conductance responsiveness was again observed in anhedonics. Bernstein and Reidel (1987) found anhedonics to be hyporesponsive in both skin conductance and finger-pulse volume during habituation. As Sponheim et al. argue in Chapter 8, such cross-measure convergence suggests a central rather than a peripheral anomaly.

Using a tone-orienting paradigm in our laboratory, we have seen this same pattern of reduced skin conductance responsiveness in anhedonics. Preliminary analysis of subjects provided clear but statistically nonsignificant differences (Ebert, 1987). Anhedonics' mean response was smaller than controls' on every one of 24 trials. Additionally, anhedonics habituated faster to a series of 65-dB tones and then to a series of 95-dB tones. In line with the 40% to 50% of schizophrenics who fail to respond on the first trial in standard habituation paradigms (Bernstein, 1992; Dawson & Nuechterlein, 1984; Öhman, 1981) and the 37.5% rate in Bernstein and Reidel's (1987) anhedonics, 45% of our anhedonics were trial-1 nonresponders. Having recently analyzed an enlarged sample, we still find a trend for hyporesponsiveness in anhedonics. However, statistical support is not strong, being dependent on the skin conductance measure assessed. Nevertheless, these findings across several laboratories provide clear evidence of an autonomic orienting deficit in anhedonics.

Heart rate differences in the Ebert (1987) study were nonsignificant but suggestive of a mechanism for anhedonics' orienting deficit. Anhedonics failed to show any clear response on the first trial. By trial 3, however, their deceleration, interpretable as standard cardiac orienting (Graham, 1979), was comparable to that obtained for normal control subjects on trial 1. These data suggest a more complex deficit than simply a failure to orient. On the other hand, a speculation about some simple temporal factor delaying anhedonics' development of orienting is belied by the absence of a similar pattern in skin conductance. We are enlarging the sample size to see if these data prove reliable. A larger sample may also permit us to determine whether a subset of the group is carrying the effect.

Memory Template Function

Complementing the evidence for an autonomic orienting deficit is a follow-up of the N200 finding of G.A. Miller (1986). N200 was larger for anhedonics than for controls in that study. In addition, mean N200 was apparently larger for anhedonics than for controls in a study by Simons (1982), although the N200 was not quantified in that ERP study. Group effects for auditory N200 were either not quantified or not obtained in the other ERP studies reviewed so far (however, cf. Simons and Russo, 1987, for group effects with visual stimuli). Studies of auditory ERPs in unselected normals have revealed multiple subcomponents in the latency range of N200 (Hicks & Miller, submitted; Näätänen 1990, 1992; Pritchard, Shappell, & Brandt, 1991). The studies reviewed here did not allow the separation of these subcomponents and, indeed, would not normally have been considered "N200 paradigms." Yet the N200 and other psychophysiological findings in the 1986 study suggested a specific hypothesis about anomalous information processing in anhedonics. In that pitch discrimination task, anhedonics showed a sequence of physiological responses on all trials matching those that controls produced on difficult trials. Anhedonics' overt performance matched that of controls, but on the basis of the psychophysiological data, it appeared that anhedonics were accomplishing normal performance by using unusual strategies. Determining which of several N200 subcomponents was exaggerated in anhedonics could help identify this alternative strategy.

Giese-Davis et al. (1993) designed a study to address anhedonics' use of alternative strategies via the N200. By expanding on the methodology of Sams, Alho, and Näätänen (1983), within-subjects manipulations in this paradigm allowed differentiation of two auditory templates. Each template was conceived as a representation in short-term memory of recent stimulus regularities (Näätänen, 1990; Pritchard et al., 1991). It is believed that in the context of rapidly repeated tones one type of template arises involuntarily (i.e., independent of the subject's attentional behavior). Magnetoencephalography and source analysis have shown this template to be implemented in primary auditory cortex atop the temporal lobe (reviewed by Näätänen, 1992). Another template (location as yet unknown) arises only for voluntarily attended stimuli. Stimuli mismatching these templates elicit an N200, but the two templates are associated with different N200 subcomponents, differing in latency and scalp topography. In both cases, the templates are conceptualized as neuronal representations of stimulus features. Experimental manipulation and waveform subtraction allow the two N200 subcomponents to be differentiated.

Giese-Davis et al. (1993) confirmed the sensitivity of the N2a subcomponent (also known as mismatch negativity) to mismatches even when the subject ignores the stimuli. Specifically, N2a amplitude was proportional to the degree of mismatch between the current stimulus and the available template based on recent trials. Anhedonics showed the same within-subjects effects for N2a as controls, indicating that the involuntary-template processor was capable of normal functioning. In contrast were the data for the voluntary-attention component (N2b). Control subjects' N2b was enhanced only on trials with a mismatch to a strong template, as expected from research on unselected normals (Sams et al., 1983). In contrast, anhedonics showed the exaggerated N2b on trials when a strong template was presumably in place, regardless of whether the incoming stimulus matched or mismatched the template. Preliminary analyses subsequently conducted on a larger sample of questionnaire-selected, nonpatient subjects, categorized on the basis of clinical interview diagnosis instead of questionnaire scores, found similar results. N2a was not sensitive to diagnosis, whereas N2b was exaggerated in diagnosed nonpatients (Fernandes, Giese-Davis, Hicks, Klein, & Miller, submitted).

The implication is that anhedonics found objectively familiar stimuli to be subjectively novel. These data, which replicate the large N200 reported by G.A. Miller (1986)—despite numerous differences in the paradigms—point to a specific information processing deficit in anhedonics, involving the comparison of voluntarily attended stimuli with templates stored in working memory.

The parallel between the present N2a/N2b findings and Knight's (1984, 1992, 1993) concept of early information processing deficits in schizophrenia is encouraging. Knight proposed two processing stages relevant to schizophrenics' attentional and input dysfunctions. The first stage accomplishes a sensory representation of physical stimulus qualities. It has high capacity and does not require attentional resources. The second stage is a limited-capacity, short-term memory, which processes the representation produced in the first stage. Items of stimulus information are transferred into stage 2 in parallel, but its output is serial. Thus, it involves attentional switching from one item to another. Knight (1992, 1993) suggested that stage 2 is the likely site of some cognitive deficiencies in schizophrenics. Specifically, either the allocation of resources to support stage 2 or the consolidation of information in stage 2 is problematic. Bernstein (1992) reviewed other concepts of information processing deficits in schizophrenia that are broadly consistent with Knight's focus on a second-stage, limited-capacity process (though see Gray, Hemsley, Feldon, Gray, & Rawlins, 1991, for a contrary view).

Pritchard and colleagues' (1991) review of the normative N200 literature proposed a subdivision of N200 components into N2a, N2b, and N2c, similar to Knight's (1992) cognitive model and Giese-Davis and co-workers' (1993) findings. Pritchard's N2a and N2b matches Giese-Davis's closely (the paradigm used by Sams et al., 1983, does not bring out a separate N2c). Importantly, Pritchard's interpretation of the functional significance of N200 subcomponents appears to match well with Knight's stages. N2a appears to be an ERP manifestation of stage 1, and N2b is associated with stage 2.

Although anhedonics' N200 enhancement fits Knight's (1992) model of overt performance deficits in schizophrenics, three issues remain unresolved. First, a similar N200 enhancement (specifically, N2b in the absence of any N2a differences) has not been found to be specific to schizophrenia. Second, it is not clear how to reconcile anhedonics' N200 *enhancement* with the *low* autonomic responsiveness seen in anhedonics, schizophrenics, and mood-disorder patients or with the CNV hyporesponsiveness discussed earlier for anhedonics. Third, although we concluded that there is ample psychophysiological evidence for a pleasure deficit in anhedonia, it is not clear how that conclusion relates to autonomic hyporesponsiveness for neutral, no-task stimuli or to N200 enhancement for neutral task-related stimuli.

It is possible to speculate that some of the observed responses reflect some process compensating for another process that is functioning abnormally. We have recently found that skin conductance nonresponders have exaggerated N2b (Fernandes & Miller, 1993). Thus, the process reflected in N2b may be compensating for the process reflected in skin conductance. We have proposed (G.A. Miller & Yee, 1994) (a) that the autonomic data reflect the core deficit, (b) that N2b reflects the compensatory mechanism, and (c) that it is the eventual failure of that mechanism that moves the nonpatient anhedonic into clinical decompensation.

Several aspects of this hypothesis are testable. First, active schizophrenics should not show an enhanced N200, at least not an N2b in the paradigm used by Sams et al. (1983) and Giese-Davis et al. (1993) because, presumably, they lack the compensatory mechanism. Second, variations in clinical status following initial diagnosis could covary with N200 enhancement. Specifically, improvement in clinical symptoms should witness the emergence of an exaggerated N200, if the improvement is due to restoration of the compensatory mechanism.

If this hypothesis is borne out, it would then be important to identify the determinants of N200 amplitude. Of special interest would be mechanisms or strategies subject to voluntary control or to training. They could be the focus of preventive interventions with individuals at risk. In this story, the N200 enhancement is not a deficit but a compensatory strength, which clinically we may wish to foster.

General and Regional EEG Findings

Although no specific EEG abnormality has been consistently identified in schizophrenia, frequency analyses have demonstrated similar EEG findings for schizophrenics and anhedonics. Pierson and co-workers' (1987) EEG power spectrum analysis indicated reduced alpha and enhanced beta power among anhedonics, as has been reported in schizophrenics (Bernstein, 1987; Iacono, 1982; Itil, 1977; Shagass, 1976; Sponheim, Clementz, Iacono, & Beiser, 1994). G.A. Miller (1986) reported that anhedonics displayed reduced EEG alpha and theta power at rest. Similarly, anhedonics displayed less alpha than controls in two recent studies of resting alpha in our laboratory (Etienne, Deldin, Giese-Davis, & Miller, 1990; Fernandes, 1992), although the mean differences in both studies were not statistically reliable. Moreover, converging evidence exists for regional activity differences in schizophrenia and perhaps in anhedonia. These regional differences, however, are less consistent than those suggesting general activity differences. Although it is well-known that the exact location of the source of electrical activity in the brain cannot be determined unequivocally from scalp electrode placements (e.g., Lutzenberger, Elbert, & Rockstroh, 1987), neuropsychological studies (see Davidson, Chapman, Chapman, & Henriques, 1990) lend support to analyzing EEG activity in terms of broad regions. Typically, the findings suggest greater left-hemisphere activity in schizophrenia, either posteriorly (Morihisa, Duffy, & Wyatt, 1983) or temporally (Salisbury et al., 1994). Greater reactivity in schizophrenics' skin conductance orienting responses from the right than the left hand provides additional support for left-hemisphere activity. Some studies, however, have found greater right-hemisphere activity in schizophrenia (e.g., Volavka, Abrams, Taylor, & Recker, 1981). Moreover, regional findings may be affected by subtype of schizophrenia (Serafetinides, Coger, Martin, & Dymond, 1981), comorbidity with other disorders (Coryell, Keller, Lavori, & Endicott, 1990), gender (Frazier, Silverstein, & Fogg, 1989), or anxiety (Heller, 1990; Wale & Carr, 1990).

Interestingly, the N200 data discussed previously support greater left-hemisphere activity in schizophrenia. Based on the assumption that the left and right cerebral hemispheres are normally specialized for sequential-analytic and parallel-holistic types of processing, respectively, Venables (1984) proposed that schizophrenics have some deficiency in right-hemisphere processing that requires them to rely on sequential processes when normals would rely on holistic processes. A deficiency in task performance would result that would be handled best by the type of processing associated with consolidation of information manifested in N2b, precisely what was found in our anhedonic subjects. Thus, it is possible that anhedonics might display greater left-hemisphere activity than controls. Bruder and colleagues (1989), in a study of depressed subjects with and without anhedonia, found support for left-hemisphere involvement in anhedonia. Specifically, melancholic subjects (with anhedonia) demonstrated larger right-ear (left-hemisphere) advantages on a dichotic listening task than atypical depressives (without anhedonia). If one assumes that the two depressed groups did not differ in degree of depression and therefore right-hemisphere function, then the presence of anhedonia may have been related to greater left-hemisphere involvement.

Eye-Tracking Problems

A number of studies have established eye-tracking abnormalities in schizophrenics (for reviews see Clementz & Sweeney, 1990; Holzman, 1987; Iacono, 1988). Simons and Katkin (1985) reported two studies of smooth-pursuit eye-movement (SPEM) in anhedonics. Although no group mean difference in tracking performance was observed, anhedonic and control groups did differ significantly in eye-movement variance in both studies. That is, the anhedonic group in both studies included a subset of subjects with markedly deviant SPEM. Considerable controversy has arisen regarding methods of quantifying eye-tracking error (Clementz & Sweeney, 1990), and the studies by Simons and Katkin were not intended to investigate methodological issues. Their initially promising result deserves further study, both in scoring methods and in characterizations of that subset of subjects who show the errors.

Blink Rate

Another interesting finding among schizophrenics is their high blink rate (see Karson, Dykmam, & Paige, 1990, for a review). There is no published study of blink rate in anhedonics to date. However, the positive findings of studies examining eye-tracking anomalies as well as Lutzenberger and co-workers' (1981) report of systematic eye movement artifact in anhedonics suggest posible blink rate anomalies in anhedonia. We plan to examine data from several studies to determine whether anhedonics have a higher blink rate than normal control subjects. Preliminary analyses (Fernandes et al., submitted) of data from the paradigm of Giese-Davis et al. (1993) found no group differences in blink rate.

Because EEG alpha levels typically change as a function of whether the eyes are open or closed, a comprehensive account of eye movement and blink anomalies in schizophrenia would also need to address the low overall alpha levels reported in this disorder. Karson et al. (1990) have come closest to proposing such a theory. Based on evidence for reduced activity in the visual cortex during blinks and for alpha frequency induction preceding blinks, they suggest that an increased blink rate in schizophrenia reflects abnormal activity in a circuit involving subcortical structures (pontine tegmentum, cerebellum, substantia nigra, midbrain tectum, and lateral geniculate bodies) and the occipital cortex. Indeed, this theory could account for the deficient inhibition of the startle blink reflex observed in schizophrenics (Braff et al., 1978; for review, see Dawson, Schell, Hazlett, Filion, & Nuechterlein, 1995). In any case, additional study is warranted to understand the mechanisms associated with blink rate and how it may be related to psychosis vulnerability.

Context Updating and Resource Allocation in Anhedonia

By far the largest number of psychophysiological studies have looked for P300 effects in anhedonia. The reduction of P300 in schizophrenia is well established (Pritchard, 1986), although it has been observed in a number of other disorders as well. The interpretation of the functional significance of P300 is controversial (Johnson, 1988, 1993), but the dominant position is that it reflects some aspect of working memory and perhaps specifically the updating of the subject's current model of the world (Donchin & Coles, 1988). Viewed broadly, this notion appears relevant to orienting (Roth, 1983), but the relationship between P300 and autonomic orienting is unclear (e.g., Bernstein, 1987, 1992).

However, results of two studies of P300 in schizophrenia encourage careful study of P300 effects in anhedonia. Pfefferbaum, Ford, White, and Roth (1989) and Ward, Catts, Fox, Michie, and McConaghy (1991) found that schizophrenics' P300 correlated significantly with negative symptoms. Typically, in the clinical picture of schizophrenia, negative symptoms (including anhedonia) are the more durable, whereas positive symptoms are more episodic and more responsive to medication. Furthermore, negative symptoms are more typical of the premorbid course of schizophrenia (Manschreck & Maher, 1991). Thus, a thorough evaluation of possible P300 anomalies in nonpatient anhedonia is particularly important, because there are grounds to expect a P300 anomaly to be a risk marker in nonpatients.

Simons (1982, experiment 2) first demonstrated that anhedonics showed the same general decrement in P300 amplitude as schizophrenics. Subsequent studies, considered individually, have provided somewhat mixed support for that conclusion. Simons and Russo (1987) noted that most studies not showing a significantly reduced P300 used tasks or recording sites that are not optimal for observing P300. Given the impressive consistency in mean group differences (at least 15 of 16 known studies reviewed by G.A. Miller & Yee, 1994) using a variety of paradigms across several laboratories, the conclusion that anhedonics show some P300 attenuation, like schizophrenics, appears solidly established. The effect size in any given study varies with task and details of recording and scoring. Interestingly, the studies that showed a statistically reliable effect employed relatively demanding tasks.

Proponents of the context-updating view of P300 (Donchin & Coles, 1988) have found it to be sensitive to task demands conceptualized around the notion of cognitive resources (e.g., Strayer & Kramer, 1990; Wickens, Kramer, Vanasse, & Donchin, 1983). A systematic study of resource allocation in a dual-task paradigm (Yee & Miller, 1994) found not only reduced P300 but insensitivity of P300 to a manipulation of task priority in anhedonics. The latter result implies an inflexibility in, or alternative strategy for, alterations of resource allocation among competing tasks. In the laboratory, resource allocation is commonly studied by requiring overt performance on two tasks performed concurrently.

Simons and Russo (1987) and Nuechterlein (1990) speculated that anhedonics' decreased capacity for pleasure may be a function of a problem in the allocation of processing resources. This view is consistent with a proposal of a fundamental deficit in resource quantity or allocation in schizophrenia (Callaway & Naghdi, 1982; Nuechterlein & Dawson, 1984; Nuechterlein & Green, 1991; see also Bernstein, 1987; Mirsky & Duncan, 1986; Pogue-Geile & Oltmanns, 1980; Bernstein, 1992, reviewed several similar proposals from others). The Yee and Miller (1994) study provides the clearest evidence to date of a resource allocation problem in anhedonia and therefore supports this view. Nuechterlein and Dawson (1984) emphasized that processing deficits tend to be observed in schizophrenics and children of schizophrenics when task processing demands are high (except in the most severely disturbed patients, who show difficulties even for relatively simple tasks). Thus, it is not surprising that, in P300 studies of anhedonia, it is those paradigms requiring active processing that have provided P300 reductions in anhedonics that reached statistical significance.

Although the evidence for a P300-amplitude decrement in anhedonia is clear, the psychological interpretation of this evidence is constrained by the cogency of theories of P300 amplitude in normals. As noted previously, an orienting view of P300 leaves much to be desired. Thus, it is not immediately clear how to link the P300 and orienting findings in anhedonia (or in schizophrenia). The context-updating view of P300 is quite broad and suggests that anhedonics have some problem in the updating of working memory. The suspicion that P300 more specifically reflects the allocation of resources for such updating leads to a focus on possible resource allocation problems in anhedonia.

Summary of Psychophysiological Studies With Anhedonics

Descriptively, there is decisive evidence of P300 reduction in anhedonia. Early ERP components and overt performance are consistently intact. There has been either less research or less consistency in, but some replication of, N200 and PINV enhancement, EEG overall or regional activity, E-wave and skin conductance reduction, and eye-tracking deviation. Contingent negative variation has been reported to be normal, selectively reduced, or selectively enhanced, depending on the stimuli, although sexual stimuli prompted reduced CNV in both studies that used such stimuli. Simons (1988) has argued that the E-wave component of the CNV in the absence of an RT task depends on expectancy of emotional stimuli. Most CNV studies of anhedonia have used nonemotional imperative stimuli, so the inconsistent effects for CNV may reflect a nonoptimal choice of stimuli. There have also been several reports of heart rate effects, with anhedonics consistently being less responsive. Finally, the studies described by Simons et al. (1993) and others collectively establish that processing of pleasant emotional stimuli is abnormal in some way in anhedonia. In contrast, in a number of studies of basic perceptual processing, using a variety of paradigms and psychophysiological measures, no evidence of aberrant function in anhedonics has been published. This distinction between measures sensitive to active cognition and passive perception is most clear in Schell et al. (1995). As noted, anhedonics resembled controls in startle eye-blink reflex modification during a passive task (replicating Simons and Giardina, 1992). However, when the same subjects were given active performance instructions, anhedonics showed deficient attentional modulation of the eye-blink reflex.

Although research with nonpatients at risk avoids a host of interpretative problems that are due to gross impairment, medication, and so on encountered in research with patients, we now face the converse problem: a host of psychophysiological anomalies in the absence of coincident overt impairment in anhedonics. Unlike a patient population, however, an at-risk population would be *expected* to show just such a pattern—subtle anomalies in nonessential manifestations of information processing. To show more severe, pervasive anomalies would move one from at-risk to patient status. Thus, the interpretative task is to account for the numerous psychophysiological findings in anhedonia in a way that permits us to explain why such individuals function normally on overt measures.

We have proposed (Miller & Yee, 1994) that anhedonics' enhanced N2b reflects the exaggerated operation of a voluntary cognitive mechanism which is compensating (successfully so far, in these nonpatients) for some other processing anomaly. The latter cannot be a gross deficiency in orienting, because N2a effects are normal in anhedonia; and in Näätänen's (1990) analysis of N200 subcomponents, it is N2a that is closely linked to orienting.

Within the ERP literature on anhedonia, we have normal exogenous components, *obligatory* sensory processing; normal N2a, *automatic* stimulus template development and comparison; exaggerated N2b, *voluntary* attention to and registration of unexpected stimuli; reduced

P300, *voluntary* resource allocation for the updating of working memory; and reduced CNV and enhanced PINV, suggesting a problem in the *voluntary* control of sustained attention. Thus, a consistent pattern is seen of voluntary rather than automatic processes being disrupted in anhedonia, which is in line with Knight's (1992, 1993) model for schizophrenia and the more general proposal of Nuechterlein and Dawson (1984).

The autonomic data, viewed in an orienting framework, cannot be as readily characterized as suggesting a problem in just automatic or just voluntary processing. Orienting has been conceived variously as involving both types of processes: an automatic detection (at some initial level in the system) of stimulus change, followed *under some circumstances* by what has been called an interrupt (Graham, 1979) of ongoing voluntary or controlled processing, leading to a reallocation of voluntary processing capacity to the stimulus (Öhman, 1979).

We have proposed (Miller & Yee, 1994) that the ERP and autonomic data on anhedonia may be integrated as follows. Capabilities for sensory registration of stimuli and detection of stimulus change are normal. However, something about the normally routine step of interrupting ongoing processing and allocating limited resources to memory-dependent evaluation of a given stimulus is awry. Thus, initial but not later stages of orienting are intact. Anhedonics can compensate for problems in some later stage(s) via an overreaction, an oversensitivity, in that portion of voluntary attentional processes that are manifested in N2b. Put more simply, anhedonics achieve their compensation by relying on a different strategy than normals do. Furthermore, given that anhedonics' psychophysiology is within the range observed for normals under some circumstances, it is parsimonious to assume that anhedonics' strategies are selected from a normal repertoire.

This line of reasoning can be made more specific. Yee and Miller (1994) found an interaction for P300 during three versions of a visual running-memory task. The anhedonics produced the same P300 amplitude for all conditions that controls did for the most difficult condition (significant Group × Condition effect). This suggests that anhedonics adopted the information processing strategy for all conditions that controls used for the most difficult condition. In other words, anhedonics found the easy tasks to be more difficult than controls did. Given such subjective difficulty, the anhedonics compensated by adopting a strategy that is quite normal for difficult circumstances. The result was abnormal psychophysiology and normal overt performance. As noted above, this pattern is to be expected in at-risk populations. Should this compensatory strategy ever prove inadequate to the task at hand, overt performance would deteriorate. If severe enough, the overt clinical picture could warrant a formal diagnosis.

Psychophysiological Evidence for a Specific Processing Deficit in Perceptual Aberration

Simons and colleagues and the Tübingen group have done most of the published psychophysiological research on subjects identified with the Chapmans' perceptual aberration and magical ideation scales. These subjects have usually been run as an additional group in studies comparing anhedonics and controls. Thus, the available paradigms are largely a subset of those reviewed above. The smaller number of published studies and the wide array of paradigms make generalizations difficult. However, several patterns do emerge from this literature.

In contrast to anhedonics' record of consistently small P300, perceptual aberrators have produced a statistically small P300 in only one condition of one study (Rockstroh et al., 1982). Simons (1980) observed a failure of perceptual aberrators to differentiate two warning stimuli as measured by P300, but their overall P300 was not smaller than that of controls. A number of studies have found no differences in P300 for perceptual aberrators versus controls. At present, there is little basis for believing that this measure differentiates this group from controls. Also clear is the absence of skin conductance effects for perceptual aberrators in any study to date (for review, see G.A. Miller & Yee, 1994). Thus, a general orienting deficit can be ruled out for this group on the basis of available data.

Several scattered findings involving several other psychophysiological measures have appeared in individual studies, none yet replicated and none markedly distinct from those obtained in anhedonics. Simons and Katkin (1985) reported more between-subjects variance in perceptual aberrator's eye-tracking than in anhedonics', which as already noted was more variable than controls'. (Simons, personal communication, August 21, 1989, was not able to replicate this effect in a third sample.) G.A. Miller (1986) observed less resting EEG theta and larger O-wave during pitch discrimination for perceptual aberrators (and anhedonics) relative to controls. Rockstroh et al. (1982) reported a failure of perceptual aberrators to learn self-control of O-wave via biofeedback, although control of E-wave was normal. In the PINV loss-of-control paradigm, Lutzenberger et al. (1981) found PINV enhancement and systematic eye-movement artifact in perceptual aberrators, similar to what they found in anhedonics. In contrast, Ward, Catts, Armstrong, and McConaghy (1984) reported a smaller PINV in perceptual aberrators. Because theirs was not a standard PINV paradigm, it is difficult to determine whether this finding contradicts the results of Lutzenberger et al.

A clearer picture emerges, however, regarding heart rate data across studies. In G.A. Miller's (1986) pitch-discrimination study, involving no emotional stimuli, perceptual aberrators (and anhedonics) produced more cardiac deceleration than controls. Several studies have observed enhanced cardiac acceleration for perceptual aberrators in the face of potentially aversive stimuli, and in each case they differed from anhedonics as well as from controls. Simons (1980) reported supranormal acceleration during habituation to baby cries but not simple tones or telephone rings. Lutzenberger et al. (1981) reported obtaining it to an aversive imperative stimulus after loss of control. Rockstroh et al. (1982) found it in response to negative reinforcement in the CNV biofeedback paradigm. In two CNV studies using either nude slides or aversive tones, perceptual aberrators failed to differentiate the warning stimuli in heart rate acceleration, in contrast to controls, who produced larger responses than those before neutral stimuli (Simons, 1980). The one exception to this pattern for heart rate is an unpublished study by Simons (personal communication, July 22, 1991), in which perceptual aberrators' heart rate resembled controls during an emotional imagery task.

Interpretation of these diverse findings is difficult. However, the skin conductance and heart rate data do suggest one possible characterization of perceptual aberrators. Lang, Bradley, and Cuthbert (1990) proposed that emotion may be viewed in dimensional terms, involving arousal and valence, with skin conductance a measure of the former and heart rate a measure of the latter. Applying this view in light of the normal skin conductance but exaggerated heart rate findings for perceptual aberrators, we proposed that this group differs from controls in their handling of the emotional valence dimension (G.A. Miller & Yee, 1994). Specifically, they appear to experience normally aversive stimuli as even more aversive than controls do. It is possible, if not straightforward, to align this view with the basic fact that this at-risk group reports an unusual number of experiences of perceptual distortions and related unusual experiences and beliefs. This view is also consistent with the relatively large number of clinical symptoms such subjects report during diagnostic interviews (Chapman & Chapman, 1987).

A different characterization of perceptual aberrators emerges from two studies from Simons's laboratory using a startle-probe paradigm. As noted earlier, the procedure (Graham, 1979) produces backward-masking phenomena that have been used in numerous studies of schizophrenics to examine possible problems in stimulus filtering and short-term memory (Braff et al., 1978). Simons and Giardina (1992) found inhibition of the startle-blink reflex to be similar in anhedonics and controls, but perceptual aberrators showed reduced reflex inhibition at short interstimulus intervals (ISIs). Although one study using a more limited set of ISIs (Blumenthal, Schicatano, Norris, & Chapman, 1993) found no group differences (replicating Simons & Giardina, 1992, at those ISIs), using a wider range of ISIs, Schell et al. (1995) replicated Simons and Giardina's positive findings in a similar passive task and extended them to an active, attentional task. Perlstein, Fiorito, Simons, and Graham (1989) replicated the passive-task finding and also found that, like schizophrenics (Boutros, Zouridakis, & Overall, 1991; Deldin, Miller, & Gergen, 1991; Freedman, Adler, & Waldo, 1987; Judd, McAdams, Budnick, & Braff, 1992; though see Kathmann & Engel, 1990), perceptual aberrators failed to suppress the amplitude of the P50 component of the ERP. Simons et al. (1993) interpreted their data in terms of long-standing notions of sensory overload that is due to inadequate filtering in schizophrenia. Whereas they suggested that anhedonia is characterized by a problem in emotional processing—specifically, the coherence of the representation of emotion in memory—they suggested that perceptual aberrators have a problem in perceptual information processing, specifically, the filtering of transient input. The parameters of this filtering problem deserve more research. Blumental et al. (1993) did not replicate the reduced reflex inhibition in perceptual aberration, and Chmielewski (1994) did not replicate the P50 suppression effect. A new study by Simons and colleagues (personal communication, August 3, 1994) suggests that the startle reflex findings for perceptual aberrators depend on mode of stimulus presentation or type of task. Preliminary analyses indicated that startle stimuli during imagery of emotional slides demonstrate group differences more clearly than the startle stimuli during actual presentation of the emotional slides later imagined.

This picture is somewhat complicated by the recent report that perceptual aberrators show an enhanced affective modulation of the startle reflex (Cook, personal communication, May 28, 1994). Presently available data do not allow a choice between or an integration among the hypothesis that perceptual aberrators are overly responsive to aversive stimuli, the hypothesis that they suffer from a defect in input filtering, or the possibility that one of these phenomena is a result of the other (e.g., that stimuli are aversive because of inadequate filtering mechanisms). Further studies are needed to pursue these interpretations.

The Wisconsin Scales: Summary and Future Directions

Although the evidence for overt behavioral performance differences is suggestive, the psychophysiological differences have more clearly distinguished questionnaire-defined, nonpatient anhedonics and perceptual aberrators from controls. The studies that produced these findings have been diverse in paradigm, task, and stimuli. A variety of psychophysiological measures have been used, and standard interpretations of their functional significance differ in content, specificity, and degree of consensus. Such a diversity of findings inhibits the emergence of a single, fundamental interpretation covering all findings to date. However, it is striking that anhedonics' and perceptual aberrators' overt performance has consistently been normal in the psychophysiology literature and that at least some of the psychophysiological findings appear to be task dependent. These individuals are generally able to perform normally in the laboratory, consistent with their ability to function adequately in daily life, unlike seriously disturbed patients.

The few positive laboratory performance findings deserve further attention. In each case, the psychosis-prone groups differed from controls in the way that schizophrenics do. Thus, these findings must be taken seriously. Importantly, the normal overt performance of these at-risk individuals across the wide range of psychophysiological studies clearly argues against a generalized performance deficit.

The psychophysiological data on anhedonics suggest they (a) have a pleasure deficit, (b) are relatively unresponsive to stimuli that normally elicit an orienting response, and (c) are nevertheless able to process such stimuli by some means that allows them to produce normal overt behavior. An anomaly in anhedonics' orienting and maintenance of working memory (in one form or another) is especially interesting given the view that these phenomena in normals require availability and appropriate allocation of limited processing resources (e.g., Dawson, 1990; Öhman, 1979; Siddle, 1991). Öhman (1981; Öhman, Nordby, & d'Elia, 1986) has proposed specifically that orienting anomalies arise two ways in schizophrenia: both hyporesponding to task-relevant stimuli (see also Nuechterlein & Dawson, 1984) and hyperresponding (distractibility) to task-irrelevant stimuli. Continued research is needed to explore the apparent resource allocation deficit and possible N2b-indexed compensatory mechanism in anhedonia. This could lead to prevention efforts, if training methods can be developed to foster more efficient allocation in subjects at risk.

Further exploration is also warranted concerning anhedonics' abnormal processing of emotional stimuli. As discussed earlier, Simons et al. (1993) have suggested that the problem is not simply an attenuation of response amplitude but a disorganization of the various response components in emotion. Specifically, they proposed that the difficulty experienced by anhedonics concerns the representation of positive emotion in memory. This view complements the orienting perspective on information processing deficits in short-term memory described previously, because Simons's proposal for emotional processing deficits involves longer-term memory.

Much less can be said about the perceptual aberrators at present. Psychophysiological study of this group has largely been pursued with general notions of what to look for but without the specific hypotheses (G.A. Miller, 1986) or face-valid findings (G.A. Miller et al., 1981; Simons et al., 1982a) that emerged early on for the anhedonics. Thus, studies with these groups have been more exploratory. Several laboratories are pursuing a variety of studies focusing on perceptual processes in the perceptual aberration group. Based on both face validity and recent empirical success, this direction appears to be the most promising for characterizing this group.

Remaining Questions

The literature on psychometrically selected subjects potentially at risk for schizophrenia has begun to progress from asking *whether* there is a difference to *why* there is a difference for the at-risk groups. Some combination of psychometric, behavioral performance, and psychophysiological measures may offer much to psychopathology research, intervention, and prevention. Still, at least five related questions remain unanswered.

First, the predictive validity and specificity of anhedonia and perceptual aberration for future psychopathology are not yet fully established. Prospective follow-up results for perceptual aberrators are very promising, and as noted earlier, on nearly all measures, anhedonics fell between perceptual aberrators and controls, though the results were not significantly different from controls on any individual measure; differences between anhedonics and perceptual aberrators were not tested (Chapman et al., 1994).

Second, it is not known whether self-report on these scales will bear on psychopathology in the same manner across socially and ethnically diverse groups. Racial minorities have been systematically excluded in most of the research with these scales because of anecdotal evidence of ethnic-group differences on the scales and the lack of within-group norms. The latter have only recently become available (Chmielewski, Fernandes, Yee, & Miller, in press).

Third, it is not clear whether anhedonics' or perceptual aberrators' psychophysiological abnormalities are associated with specific types of psychopathology or only with nonspecific risk for a range of disorders. Longitudinal follow-up of psychophysiologically assessed anhedonics and perceptual aberrators is needed.

Fourth, the sensitivity of behavioral and psychophysiological measures used to assess these subjects is not known. Because the Chapman scales assess a continuous distribution of putatively psychosis-related features within a population, categorically labeling an individual as an anhedonic or perceptual aberrator can be arbitrary in some cases. This dimensional quality of the Chapman scales further complicates the interpretation of results from behavioral or psychophysiological assessment in these individuals.

Fifth, it is unclear whether individuals' scores on the Chapman scales change as a function of developmental stage or how such changes may relate to the behavioral performance or psychophysiological findings. While the majority of work with the Chapman scales has been completed with late-adolescent and early adult subjects, earlier identification of subjects at risk for psychosis and longitudinal follow-up are necessary before a comprehensive theory about such individuals can be proposed.

Acknowledgments. Portions of this work were supported by National Institute of Mental Health grants MH39628 and MH49296 and by the University of Illinois Research Board. We gratefully acknowledge consultation with Phillip M. Chmielewski, Michael G.H. Coles, Emanuel Donchin, Raymond A. Knight, Arthur F. Kramer, Daniel N. Klein, Risto Näätänen, Robert F. Simons, and Cindy M. Yee.

References

Allen, J., & Schuldberg, D. (1989). Positive thought disorder in a hypothetically psychosis-prone population. *Journal of Abnormal Psychology, 98*, 491–494.

Allen, J.J., Chapman, L.J., & Chapman, J.P. (1987). Cognitive slippage and depression in hypothetically psychosis-prone college students. *The Journal of Nervous and Mental Disease, 175*, 347–353.

Allen, J.J., Chapman, L.J., Chapman, J.P., Vuchetich, J.P., & Frost, L.A. (1987). Prediction of psychoticlike symptoms in hypothetically psychosis-prone college students. *Journal of Abnormal Psychology, 96*, 83–88.

Asarnow, R.F., Nuechterlein, K.H., & Marder, S.R. (1983). Span of apprehension performance, neuropsychological functioning, and indices of psychosis-proneness. *The Journal of Nervous and Mental Disease, 171*, 662–669.

Balogh, D.W., & Merritt, R.D. (1985). Susceptibility to Type A backward pattern masking among hypothetically psychosis-prone college students. *Journal of Abnormal Psychology, 94*, 377–383.

Balogh, D.W., & Merritt, R.D. (1987). Visual masking and the schizophrenia spectrum. Interfacing clinical and experimental methods. *Schizophrenia Bulletin, 13*, 679–698.

Beckfield, D.F. (1985). Interpersonal competence among college men hypothesized to be at risk for schizophrenia. *Journal of Abnormal Psychology, 94*, 397–404.

Berenbaum, H., Snowhite, R., & Oltmanns, T.F. (1987). Anhedonia and emotional responses to affect evoking stimuli. *Psychological Medicine, 17*, 677–684.

Bernstein, A.S. (1987). Orienting response research in schizophrenia: Where we have gone and where we might go. *Schizophrenia Bulletin, 13*, 623–641.

Bernstein, A.S. (1992). The orienting response as an index of attentional dysfunction in schizophrenia. In B.A. Campbell, H. Hayne, & R. Richardson (Eds.), *Attention and information processing in infants and adults* (pp. 297–323). Hillsdale, NJ: Lawrence Erlbaum.

Bernstein, A.S., & Reidel, J.A. (1987). Psychophysiological response patterns in college students with high physical anhedonia: Scores appear to reflect schizotypy rather than depression. *Biological Psychiatry, 22*, 829–847.

Blumenthal, T.D., Schicatano, E.J., Norris, C.M., & Chapman, J.G. (1993). Normal startle eyeblink inhibition in psychosis-prone college students. *Psychophysiology, 30* (Suppl. 1), 18.

Boutros, N.N., Zouridakis, G., & Overall, J. (1991). Replication and extension of P50 findings in schizophrenia. *Clinical Electroencephalography, 22*, 40–45.

Braff, D., Stone, C., Callaway, E., Geyer, M., Glick, I., & Bali, L. (1978). Prestimulus effects on human startle reflex in normals and schizophrenics. *Psychophysiology, 15*, 339–343.

Brecher, M., & Begleiter, H. (1983). Event-related potentials to incentive stimuli in unmedicated schizophrenic patients. *Biological Psychiatry, 18*, 661–674.

Bruder, G.E., Quitkin, F.M., Stewart, J.W., Martin, C., Voglmaier, M.M., & Harrison, W.M. (1989). Cerebral laterality and depression: Differences in perceptual asymmetry among diagnostic subtypes. *Journal of Abnormal Psychology, 98*, 177–186.

Bull, K., & Lang, P.J. (1972). Intensity judgments and physiological response amplitude. *Psychophysiology, 9*, 428–436.

Callaway, E., & Naghdi, S. (1982). An information processing model for schizophrenia. *Archives of General Psychiatry, 33*, 339–347.

Chapman, L.J., & Chapman, J.P. (1978). *Revised physical anhedonia scale.* Unpublished test.

Chapman, L.J., & Chapman, J.P. (1985). Psychosis proneness. In M. Alpert (Ed.), *Controversies in schizophrenia* (pp. 157–172). New York: Guilford Press.

Chapman, L.J., & Chapman, J.P. (1987). The search for symptoms predictive of schizophrenia. *Schizophrenia Bulletin, 13*, 497–503.

Chapman, L.J., & Chapman, J.P. (1992, November). *Follow-up of psychosis-prone subjects.* Paper presented at the meeting of the Society for Research in Psychopathology, Palm Springs, CA.

Chapman, L.J., Chapman, J.P., Kwapil, T.R., Eckblad, M., & Zinser, M.C. (1994). Putatively psychosis-prone subjects ten years later. *Journal of Abnormal Psychology, 103*, 171–183.

Chapman, L.J., Chapman, J.P., Raulin, M.L. (1976). Scales for physical and social anhedonia. *Journal of Abnormal Psychology, 85*, 374–382.

Chapman, L.J., Chapman, J.P., Raulin, M.L. (1978). Body-image aberration in schizophrenia. *Journal of Abnormal Psychology, 87*, 399–407.

Chmielewski, P.M. (1994). *Auditory sensory gating in groups at-risk for schizophrenia and mood disorders.* Unpublished master's thesis, University of Illinois at Urbana–Champaign, Champaign, IL.

Chmielewski, P.M., Fernandes, L.O.L., Yee, C.M., & Miller, G.A. (in press). Ethnicity and gender in scales of psychosis-proneness and mood disturbances. *Journal of Abnormal Psychology.*

Clementz, B.A., Grove, W.M., Katsanis, J., & Iacono, W.G. (1991). Psychometric detection of schizotypy: Perceptual aberration and physical anhedonia in relatives of schizophrenics. *Journal of Abnormal Psychology, 100*, 607–612.

Clementz, B.A., & Sweeney, J.A. (1990). Is eye movement dysfunction a biological marker for schizophrnia? A methodological review. *Psychological Bulletin, 108*, 77–92.

Coles, M.G.H. (1989). Modern mind-brain reading: Psychophysiology, physiology, and cognition. *Psychophysiology, 26*, 251–269.

Coryell, W., Keller, M., Lavori, P., & Endicott, J. (1990). Affective syndromes, psychotic features, and prognosis: I. Depression. *Archives of General Psychiatry, 47*, 651–657.

Davidson, R.J. (1992). Anterior cerebral asymmetry and the nature of emotion. *Brain and Cognition, 20*, 125–151.

Davidson, R.J., Chapman, J.P., Chapman, L.J., & Henriques, J.B. (1990). Asymmetrical brain electrical activity discriminates between psychometrically-matched verbal and spatial cognitive tasks. *Psychophysiology, 27*, 528–543.

Dawson, M.E. (1990). Psychophysiology at the interface of clinical science, cognitive science, and neuroscience. *Psychophysiology, 27*, 243–255.

Dawson, M.E., & Nuechterlein, K.H. (1984). Psychophysiological dysfunctions in the developmental course of schizophrenic disorders. *Schizophrenia Bulletin, 10*, 204–232.

Dawson, M.E., Nuechterlein, K.H., & Schell, A.M. (1992). Electrodermal anomalies in recent-onset schizophrenia: Relationship to symptoms and prognosis. *Schizophrenia Bulletin, 18*, 295–311.

Dawson, M.E., Schell, A.M., Hazlett, E.A., Filion, D.L., & Nuechterlein, K.H. (1995). Attention, startle eye-blink modification, and psychosis proneness. In A. Raine, T. Lencz, & S.A. Mednick (Eds.), *Schizotypal personality disorder* (pp. 250–271). Cambridge, England: Cambridge University Press.

Deldin, P.J., Miller, G.A., & Gergen, J.A. (1991, December). *Partial evidence for P50 suppression in schizophrenia*. Paper presented at the meeting of the Society for Research in Psychopathology, Boston.

Donchin, E., & Coles, M.G.H. (1988). Is the P300 component a manifestation of context updating? *Brain and Behavioral Sciences, 11*, 357–374.

Drewer, H.B., & Shean, G.D. (1993). Reaction time crossover in schizotypal subjects. *Journal of Nervous and Mental Disease, 181*, 27–30.

Ebert, L. (1987). *Orienting and habituation in anhedonic and dysthymic college students*. Unpublished master's thesis, University of Illinois. Urbana, IL.

Eckblad, M., & Chapman, L.J. (1983). Magical ideation as an indicator of schizotypy. *Journal of Consulting and Chinical Psychology, 51*, 215–225.

Edell, W.S., & Chapman, L.J. (1979). Anhedonia, perceptual aberration, and the Rorschach. *Journal of Consulting and Clinical Psychology, 47*, 377–384.

Edell, W.S., & Joy, S.P. (1989, November). *Psychosis proneness in young hospitalized psychiatric patients*. Paper presented at the meeting of the Society for Research in Psychopathology, Miami.

Edell, W.S., & Kaslow, N.J. (1991). Parental perception and psychosis proneness in college students. *American Journal of Family Therapy, 19*, 195–205.

Eikmeier, G., Lodemann, E., Olbrich, H.M., Pach, J., Zerbin, D., & Gastpar, M. (1992). Altered frontocentral PINV topography and the primary negative syndrome in schizophrenia. *Schizophrenia Research, 8*, 251–256.

Etienne, M.A., Deldin, P.J., Giese-Davis, J., & Miller, G.A. (1990, October). *Differences in EEG distinguish populations at risk for psychopathology*. Paper presented at the meeting of the Society for Psychophysiological Research, Boston.

Fawcett, J., Clark, D.C., Scheftner, W.A., & Hedeker, D. (1983). Differences between anhedonic and normally hedonic depressive states. *American Journal of Psychiatry, 140*, 1027–1029.

Fernandes, L.O.L. (1992). *Brain electrical activity in late adolescent dysthymia and anhedonia*. Unpublished master's thesis, University of Illinois, Urbana, IL.

Fernandes, L.O.L., Giese-Davis, J.E., Hicks, B.D., Klein, D.N., & Miller, G.A. (submitted). Converging evidence for a cognitive anomaly in early psychopathology.

Fernandes, L.O.L., & Miller, G.A. (1993). Unpublished analyses.

Fitzgibbons, L., & Simons, R.F. (1992). Affective response to color-slide stimuli in subjects with physical anhedonia: A three-systems approach. *Psychophysiology, 29,* 613–620.

Frazier, M.F., Silverstein, M.L., & Fogg, L. (1989). Lateralized cerebral dysfunction in schizophrenia and depression: Gender and medication effects. *Archives of Clinical Neuropsychology, 4,* 33–44.

Freedman, R., Adler, L.E., & Waldo, M. (1987). Gating of the auditory-evoked potential in children and adults. *Psychophysiology, 24,* 223–227.

Frith, C. (1991). In what context is latent inhibition relevant to the symptoms of schizophrenia? *Behavioral and Brain Sciences, 14,* 28–29.

Gatchel, R.J., & Lang, P.J. (1973). Accuracy of psychophysical judgments and physiological response amplitude. *Journal of Experimental Psychology, 98,* 175–183.

Giese-Davis, J.E., Miller, G.A., & Knight, R.A. (1993). Memory template comparison processes in anhedonia and dysthymia. *Psychophysiology, 30,* 646–656.

Graham, F.K. (1979). Distinguishing among orienting, defense, and startle reflexes. In H.D. Kimmel, E.G. van Olst, & J.F. Orlebeke (Eds.), *The orienting reflex in humans* (pp. 137–167). Hillsdale, NJ: Lawrence Erlbaum.

Gray, J.A., Hemsley, D.R., Feldon, J., Gray, N.S., & Rawlins, J.N.P. (1991). Authors' response. *Behavioral and Brain Sciences, 14,* 56–71.

Grove, W.M. (1982). Psychometric detection of schizotypy. *Psychological Bulletin, 92,* 27–38.

Grove, W.M., Lebow, B.S., Clementz, B.A., Cerri, A., Medus, C., & Iacono, W.G. (1991). Familial prevalence and coaggregation of schizotypy indicators: A multitrait family study. *Journal of Abnormal Psychology, 100,* 115–121.

Haberman, M.C., Chapman, L.J., Numbers, J.S., & McFall, R.M. (1979). Relation of social competence to scores on two scales of psychosis proneness. *Journal of Abnormal Psychology, 88,* 675–677.

Heller, W. (1990). The neuropsychology of emotion: Developmental patterns and implications for psychopathology. In N.L. Stein, B.L. Leventhal, & T. Trabasso (Eds.), *Psychological and biological approaches to emotion* (pp. 167–211). Hillsdale, NJ: Lawrence Erlbaum.

Hicks, B.D., & Miller, G.A. (submitted). *Distinguishing N2b and N2c subcomponents of the event-related brain potential*.

Holzman, P.S. (1987). Recent studies of psychophysiology in schizophrenia. *Schizophrenia Bulletin, 13,* 49–75.

Iacono, W.G. (1982). Bilateral electrodermal habituation–dishabituation and resting EEG in remitted schizophrenics. *Journal of Nervous and Mental Disease, 170,* 91–101.

Iacono, W.G. (1988). Eye movement abnormalities in schizophrenic and affective disorders. In C.W. Johnson & F.J. Pirozzolo (Eds.), *Neuropsychology of eye movements* (pp. 117–143). Hillsdale, NJ: Lawrence Erlbaum.

Itil, T.M. (1977). Qualitative and quantitative EEG findings in schizophrenia. *Schizophrenia Bulletin, 3*, 61–79.

Johnson, R., Jr. (1988). The amplitude of the P300 component of the event-related potential: Review and synthesis. In P.K. Ackles, J.R. Jennings, & M.G.H. Coles (Eds.), *Advances in psychophysiology,* (Vol. 3, pp. 69–137). Greenwich, CT: JAI Press.

Johnson, R., Jr. (1993). On the neural generators of the P300 component of the event-related potential. *Psychophysiology, 30*, 90–97.

Judd, L.L., McAdams, L., Budnick, B., & Braff, D.L. (1992). Sensory gating deficits in schizophrenia: New results. *American Journal of Psychiatry, 149*, 488–493.

Jutai, J.W. (1989). Spatial attention in hypothetically psychosis-prone college students. *Psychiatry Research, 27*, 207–215.

Karson, C.N., Dykman, R.A., & Paige, S.R. (1990). Blink rates in schizophrenia. *Schizophrenia Bulletin, 16*, 345–354.

Kathmann, N., & Engel, R.R. (1990). Sensory gating in normals and schizophrenics: A failure to find strong P50 suppression in normals. *Biological Psychiatry, 27*, 1216–1226.

Katsanis, J., Iacono, W.G., & Beiser, M. (1990). Anhedonia and perceptual aberration in first-episode psychotic patients and their relatives. *Journal of Abnormal Psychology, 99*, 202–206.

Kendler, K.S., & Hewitt, J. (1992). The structure of self-report schizotypy in twins. *Journal of Personality Disorders, 6*, 1–17.

Klein, D.F. (1974). Endogenomorphic depression: A conceptual and terminological revision. *Archives of General Psychiatry, 31*, 447–454.

Knight, R.A. (1984). Model of cognitive deficit in schizophrenia. In R.A. Dieusthier, W.D. Spaulding, & J.K. Cole (Eds.), *Current thory and research in Schizophrenia* (Vol. 31, pp. 93–156). Lincoln, NE: University of Nebraska Press.

Knight, R.A. (1992). Specifying cognitive deficiencies in poor premorbid schizophrenics. In E.F. Walker, R. Dworkin, & B. Cornblatt (Eds.), *Progress in experimental psychology and psychopathology research* (Vol. 15, pp. 252–289). New York: Springer-Verlag.

Knight, R.A. (1993). Comparing cognitive models of schizophrenics' input dysfunction. In R.L. Cromwell & C.R. Snyder (Eds.), *Schizophrenia: Origins, processes, treatment and outcome.* Oxford, England: Oxford University Press.

Lang, P.J., Bradley, M.M., & Cuthbert, B.N. (1990). Emotion, attention, and the startle reflex. *Psychological Review, 97*, 377–395.

Lang, P.J., Gatchel, R.J., & Simons, R.F. (1975). Electro-cortical and cardiac correlates of psychophysical judgment. *Psychophysiology, 12*, 649–655.

Lang, P.J., Öhman, A., & Vaitl, D. (1988). *The international affective picture system (photographic slides).* Gainesville, FL: Center for Research in Psychophysiology, University of Florida.

Lenzenweger, M.F. (1993). Explorations in schizotypy and the psychometric high-risk paradigm. In L.J. Chapman, J.P. Chapman, & D.C. Fowles (Eds.), *Progress in personality and psychopathology research (Vol. 16, pp. 66-116). New York: Springer-Verlag.

Lenzenweger, M.F., Cornblatt, B.A., & Putnick, M. (1991). Schizotypy and sustained attention. *Journal of Abnormal Psychology, 100*, 84-89.

Lenzenweger, M.F., & Loranger, A.W. (1989a). Detection of familial schizophrenia using a psychometric measure of schizotypy. *Archives of General Psychiatry, 46*, 902-907.

Lenzenweger, M.F., & Loranger, A.W. (1989b). Psychosis proneness and clinical psychopathology: Examination of the correlates of schizotypy. *Journal of Abnormal Psychology, 98*, 3-8.

Levenson, R.W. (1992). Autonomic nervous system differences among emotions. *Psychological Science, 3*, 23-27.

Levin, R., & Raulin, M.L. (1991). Preliminary evidence for the proposed relationship between frequent nightmares and schizotypal symptomatology. *Journal of Personality Disorders, 5*, 8-14.

Loveless, N.E. (1979). Event-related slow potentials of the brain as expressions of orienting function. In H.D. Kimmel, E.G. van Olst, & J.F. Orlebeke (Eds.), *The orienting reflex in humans* (pp. 77-100). Hillsdale, NJ: Lawrence Erlbaum.

Lutzenberger, W., Elbert, T., & Rockstroh, B. (1987). A brief tutorial on the implications of volume conduction for the interpretation of the EEG. *Journal of Psychophysiology, 1*, 81-89.

Lutzenberger, W., Elbert, T., Rockstroh, B., Birbaumer, N., & Stegagno, L. (1981). Slow cortical potentials in subjects with high or low scores on a questionnaire measuring physical anhedonia and body image distortion. *Psychophysiology, 18*, 371-380.

Lutzenberger, W., Birbaumer, N., Rockstroh, B., & Elbert, T. (1983). Evaluation of contingencies and conditional probabilities: A psychophysiological approach to anhedonia. *Archives of Psychiatry and Neurological Sciences, 233*, 471-488.

Manschreck, T.C., & Maher, B.A. (1991). Approximations to a neuropsychological model of schizophrenia. *Behavioral and Brain Sciences, 14*, 36-37.

Martin, E.M., & Chapman, L.J. (1982). Communication effectiveness in psychosis-prone college students. *Journal of Abnormal Psychology, 91*, 420-425.

Meehl, P. (1962). Schizotaxia, schizotypy, schizophrenia. *American Psychologist, 17*, 827-838.

Meehl, P. (1990). Toward an integrated theory of schizotaxia, schizotypy, and schizophrenia. *Journal of Personality Disorders, 4*, 1-99.

Merritt, R.D., & Balogh, D.W. (1989). Backward masking spatial frequency effects among hypotehetically schizotypal individuals. *Schizophrenia Bulletin, 15*, 573-583.

Miller, E.N., & Chapman, L.J. (1983). Continued word association in hypothetically psychosis-prone college students. *Journal of Abnormal Psychology, 92*, 468-478.

Miller, G.A. (1986). Information processing deficits in anhedonia and perceptual aberration: A psychophysiological analysis. *Biological Psychiatry, 21*, 100-115.

Miller, G.A., Simons, R.F., & Lang, P.J. (1981, June). *Electrocortical measures of information processing deficits in anhedonia*. Paper presented at EPIC VI, the Sixth International Conference on Event-Related Slow Potentials of the Brain, Lake Forest, IL.

Miller, G.A., & Yee, C.M. (1985, October). *Affective responsiveness in anhedonia and dysthymia.* Paper presented at the meeting of the Society for Psychophysiological Research, Houston, TX.
Miller, G.A., & Yee, C.M. (1994). Risk for severe psychopathology: Psychometric screening and psychophysiological assessment. In P.K. Ackles, J.R. Jennings, & M.G.H. Coles (Eds.), *Advances in psychophysiology* (Vol. 5, pp. 1–54). London: Jessica Kingsley.
Miller, G.A., Yee, C.M., Anhalt, J.M. (submitted). *ERP incentive effects and potential risk for psychopathology.*
Mirsky, A.F., & Duncan, C.C. (1986). Etiology and expression of schizophrenia: Neurobiological and psychosocial factors. *Annual Review of Psychology, 37,* 291–319.
Mo, S.S., & Chavez, M.R. (1986). Perceptual aberration and brain hemisphere reversal of foreperiod effect on time estimation. *Journal of Clinical Psychology, 42,* 787–792.
Morihisa, J.M., Duffy, F.H., & Wyatt, R.J. (1983). Brain electrical activity mapping (BEAM) in schizophrenic patients. *Archives of General Psychiatry, 40,* 719–728.
Näätänen, R. (1990). The role of attention in auditory information processing as revealed by event-related potentials and other brain measures of cognitive function. *Behavioral and Brain Sciences, 13,* 201–232.
Näätänen, R. (1992). *Attention and brain function.* Hillsdale, NJ: Lawrence Erlbaum.
Neale, J.M. (1982). Information processing and vulnerability: High-risk research. In M.J. Goldstein (Ed.), *Preventing intervention in schizophrenia: Are we ready?* (pp. 78–89). Rockville, MD: National Institute of Mental Health.
Nuechterlein, K.H. (1977). Reaction time and attention in schizophrenia: A critical evaluation of the data and theories. *Schizophrenia Bulletin, 3,* 373–428.
Nuechterlein, K.H. (1990). Methodological considerations in the search for indicators of vulnerability to severe psychopathology. In J.W. Rohrbaugh, R. Parasuraman, & R. Johnson, Jr. (Eds.), *Event-related potentials: Basic issues and applications* (pp. 364–373). New York: Oxford University Press.
Nuechterlein, K.H., & Dawson, M.E. (1984). Information processing and attentional functioning in the developmental course of schizophrenic disorders. *Schizophrenia Bulletin, 10,* 160–203.
Nuechterlein, K.H., & Green, M.F. (1991). Neuropsychological vulnerability or episode factors in schizophrenia? *Behavioral and Brain Sciences, 14,* 37–38.
Numbers, J.S., & Chapman, L.J. (1982). Social deficits in hypothetically psychosis-prone college women. *Journal of Abnormal Psychology, 91,* 255–260.
Öhman, A. (1979). The orienting response, attention, and learning: An information processing perspective. In H.D. Kimmel, E.G. van Olst, & J.F. Orlebeke (Eds.), *The orienting reflex in humans* (pp. 443–471). Hillsdale, NJ: Lawrence Erlbaum.
Öhman, A. (1981). Electrodermal activity and vulnerability to schizophrenia: A review. *Biological Psychology, 12,* 87–145.
Öhman, A., Nordby, H., & d'Elia, G. (1986). Orienting and schizophrenia: Stimulus significance, attention, and distraction in a signaled reaction time task. *Journal of Abnormal Psychology, 95,* 326–334.

Overby, L.A. (1992). Perceptual asymmetry in psychosis-prone college students: Evidence for left-hemisphere overactivation. *Journal of Abnormal Psychology, 101*, 96–103.

Perlstein, W.M., Fiorito, E., Simons, R.F., & Graham, F.K. (1989). Prestimulation effects on reflex blink and EPs in normal and schizotypal subjects. *Psychophysiology, 26*, S48.

Perosa, L.M., & Simons, R.F. (1991, August). *Family perceptions of adolescents at risk for schizophrenia.* Paper presented at the meeting of the American Psychological Association, San Francisco.

Pfefferbaum, A., Ford, J., White, P., & Roth, W.T. (1989). P3 in schizophrenia is affected by stimulus modality, response requirements, medication status and negative symptoms. *Archives of General Psychiatry, 46*, 1035–1044.

Pierson, A., Loas, G., & Lesevre, N. (1990). Étude de potentiels evoques cognitifs en fonction de la valence affective et de la signification des stimulus chez des sujets sains anhédoniques avec attitudes dysfonctionnelles. *L'Encephale, 16*, 209–216.

Pierson, A., Ragot, R., Ripoche, A., & Lesevre, N. (1987). Electrophysiological changes elicited by auditory stimuli given a positive or negative value: A study comparing anhedonic with hedonic subjects. *International Journal of Psychophysiology, 5*, 107–123.

Pierson, A., Ragot, R., Ripoche, A., & Lesevre, N. (1988). Modifications d'indices d'activation varies en fonction de la valence acquise par un stimulus chez des sujets anhédoniques et dépressogenes. *Neurophysiologie Clinique, 18*, 33–49.

Pogue-Geile, M.F., & Oltmanns, T.F. (1980). Sentence perception and distractibility in schizophrenic, manic, and depressed patients. *Journal of Abnormal Psychology, 89*, 115–124.

Pritchard, W.S. (1986). Cognitive event-related potential correlates of schizophrenia. *Psychological Bulletin, 100*, 43–66.

Pritchard, W.S., Shappell, S.A., & Brandt, M.E. (1991). Psychophysiology of N200/N400: A review and classification scheme. In J.R. Jennings, P.K. Ackles, & M.G.H. Coles (Eds.), *Advances in psychophysiology* (Vol. 4, pp. 43–106). London: Jessica Kingsley.

Raulin, M.L., & Henderson, C.A. (1987). Perception of implicit relationships between personality traits by schizotypic college subjects: A pilot study. *Journal of Clinical Psychology, 43*, 463–467.

Rockstroh, B., Elbert, T., Birbaumer, N., & Lutzenberger, W. (1982). *Slow brain potentials and behavior.* Baltimore: Urban & Schwarzenberg.

Rohrbaugh, J.W., & Gaillard, A.W.K. (1983). Sensory and motor aspects of the contingent negative variation. In A.W.K. Gaillard & W. Ritter (Eds.), *Tutorials in event-related potential research: Endogenous components* (pp. 269–310). Amsterdam, the Netherlands: North-Holland.

Roth, W.T. (1983). A comparison of P300 and skin conductance response. In A.W.K. Gaillard & W. Ritter (Eds.), *Tutorials in event-related potential research: Endogenous components* (pp. 177–199). Amsterdam, the Netherlands: North-Holland.

Salisbury, D.F., O'Donnell, B.F., McCarley, R.W., Nestor, P.G., Faux, S.F., & Smith, R.S. (1994). Parametric manipulations of auditory stimuli differentially

affect P3 amplitude in schizophrenics and controls. *Psychophysiology, 31*, 29-36.
Sams, M., Alho, K., & Näätänen, R. (1983). Sequential effects on the ERP in discriminating two stimuli. *Biological Psychology, 17*, 41-58.
Schell, A.M., Dawson, M.E., Hazlett, E.A., & Filion, D.L. (1995). Attentional modulation of startle in psychosis-prone college students. *Psychophysiology, 32*, 266-273.
Schuldberg, D., & London, A. (1989). Psychological differentiation and schizotypal traits: Negative results with the Group Embedded Figures Test. *Perceptual and Motor Skills, 68*, 1219-1226.
Serafetinides, E.A., Coger, R.W., Martin, J., & Dymond, A.M. (1981). Schizophrenic symptomatology and cerebral dominance patterns: A comparison of EEG, AER, and BPRS measures. *Comprehensive Psychiatry, 22*, 218-225.
Shagass, C. (1976). An electrophysiological view of schizophrenia. *Biological Psychiatry, 11*, 3-30.
Siddle, D.A.T. (1991). Orienting, habituation, and resource allocation: An associative analysis. *Psychophysiology, 28*, 245-259.
Silverstein, S.M., Raulin, M.L., Pristach, E.A., & Pomerantz, J.R. (1992). Perceptual organization and schizotypy. *Journal of Abnormal Psychology, 101*, 265-270.
Simons, R.F. (1980). *The psychophysiology of the schizotype: Electodermal, electrocortical, heart rate, and eye tracking characteristics of subjects reporting physical anhedonia or body image aberration.* Unpublished doctoral dissertation, University of Wisconsin-Madison.
Simons, R.F. (1981). Electrodermal and cardiac orienting in psychometrically defined high-risk subjects. *Psychiatry Research, 4*, 347-356.
Simons, R.F. (1982). Physical anhedonia and future psychopathology: A possible electrocortical continuity. *Psychophysiology, 19*, 433-441.
Simons, R.F. (1988). Event-related slow brain potentials: A perspective from ANS psychophysiology. In *Advances in Psychophysiology* (Vol. 3, pp. 223-267). Greenwich, CT: JAI Press.
Simons, R.F., Fitzgibbons, L., & Fiorito, E. (1993). Emotion-processing in anhedonia. In N. Birbaumer & A. Öhman (Eds.), *The organization of emotion* (pp. 288-306). Toronto: Hogrefe.
Simons, R.F., & Giardina, B.D. (1992). Reflex modification in psychosis-prone young adults. *Psychophysiology, 29*, 8-16.
Simons, R.F., & Katkin, W. (1985). Smooth pursuit eye movements in subjects reporting physical anhedonia and perceptual aberrations. *Psychiatry Research, 14*, 275-289.
Simons, R.F., & Lang, P.J. (1976). Psychophysical judgment: Electro-cortical and heart rate correlates of accuracy and uncertainty. *Biological Psychology, 4*, 51-64.
Simons, R.F., MacMillan, F.W., & Ireland, F.B. (1982a). Anticipatory pleasure deficit in subjects reporting physical anhedonia: Slow cortical evidence. *Biological Psychology, 14*, 297-310.
Simons, R.F., MacMillan, F.W., & Ireland, F.B. (1982b). Reaction-time crossover in preselected schizotypic subjects. *Journal of Abnormal Psychology, 91*, 414-419.

Simons, R.F., Öhman, A., & Lang, P.J. (1979). Anticipation and response set: Cortical, cardiac, and electrodermal correlates. *Psychophysiology, 16*, 222–233.

Simons, R.F., & Russo, K.R. (1987). Event-related potentials and continuous performance in subjects with physical anhedonia or perceptual aberrations. *Journal of Psychophysiology, 4*, 401–410.

Simons, R.F., Russo, K.R., & Hoffman, J.E. (1988). Event-related potentials and eye-movement relationships during psychophysical judgments: The biasing effect of rejected trials. *Journal of Psychophysiology, 2*, 27–37.

Sponheim, S.R., Clementz, B.A., Iacono, W.G., & Beiser, M. (1994). Resting EEG in first-episode and chronic schizophrenia. *Psychphysiology, 31*, 37–43.

Spring, B., Lemon, M., Weinstein, L., & Haskell, A. (1989). Distractibility in schizophrenia: State and trait aspects. *British Journal of Psychiatry, 155*, 63–68.

Strayer, D.L., & Kramer, A.F. (1990). Attentional requirements of automatic and controlled processing. *Journal of Experimental Psychology: Learning, Memory, and Cognition, 16*, 67–82.

Venables, P.H. (1984). Cerebral mechanisms, autonomic responsiveness, and attention in schizophrenia. In R.A. Diesuthier, W.D. Spaulding, & J.K. Cole (Eds.), *Current theory and research in schizophrenia* (Vol. 31, pp. 47–91). Lincoln, NE: University of Nebraska Press.

Venables, P., & Bernstein, A. (1983). The orienting response and psychopathology: Schizophrenia. In D. Siddle (Ed.), *Orienting and habituation: Perspectives in human research* (pp. 475–504). New York: Wiley.

Volavka, J., Abrams, R., Taylor, M.A., & Recker, D. (1981). Hemispheric lateralization of fast EEG activity in schizophrenia and endogenous depression. *Advances in Biological Psychiatry, 6*, 72–75.

Wale, J., & Carr, V. (1990). Differences in dichotic listening asymmetries in depression according to symptomatology. *Journal of Affective Disorders, 18*, 1–9.

Walter, W.G., Cooper, R., McCallum, W.C., & Winter, A.L. (1964). Contingent negative variation: An electric sign of sensory motor association and expectancy in the human brain. *Nature, 203*, 380–384.

Ward P.B., Catts, S.V., Armstrong, M.S., & McConaghy, N. (1984). P300 and psychiatric vulnerability in university students. *Annals of the New York Academy of Sciences, 425*, 645–652.

Ward, P.B., Catts, S.V., Fox, A.M., Michie, P.T., & McConaghy, N. (1991). Auditory selective attention and event-related potentials in schizophrenia. *British Journal of Psychiatry, 158*, 534–539.

Ward, P.B., McConaghy, N., & Catts, S.V. (1991). Word association and measures of psychoisis proneness in university students. *Personality and Individual Differences, 12*, 473–480.

Weerts, T.C., & Lang, P.J. (1973). The effects of eye fixation and stimulus and response location on the contingent negative variation (CNV). *Biological Psychology, 1*, 1–19.

Wickens, C.D., Kramer, A., Vanasse, L., & Donchin, E. (1983). The performance of concurrent tasks: A psychophysiological analysis of the reciprocity of information processing resources. *Science, 221*, 1080–1082.

Wilkins, S., & Venables, P.H. (1992). Disorder of attention in individuals with schizotypal personality. *Schizophrenia Bulletin, 18*, 717–723.

Yee, C.M., Deldin, P.J., & Miller, G.A. (1992). Stimulus intensity effects in dysthymia and anhedonia. *Journal of Abnormal Psychology, 101*, 230–233.

Yee, C.M., & Miller, G.A. (1988). Emotional information processing: Modulation of fear in normal and dysthymic subjects. *Journal of Abnormal Psychology, 97*, 54–63.

Yee, C.M., & Miller, G.A. (1994). A dual-task analysis of resource allocation in dysthymia and anhedonia. *Journal of Abnormal Psychology, 103*, 625–636.

Zahn, T.P. (1986). Psychophysiological approaches to psychopathology. In M.G.H. Coles, E. Donchin, & S.W. Porges (Eds.), *Psychophysiology: Systems, processes and applications* (pp. 508–610). New York: Guilford Press.

Zahn, T.P. (1988). Studies of autonomic psychophysiology and attention in schizophrenia. *Schizophrenia Bulletin, 14*, 205–208.

Zborowski, M.J., & Garske, J.P. (1993). Interpersonal deviance and consequent social impact in hypothetically schizophrenia-prone men. *Journal of Abnormal Psychology, 102*, 482–489.

3
Expressed Emotion: Toward Clarification of a Critical Construct

JILL M. HOOLEY, LAURA R. ROSEN, AND JOHN E. RICHTERS

The most useful and productive concepts in psychopathology are often the ones we understand the least well. This has always been true of such concepts as stress, resilience, and protective factors, and it is no less true of the expressed emotion (EE) construct, a relatively recent addition to the list. Operationally, EE is a measure of the extent to which the relative of a psychiatric patient talks about the patient in a critical, hostile, or emotionally overinvolved way during a semistructured clinical interview. Although the precise nature of the EE construct is not well understood, it is widely believed to reflect an underlying critical and/or negative attitude of the family member toward the patient that expresses itself in daily interactions. Empirical support for this assumption is provided by the results of several laboratory-based studies showing that high levels of EE are associated with more negative patient–relative interactions (Hahlweg et al., 1989; Hooley, 1986; Hooley, 1990; Hooley & Hahlweg, 1986; Kuipers, Sturgeon, Berkowitz, & Leff, 1983; Miklowitz, Goldstein, Falloon, & Doane, 1984; Mueser et al., 1993; Strachan, Leff, Goldstein, Doane, & Burtt, 1986).

The popularity of EE among researchers, however, stems not merely from its usefulness as an indirect index of interactions between patients and family members, but from its usefulness in predicting psychiatric relapse. Numerous studies have shown that psychiatric patients who return home following hospitalization to live with family members who are rated as high in EE suffer relapse rates that are significantly higher than the rates suffered by patients who live with less critical family members (e.g., Brown, Birley, & Wing, 1972; Hooley, Orley, & Teasdale, 1986; Miklowitz, Goldstein, Nuechterlein, Snyder, & Mintz, 1988; Vaughn & Leff, 1976a; Vaughn, Snyder, Jones, Freeman, & Falloon, 1984). Moreover, interventions designed to reduce EE levels in family members have been shown to reduce the likelihood of early psychiatric relapse in discharged patients (Hogarty et al., 1986; Leff, Kuipers, Berkowitz, Eberlein-Fries, & Sturgeon, 1982; Leff et al., 1989; Tarrier et al., 1988).

By the early 1980s, however, the EE construct was in danger of falling victim to its own success. Researchers began to bring premature closure to the issue of causality by assuming that high levels of EE (a) reflected trait-like characteristics of family members and (b) were causally important in bringing about psychiatric relapse in patients. Some nonreplications of the EE–relapse link, however, have generated new concerns about these assumptions. Psychosocial interventions have also been criticized because of their implicit assumption that families influence the course of schizophrenia. And outside the scientific community, concerns were raised that "high expressed emotion" was being used as a synonym for the discredited "schizophrenogenic family" model of an earlier generation (e.g., Lefley, 1992). Not surprisingly, many families who carry the burden of care for severely impaired psychiatric patients have become angered by the implication that they are somehow responsible for the poor prognosis of their psychiatrically ill family members. For all of these reasons, earlier uncritical assumptions about the EE construct have given way to renewed scientific and political controversy.

In light of these developments, it seems useful and timely to step back from the research enterprise and take stock of what we know and do not know about the EE construct. In this chapter, we provide a selective review of recent developments in the EE literature in an attempt to extend beyond earlier reviews (Hooley, 1985; Koenigsberg & Handley, 1986; Kuipers, 1979; Leff & Vaughn, 1985). We begin by describing the measurement of EE and the contexts in which high levels of EE have been found. Next we examine the evidence linking high levels of EE to increased risk of psychiatric relapse. Finally, we consider three competing explanatory models of the EE–relapse link, evaluate the plausibility of those models in the light of existing data, and suggest avenues of inquiry that hold promise for increasing our understanding of what is clearly a much more complex construct than many have previously believed.

Expressed Emotion as a Measure of Family Environment

Expressed emotion is a summary rating of the content and affective tone of comments made by a relative about a specific family member during a private 1- to 2-h semistructured interview about that family member. This interview, the Camberwell Family Interview (CFI; Vaughn & Leff, 1976b), has most often been used to query the family members of psychiatric patients about events leading to the patient's current hospitalization. Within the structure of the interview, relatives are allowed to talk freely about the patient and his or her symptoms. The interview is audiotaped for later coding by a trained rater.[1]

The EE rating is based primarily on the number of critical remarks the relative makes about the patient during the course of the interview. Critical remarks are those that, based on either content or voice tone, indicate dislike or disapproval of some characteristic of the patient or his or her behavior. According to convention, relatives of schizophrenic patients are rated as being high in EE if they make six or more critical remarks during the course of the CFI. But, even in the absence of high levels of criticism, a high EE rating can be assigned if the relative demonstrates any evidence of hostility (i.e., generalized criticism or rejection of the patient) or shows high levels of emotional overinvolvement (EOI) (a dramatic or exaggerated response to the patient's illness, overconcern, or extreme protectiveness or self-sacrifice). Relatives who are neither critical, hostile, nor emotionally overinvolved are rated as being low in EE. More detailed information about the assessment of EE can be found in Leff and Vaughn (1985).

It should be clear from the preceding description that the term *expressed emotion* is something of a misnomer. The EE index represents a heterogenous mix of criticism/hostility and EOI. Early EE research demonstrated that both of these variables predicted patients' relapses (e.g., Brown et al., 1972). In all other respects, however, criticism and EOI appear to have little in common, either conceptually or statistically. They are not typically correlated and may themselves have different sets of correlates (e.g., Miklowitz, Goldstein, & Falloon, 1983). Furthermore, given the empirical evidence that the most important element of EE is the measure of criticism, EE researchers might do worse than take this fact into consideration and rename the construct. Substituting the term "negative affect" or simply "criticism" would surely remove some of the construct's mystery. It might also reduce the confusion that the current name engenders and make the construct more accessible to those who work with distressed relationships more generally.

It is also worthwhile noting that the majority of studies that have examined links between EE and patient relapse never actually observe negative interactions between the family members and the patients in their samples. In the CFI, the emotion that is "expressed" is expressed to a researcher, not to a patient directly. The underlying assumption of these studies is that the criticisms and negative affect expressed by family members about patients during the CFI are also reflected in their day-to-day interactions. As indicated earlier, a number of studies have provided empirical support for this assumption.

---

[1] Training in the rating of EE takes approximately 1 month. Coders are required to achieve reliabilities of .80 (intraclass correlation) or better on the major EE subscales of criticism, hostility, and emotional overinvolvement.

The Pervasiveness of High Expressed Emotion

As with most other human characteristics assessed in psychology and psychiatry, it is easier to describe the measurement procedures for EE than it is to define the underlying construct. The construct was first assessed nearly 30 years ago in the service of understanding why some schizophrenic patients relapsed repeatedly after being discharged from the hospital while others did not. More recent research, however, has shown that high levels of EE are not unique to families of those diagnosed with schizophrenia. High levels of EE have been demonstrated in the relatives of patients suffering from a wide range of psychiatric and medical problems, including depression, mania, anorexia, agoraphobia, dementia, Alzheimer's disease, obesity, and diabetes (e.g., Bledin, MacCarthy, Kuipers, & Woods, 1990; Fischmann-Havstad & Marston, 1984; Hooley, Orley, & Teasdale, 1986; Koenigsberg, Klausner, Pellino, Rosnick, & Campbell, 1993; Miklowitz, Goldstein, Nuechterlein, Snyder, & Mintz, 1988; Peter & Hand, 1988; Szmuckler, Eisler, Russell, & Dare, 1985; Vaughn & Leff, 1976a; Vitaliano, Becker, Russo, Magaña-Amato, & Maiuro, 1988). High levels of EE have also been found in normal mothers of healthy 3-year-old boys (Heckelman & Hooley, 1993) and in psychiatric staff members working with chronically impaired psychiatric patients (Heinssen, Hooley, Minarik, Israel, & Fenton, 1994; Moore, Ball, & Kuipers, 1992).

So what is EE? Goldstein et al. (1992) have suggested that criticalness and negativity on the part of the relative may reflect a subclinical manifestation of the schizophrenic genotype in the family member. Viewed in this manner, high levels of family EE merely serve to identify patients who are most at risk for relapse because of their high genetic vulnerability. Unfortunately, the explanation for high EE is not likely to be so simple. Everyday experience tells us that criticism (empirically, the most important element of the EE index) exists to varying degrees in all human interactions. Moreover, high levels of EE are not restricted to the biological relatives of schizophrenic patients. A genetic model is thus unlikely to help us much in understanding the EE–relapse link.[2] Although most often studied in research concerning relapse in schizophrenic patients, EE is clearly not a construct that respects either diagnostic, familial, or genetic boundaries.

---

[2] That high levels of EE in spouses predict relapse in patients decreases the tenability of a shared genetic variance model. However, the notion of assortative mating makes it impossible to rule out this explanation in the absence of (as yet unavailable) data demonstrating that high levels of EE in individuals who are unrelated to the patient either by biology or marriage also predict negative outcomes in patients.

Expressed Emotion as a Risk Factor

The most reliable finding concerning high EE has been its predictive utility as a risk factor for relapse in patients with schizophrenia. Numerous studies conducted in several different countries attest to this. Using aggregated data from 12 studies published through 1988, Parker and Hadzi-Pavlovic (1990) report an overall relapse rate of 54% for patients living in high EE households. This is more than twice the (24%) rate found for patients living in low-EE home environments.

In addition to its well-documented association with poor prognosis in schizophrenia, high EE has also been shown to predict relapse in depressed patients. Both Vaughn and Leff (1976a) and Hooley et al. (1986) have demonstrated that unipolar patients who return home from the hospital to live with high-EE family members are at significantly greater risk of relapse than those who return to low-EE home environments. Across the two studies, the combined relapse rate associated with high EE was 65%; the relapse rate associated with low-EE was 12%. Extending EE research to manic patients, Miklowitz Goldstein, and Nuechterlein et al. (1986) reported a relapse rate of 90% in patients returning from the hospital to live with high-EE family members, compared with a relapse rate of 54% in the low-EE group. Early data from a second sample of bipolar patients (Miklowitz, 1992) appear to replicate these findings. Moreover, results from a recent Egyptian study (Okasha et al. 1994), although more difficult to interpret, indicate that the association between family criticism and depressive relapse may replicate cross-culturally. Collectively, these studies demonstrate that the EE construct has predictive validity for mood-disordered as well as schizophrenic patients.[3]

Although little studied to date, there is also reason to believe that high levels of EE in the home are associated with negative outcomes other than psychiatric relapse. Heckelman and Hooley (1993), for example, have linked higher levels of maternal criticism in normal mothers to higher levels of insecure attachment in their 3-year-old sons. Working in an entirely different area, Koenigsberg and colleagues (1993) have found that higher levels of family criticism predict poorer glycemic control in diabetic patients. What is thus most interesting about high EE is that it appears to be a general risk factor for a variety of negative psychiatric and nonpsychiatric outcomes. However, because it has been studied most

---

[3] A cutoff of six critical remarks is associated with relapse in bipolar inpatients (Miklowitz et al., 1988). In unipolar depressed inpatients, a cutoff of two or three critical comments has been found to be most predictive of outcome (Hooley et al., 1986; Vaughn & Leff, 1976a). Because the Egyptian sample comprised unipolar *and* bipolar patients who were in *remission*, Okasha and co-workers' (1994) results concerning the level of criticism associated with relapse are hard to interpret.

extensively within families containing a schizophrenic patient, it is to a more detailed discussion of this literature that we now turn.

Expressed Emotion and Relapse in Schizophrenia

In 1990, Parker and Hadzi-Pavlovic summarized the results of all EE-schizophrenic relapse studies published up to 1988 in the form of odds ratios (i.e., odds of relapse associated with high EE divided by the odds of relapse associated with low EE). Aggregating across 12 studies, they concluded that the risk of relapse associated with high EE was 3.7 times the risk of relapse associated with low EE. Remarkably, since Parker and Hadzi-Pavlovic's 1990 report, nine additional prospective studies have appeared in the literature. Nuechterlein and his colleagues have also updated their previous (1986) report (see Nuechterlein, Snyder, & Mintz, 1992) and added several more subjects to their outcome analyses. These additions have prompted us to update Parker and Hazdzi-Pavlovic's previous analysis. The results are sumarized in Table 3.1.

Even though the number of included studies[4] has almost doubled, the overall findings have changed very little. Across all of the 21 outcome studies, the 9- to 12-month relapse rate associated with high EE is 54% (344/637 patients). The relapse rate associated with low EE is 23% (133/568 patients). Following Parker and Hadzi-Pavlovic (1990), analysis of the aggregated data reveals an odds ratio for high EE of 3.8. It should also be noted that across all of the outcome studies, the weighted mean odds ratio (as opposed to the aggregated odds ratio) for high-EE is even higher, at 6.8.[5] Simply put, compared to patients returning to low-EE families, schizophrenic patients who return home to high-EE home environments have a four- to seven-fold increased likelihood of relapse. Remarkably, the odds ratio for relapse for patients with unipolar depression seems to be even higher, although here it must be acknowledged that

---

[4] The study by Buchkremer et al. (1991) has not been included, because too little information is provided in the original article. We also excluded studies such as Guttierez et al. (1988) and McCreadie and Phillips (1988) because these involved patients already in remission (i.e., not currently hospitalized for an index episode). Finally, because the study by Bertrando et al. (1992) incorporates the sample described by Cazzullo et al. (1989) (Bertrando, personal communication, November 1994), only the former study is listed.

[5] Because it takes into account the sample size for each study, the weighted mean odds ratio provides a more reliable assessment of the odds ratio across all of the studies than does the aggregated odds ratio used by Parker and Hadzi-Pavlovic (1990). However, we are also aware that even summarizing the results in this manner is subject to criticism (e.g., Rosenthal, 1991). For this reason we have also completed a meta-analysis of the EE literature (Butzlaff & Hooley, 1995). The results of this will be published elsewhere.

TABLE 3.1. Relapse rates and odds ratios for EE outcome studies.

Sample	Patients relapsed/ Sample size Low EE	High EE	Relapse rates % Low EE	% High EE	Odds ratio
Schizophrenia					
Arévalo & Vizcarro (1989)	5/13	8/18	38	44	1.28
Barrelet et al. (1990)	0/12	8/24	0	33	12.75
Bertrando et al. (1992)	4/18	14/24	22	58	4.90
Brown et al. (1962)[a]	13/47	38/50	28	76	8.28
Brown et al. (1972)	9/56	26/45	16	58	7.15
Ivanović & Vuletić (1989)	2/31	19/29	6	59	27.55
Karno et al. (1987)	7/27	10/17	26	50	4.08
Köttgen et al. (1984)[b]	11/20	7/14	55	31	1.00
Leff et al. (1987)[c]	5/54	5/16	9	68	3.20
MacMillan et al. (1986)	14/34	26/38	41	91	2.72
Moline et al. (1985)[d]	4/13	10/11	31	31	22.50
Montero et al. (1992)[e]	5/28	10/32	18	59	6.00
Možný & Votýpková (1992)	13/56	41/69	23		4.84
Niedermeier et al. (1992)	6/21	16/28	29	57	3.33
Nuechterlein et al. (1992)	0/12	12/31	0	39	15.79
Parker et al. (1988)[f]	9/15	20/42	60	48	0.61
Stirling et al. (1991)	8/17	5/16	47	31	0.51
Tarrier et al. (1988)[g]	4/19	14/29	21	48	3.50
Vaughn & Leff (1976a)	1/16	10/21	6	48	13.60
Vaughn et al. (1984)	3/18	20/36	17	56	6.25
Vaughan et al. (1992)	10/41	25/47	24	53	3.52
Aggregate	133/568	344/637	23	54	3.84

Weighted mean odds ratio = 6.83

Unipolar depression					
Hooley et al. (1986)	0/8	20/31	0	65	30.30
Vaughn & Leff (1976)	2/9	14/21	22	67	7.00
Aggregate	2/17	34/52	12	65	14.52

Weighted mean odds ratio = 20.17

Bipolar affective disorder					
Miklowitz et al. (1988)	7/13	9/10	54	90	7.71

Note. In computing an odds ratio, 0.5 was added to values of 0 to permit computation.
[a] Figures refer to deteriorated patients.
[b] Reported in Dulz & Hand (1986).
[c] Figures based on Present State Exam/Catego diagnoses.
[d] Figures based on modified cutoff of 10 critical comments.
[e] Figures based on modified cutoff of 4 critical comments.
[f] Includes 9 households with borderline EE ratings.
[g] Figures based on "education only" and "routine treatment" groups.

TABLE 3.2. Characteristics of the EE-relapse replication and nonreplication studies.

Study	% Male	Mean CC[a]	Location; N
Schizophrenia			
		Replication studies	
Barrelet et al. (1990)	50	median = 8.0	Switzerland; $N = 36$
Bertrando et al. (1992)	76	N/A	Italy; $N = 42$
Brown et al. (1962)	100	N/A	UK; $N = 97$
Brown et al. (1972)	52	7.86[b]	UK; $N = 101$
Ivanovic & Vuletic (1989)	72	3.5	Yugoslavia; $N = 60$
Karno et al. (1987)	57	3.33[c]	USA; $N = 70$ Mexican Americans
Leff et al. (1987)	N/A	1.9[d]	India; $N = 93$
Moline et al. (1985)	N/A	N/A	USA; $N = 24$ (67% African American)
Montero et al. (1992)	53	3.1	Spain; $N = 60$
Možný & Votýpková (1992)	45	N/A	Czechoslovakia; $N = 125$
Niedermeier et al. (1992)	55	N/A	Germany; $N = 49$
Nuechterlein et al. (1992)	81	5.6[e]	USA; $N = 43$
Tarrier et al. (1988)	35	N/A	UK; $N = 83$
Vaughn & Leff (1976a)	41	8.22	UK; $N = 37$
Vaughn et al. (1984)	77	6.86	USA; $N = 69$
Vaughan et al.(1992)	63	mothers = 5.8 fathers = 5.4 husbands = 6.4	Australia; $N = 91$
		Nonreplication studies	
Arévalo & Vizcarro (1989)	58	N/A	Spain; $N = 31$
Köttgen et al. (1984)[f]	65	N/A	Germany; $N = 52$
MacMillan et al. (1986)[g]	60	median = 6.9	UK; $N = 77$
Parker et al.(1988)	58	5.2[h]	Australia; $N = 57$
Stirling et al. (1991)	52	3.4	UK; $N = 33$
Unipolar depression			
Hooley et al. (1986)	41	8.3	UK; $N = 39$
Vaughn & Leff (1976)	33	7.2	UK; $N = 30$
Bipolar disorder			
Miklowitz et al. (1988)	52	5.2[i]	USA; $N = 23$

[a] Mean number of critical comments.
[b] Reported by Vaughn & Leff (1976a).
[c] Reported by Jenkins (1991).
[d] Reported by Wig et al. (1987).
[e] K. Snyder, personal communication, February 1994.
[f] Reported in Dulz & Hand (1986).
[g] Whether this is truly an example of a nonreplication study is questionable.
[h] Reported by Parker & Johnston (1987).
[i] D. Miklowitz, personal communication, August 1994.

fewer studies contributed data to the estimate of risk. This does not diminish the importance of the findings, however, or detract from what is an inescapable conclusion: EE is highly predictive of short-term (9- to 12-month) relapse in patients with schizophrenia and severe mood disorders.

The current literature, however, provides little compelling support for Hogarty's (1985) assertion that EE is only predictive of relapse in male schizophrenic patients. Although gender differences in susceptibility to high levels of EE have certainly been reported (Hogarty et al., 1986; Montero, Gómez-Beneyto, Ruiz, Puche, & Adam, 1992; Vaughn et al., 1984), the majority of studies do not report such differences. Moreover, as Table 3.2 indicates, no evidence suggests that the nonreplication studies contained a higher proportion of female patients that male patients. It is certainly possible that male patients fare worse than women when exposed to high-EE family environments. This is consistent with recent speculation that schizophrenia is a more benign illness in women (e.g., Iacono & Beiser, 1992; Lewis, 1992). The evidence available to date, however, does not suggest that the EE–relapse association holds only for male patients.

Clearly, the majority of studies provide support for the association between high levels of EE and poor outcome not only in schizophrenic but also in mood-disordered patients. Further, evidence indicates that, at least for schizophrenia (and possibly also for mood disorders (see Okasha et al., 1994)), the EE–relapse link may transcend cultural boundaries. Table 3.2 provides details of the sample characteristics of the studies described in Table 3.1. As is evident from Table 3.2, EE has been linked to relapse in European, North American, Australian, Spanish-speaking North Americans of Mexican-American heritage, and Indian schizophrenic patient samples.[6]

As Parker and Hadzi-Pavlovic (1990) have already noted, however, the figures presented in Table 3.1 overestimate the strength of the EE–relapse association in schizophrenia. This is because in several studies, the data were presented in a manner that optimized EE's potential to predict relapse. For example, in the urban Chicago sample (67% African-American) of Moline, Singh, Morris, and Meltzer (1985), the best prediction of relapse came when a cutoff of 10 rather than 6 critical comments was used. Montero et al. (1992), working in Spain, found a cutoff of four critical remarks to be most predictive of outcome. In Leff and co-workers' (1987) replication in Chandigarh, India, hostility rather than criticism was the EE component that was significantly predictive of patient relapse.

---

[6] Recent data from Taiwan (Kleinman, personal communication, October 1992) and from China (Phillips, personal communication, March 1993) suggest that EE is not predictive of relapse in Asian samples. However, it is hard to know exactly what to make of these findings in the light of a recent replication of the EE–relapse link in a Japanese sample of schizophrenic patients (Leff, personal communication, May 1994).

Because in some cases the authors selected either the component of EE or the cutoff for critical comments that provided the optimum association with relapse, the data are biased in favor of demonstrating an EE–relapse link.

The extent to which this should be viewed as a major problem however, depends on one's conceptualization of what EE is. If EE is reified, and viewed as an error-free measure of a well-understood underlying construct, studies that fail to find a significant EE–relapse association using a standard cutoff point are clearly problematic. But EE is not measured without error, and it is clearly not a well-understood construct. Thus, the variability of the optimal EE cutoff point for predicting relapse in different populations is neither surprising nor problematic. In light of the notorious unreliability of most findings in the social sciences (Meehl, 1978), the robustness of the link between high levels of EE and relapse in schizophrenia is quite remarkable.

Nonreplications of the Expressed Emotion–Relapse Link

If we resist the temptation to reify the EE construct and assume that EE (like most other psychological variables) is only a moderately reliable measure of an imperfectly measured underlying construct, nonreplications of the EE–relapse link can be viewed as inevitable. However, provided that they are not methodologically flawed, nonreplications also hold the potential to provide valuable information about the limitations of the construct.

To date, five nonreplication studies have appeared in the EE literature.[7] The first of these (Köttgen, Sönnichsen, Mollenhauer, & Jurth, 1984; see also Dulz & Hand, 1986), was conducted in Hamburg, Germany. The results revealed a higher rate of relapse (57% vs. 41%) for patients in low- as opposed to high-EE families. When subjects receiving intervention were excluded, the relapse rates were adjusted to 50% in the high-EE

---

[7] Two other studies are also sometimes (mistakenly) interpreted as failures to replicate. McCreadie and Phillips (1988) studied EE in a sample of community living (and presumably remitted) patients. They found, like others (see Leff et al., 1990), that relatives' EE measured outside a crisis period was not predictive of patients' subsequent outcomes (but see Guttierrez et al., 1988 for a demonstration of the predictive validity of EE in a remitted sample in Spain). Hogarty et al. (1988) also found no association between EE and relapse in patients involved in a treatment study examining levels of EE and standard and minimal doses of medication. However, Hogarty et al. themselves state that "this investigation is *not* [original italics] another attempt to replicate the effects of EE on relapse, since a large number of patients who were unable to stabilize and hence at greater risk of relapse were removed from the study by design." (pp. 798). For these reasons, these investigations are not discussed in this section.

control group and 55% in the low-EE control group (see Table 7 in Köttgen et al., 1984), yielding no evidence for an EE–relapse link.

Methodological problems associated with the Hamburg study, however, have limited the extent to which EE researchers have taken Köttgen and co-workers' nonreplication seriously. Vaughn (1986) noted that the study departs from other EE studies in important ways. For example, the majority of patients in the Köttgen et al. sample did not live with their families during the follow-up period. This obviously limits the potential impact that relatives and patients may have had on each other. The Hamburg group also permitted households to be classified as low-EE even if they contained another family member who was reported by the (interviewed) low-EE relative to be critical/hostile or emotionally overinvolved. Conventionally, in EE research, an *"or"* rule is applied in assessing families, such that the family rather than a single relative is the unit of analysis. The presence of one high-EE relative in a household results in the family being classified as high EE, even if other relatives are rated as being low. The potential for misclassifying families in the Hamburg investigation therefore seems to be high and may go some way toward explaining the discrepant results.[8]

In contrast to the Köttgen et al. (1984) study, the negative results of the Northwick Park investigation (MacMillan, Gold, Crow, Johnson, & Johnstone, 1986) stimulated a great deal of interest. MacMillan and colleagues reported a significant association between high levels of EE and relapse in a sample of schizophrenics experiencing their first episodes of illness. However, when medication status and duration of illness before treatment were statistically controlled, EE failed to make a significant contribution to relapse. Leff and Vaughn (1986) however, noted than of the 60 patients living in two-parent homes, both parents were interviewed in only 6 cases. The likely misclassification of "true" high-EE families as low-EE families is therefore a potential problem in this study.

MacMillan et al. (1986) have also been critized for ignoring the fact that, in a multivariate model that included treatment and illness duration, the (one-tailed) significance level for expressed emotion ($p = .07$) was close to conventionally accepted significance levels (J. Mintz, Mintz, & Goldstein, 1987). Mintz and colleagues further note that it is not appropriate to simply assume (as the Northwick Park investigators did) that high levels of EE are caused by the effects of untreated illness. Mintz et al. highlight a number of alternative models of the relation between

---

[8] Although some (see Lefley, 1992) have criticized such a family-based form of classification, evidence suggests that patients in mixed EE households have outcomes that more closely resemble those associated with high- rather than low-EE home environments (see Kavanagh, 1992). The decision to classify at the level of the household rather than at the level of the individual is purely empirically based and does not reflect any underlying political sentiment.

EE and illness duration that are consistent with the Northwick Park data yet still assign a causal role to EE. Finally, Mintz et al. note a number of errors in the original report; these errors were subsequently corrected by the original authors (MacMillan et al., 1987).

The Northwick Park study was the first to link high levels of EE to longer durations of untreated illness and the first to question seriously the causal nature of the EE–relapse link. Subsequent research, however, indicates that controlling for duration of illness in EE analyses may be easier said than done. L.I. Mintz, Nuechterlein, Goldstein, Mintz, and Snyder (1989) have published data suggesting that parental estimates of duration of illness are confounded with levels of EE. Compared with high-EE relatives, these authors found that low-EE family members were more likely to underestimate the duration of a patient's illness relative to a "best estimate" of illness duration based on hospital records, and patients' and relatives' reports. High-EE family members, on the other hand, gave duration of illness estimates that did not significantly differ from the best estimate.

The findings of Mintz et al. (1989) are important, particularly within the context of the Northwick Park study, where relatives' estimates of duration of illness were used to clarify the link between EE and outcome. In the final analysis, however, regardless of whether the Northwick Park data are interpreted as a replication of the EE–relapse link (see Leff & Vaughn, 1986; J. Mintz et al., 1987; L.I. Mintz et al., 1989), or as a failure to replicate (MacMillan et al., 1986, 1987), one important point should not be lost. The Northwick Park team rendered EE research a valuable service when they raised the possibility that mediating variables might be more important in the understanding of the EE–relapse link than had previously been estimated.

Whereas the Hamburg study was roundly critized on methodological grounds and the interpretation of the Northwick Park data will no doubt continue to remain a source of controversy, the Australian study of Parker, Johnston, and Hayward (1988) is generally considered to be a clear-cut case of failure to replicate. Using a sample of 57 schizophrenic patients recruited in Sydney, Parker et al. reported a 9-month relapse rate of 48% for patients living with high-EE family members, compared with a relapse rate of 60% for patients living with low-EE relatives. Thus, in this study, patients in high-EE family environments actually fared *better* than their low-EE counterparts. Parker et al. also performed thorough and thoughtful analyses on their data in an effort to shed light (albeit unsuccessfully) on the factors associated with their rather unusual findings.

The Australian study is generally regarded to be less vulnerable to methodological or statistical criticisms than the nonreplication studies that preceded it. However, there is a methodological concern that has not been previously noted in the literature. Parker and his colleagues assessed relapse by administering the Present State Examination (PSE; Wing, Cooper, & Sartorius, 1974) 1 month after patients had been discharged from the hospital and again at the time of the 9-month follow-up. Determination of relapse was based on either a change in PSE status from PSE "noncase" at 1 month to PSE "case" status at 9 months, or an increase in the severity of "case" status between the two assessments as reflected in an Index of Definition score that was two or more levels higher at the 9-month assessment than at the 1-month assessment.

On first glance, such a procedure seems reasonable and essentially comparable to the procedures used previously by Vaughn and Leff (1976a). However, there appears to be one important difference. In Vaughn and Leff's study, additional information was also obtained about clinical episodes that occurred between discharge and the final month covered by the PSE. For example, in their 1976 paper, Vaughn and Leff write that two depressed patients "were well at the follow-up interview, but reported an episode of depression, persisting for two weeks or more, during the months between discharge and the final month covered by the PSE" (Vaughn & Leff, 1976a, p. 128). It is noteworthy that such a retrospective check on symptoms that may have occurred between the two PSE assessments does not appear to have been incorporated into Parker and co-workers' (1988) design. Instead, the determination of relapse was based on how the patient functioned at the 9-month assessment relative to the 1-month assessment, with no attention paid to symptomatology in the intervening period unless a patient was rehospitalized. According to the authors, "it is quite possible that a subject who was quite well at the one month follow-up might have had a significant relapse (not serious enough for hospital admission) in the next few months, and recovered by the nine-month follow-up, resulting in false classification as a "nonrelapser" (Parker et al., 1988, p. 810). Parker et al. note that, like the London studies, theirs is weakened by a failure to check the clinical status of the subjects longitudinally. What is clear, however, is that the London studies did in fact do this, although not in the manner of the later studies (e.g., Vaughn et al., 1984), for which outcome was assessed at monthly intervals.

What effect might such a procedural omission have had on Parker and colleagues' (1988) data? Hogarty et al. (1986) have argued that low levels of EE are associated with the delay rather than the prevention of relapse. If this is the case, a cross-sectional clinical snapshot taken at 9-months might be expected to produce exactly the kind of data reported by Parker et al. Patients from high-EE families might have already relapsed and improved again; patients from low-EE families might just be beginning to show indications of their (delayed) relapse.

Without more clarification from Parker and his colleagues (1988), we cannot know to what extent this aspect of their design really was a problem. It is possible that this methodological departure from the previous replication studies holds the potential to explain what otherwise looks like a strong nonreplication of the EE-relapse link. It is also possible that it has not compromised Parker and co-workers' results in any serious way and that the Australian study should be considered to be a clearcut case of nonreplication. Given the recent replication of the EE-relapse link in Australia by Vaughan et al. (1992), however, it seems clear that cultural factors do not provide a very compelling explanation for Parker and co-workers' findings.

Arévalo and Vizcarro's (1989) nonreplication of the EE-relapse link in Madrid is rather difficult to understand. Again however, cultural factors fail to provide an adequate explanation as to why the relapse rates asociated with high EE and low EE were similar in this study (44% vs. 38%). Expressed emotion has been found to be predictive of later relapse in both hospitalized and remitted Spanish patients (Guttierrez et al., 1988; Montero et al., 1992). It is possible, however, that the strong association between medication and relapse in Arévalo and Vizcarro's study (10/13 of the patients who relapsed were not receiving medication) is one factor worthy of further consideration.

The most recent addition to the nonreplication literature is the study of Stirling and colleagues (Stirling et al., 1991). Using a British sample of first- or early-onset schizophrenic patients, these authors demonstrated no significant association between high levels of EE and relapse. Because other studies (e.g., Barrelet et al., 1990; Nuechterlein et al., 1986, 1992; Leff et al., 1987) have also used first- or early-onset patients and have found results supportive of the EE-relapse link, the sample of patients used does not provide an adequate explanation of the failure to replicate. The authors were also careful to avoid many of the methodological pitfalls encountered by others (e.g., Köttgen et al., 1984; MacMillan et al., 1986), and in the case of patients living in two-parent households, the possibility of wrongly classifying a high-EE household as low EE seems to be remote. It is noteworthy that levels of criticism in the families of this study were strikingly low. The mean number of critical comments was 3.4—a much lower figure than is usually reported for British and North American samples (see Table 3.2). Moreover, criticism and hostility contributed little to EE classifications: Most relatives who received a high-EE rating did so because of high levels of emotional overinvolvement.

These results suggest that high levels of EOI in the relatives of recent-onset patients are not predictive of 1-year relapse rates. Although EOI has been linked to poor outcome in other samples (Brown et al., 1972; Vaughn & Leff, 1976a; Vaughn et al., 1984), these typically have included patients with longer durations of illness. Interestingly, Leff and co-workers' (1987) Indian replication study also involved first-onset patients and failed to find any association between EOI and subsequent outcome. Although this may simply reflect cultural differences, MacMillan et al. (1986) also reported no association between EOI and outcome—again using a recent-onset sample. This was also the case in the study by Barrelet and co-workers (1990), which involved a sample of Swiss-French patients experiencing their first hospitalizations. The only other study that employed a recent-onset sample (Nuechterlein et al., 1986; see also Nuechterlein, Snyder, & Mintz, 1992) provided no information about the association between EOI and relapse.

Collectively, these findings suggest that, although EOI may be related to poor outcome in samples of chronically ill patients (but see Hogarty, 1985), no evidence supports this link in first- or recent-onset patients. This may indicate that high levels of emotional concern represent an early and natural reaction of family members to the development of symptoms in patients and are thus of little prognostic significance—at least at first. As relatives recover from the initial shock of learning that a family member has been diagnosed with a severe mental disorder, levels of EOI may decrease (see also Stirling et al., 1993), only attaining prognostic significance in cases where EOI levels remain high. However, this is simply speculation at the present time. Stirling and co-workers' (1991) study also raises the possibility that levels of criticism may be generally lower in the families of recent-onset patients. This observation raises the possibility that high levels of criticism may reflect a reaction to psychopathology in the patient, again echoing the findings of the Northwick Park investigation. This is an issue to which we shall return later.

In summary, although the nonreplication studies discussed previously do not provide a basis for discarding the EE construct, they nonetheless suggest that we should exercise caution in our interpretation of the EE–relapse link. Many of the studies that support the association between EE and relapse can be criticized on methodological or statistical grounds, and in some cases, the data can be reanalyzed in a manner that reduces the level of significance of the results. The same is also true of some of the nonreplication studies. Taken together, the findings indicate that the association between EE and relapse, although robust, may reflect a more complex and possibly developmental process between family members and patients than previously thought. A more thorough understanding of the ways in which the patient–relative relationship is changed by the initial episode of illness and by its later behavioral and psychological sequelae is clearly needed. The EE construct is likely prove a fertile ground for researchers interested in applying the principles of developmental psychopathology (Cicchetti, 1989) to understanding the families of the mentally ill.

Models of Explanation of the Expressed Emotion–Relapse Link

The studies described previously underscore how little we actually understand about *why* high levels of criticism in family members are associated with higher rates of relapse in patients. Moreover, we know even less about why some family members are highly critical of patients and others are not, why some patients suffer relapse in the presence of criticism while others do not, and why a small number of patients living with noncritical relatives nonetheless suffer relapse. Clearly, answers to these questions not only will improve our ability to predict and perhaps prevent relapse but will also shed much needed light on the processes that give rise to the EE–relapse link.

Since the initial independent replication of Brown and co-workers' (1962, 1972) early work by Vaughn and Leff (1976a), there has been a strong bias among researchers to interpret the EE–relapse link as relatively unambiguous evidence for a negative influence of critical family members on patients. Yet much of what we know about the EE–relapse link is equally consistent with the hypothesis that high levels of relapse are a *reaction to* rather than a *cause of* patients' relapses. Moreover, in light of our imperfect ability to predict relapse from measures of EE alone, the existing data may be even more consistent with an interactive model in which characteristics of patients and family members interact to produce relapse. Perhaps because of the plausibility of the EE-as-causal hypothesis, these competing explanations for the EE–relapse link have not received the attention they deserve in empirical research, seldom receiving more than cursory attention in the discussion sections of empirical articles. Clearly, however, future gains in our understanding of the processes giving rise to psychiatric relapse will require attention to competing explanations in research designs. In the following discussion, we describe three competing models for the EE–relapse link, evaluate the plausibility of those models in light of existing data, and suggest avenues of inquiry that hold promise for increasing our understanding of the EE construct. To simplify description and discussion, and because criticism has been demonstrated to be the most important element of the EE index, the terms *EE* and *criticism* are used interchangeably.

Model 1: Relatives' Criticism is a Primary Causal Factor in Patients' Relapse

Most researchers have tended to assume (Figure 3.1) that high levels of EE reflect critical, traitlike characteristics of relatives that are causally related to patient relapse. This is perhaps an understandable bias given the consistency of the EE–relapse link with contemporary notions of

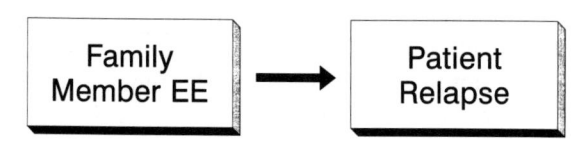

FIGURE 3.1. Model 1: Relatives' EE is a primary causal factor in patients' relapse.

diathesis–stress or vulnerability in the onset and course of psychopathology (Zubin & Spring, 1977). According to this vulnerability model, individuals with psychiatric disorders can be conceptualized as lying along a continuum of diathesis for the disorder in question. These individual differences in vulnerability represent poorly understood genetically transmitted and/or environmentally induced weaknesses within the individual that define his or her threshold for susceptibility to environmental and/or biological stressors. High levels of criticism in a close family member, according to this model, may represent a salient source of environmental stress capable of precipitating the onset of patient symptoms as a function of the patient's initial level of vulnerability. Thus, the diathesis–stress model not only accommodates the general finding of a link between EE and relapse, but it suggests the possibility that improvements in EE–relapse prediction accuracy may accrue significantly from an increase in our knowledge of where patients are initially located along the continuum of vulnerability.

Ultimately, however, the viability of the EE main effect model must be evaluated on the basis of how well its postulates are supported by data. We now consider three specific predictions of Model 1 and examine the extent to which each prediction can be considered to be supported by the available empirical data.

PREDICTION 1: HIGH LEVELS OF EXPRESSED EMOTION PREDICT ELEVATED RISK OF RELAPSE

The first postulate is that patients who return from the hospital to live with relatively critical family members will be more likely to relapse than those who return home to live with less critical family members. Clearly, this postulate is consistent with numerous reports of an EE–relapse link. As indicated in Table 3.1, a large number of studies attest to the association between EE and relapse in samples of patients diagnosed with schizophrenia and other disorders. Although, as mentioned earlier, a number of nonreplication studies do exist, the vast majority of studies have demonstrated some significant association between EE and negative outcome.

PREDICTION 2: HIGH EXPRESSED EMOTION IS A STABLE TRAIT

The second postulate of Model 1 is that criticism or high levels of EE are relatively enduring characteristics of family members, sufficiently stable to constitute a stress level capable of precipitating patient relapse. To the extent that EE levels of family members fluctuate significantly in the absence of specific intervention, it is difficult (though not impossible) to imagine how they would constitute a sufficient enough source of stress to trigger patient relapse.

The available data provide little support for the assumption of stability. As noted earlier, high levels of EE (and particularly criticism) appear to be reduced by some forms of intervention (Hogarty et al., 1986; Lam 1991; Leff et al., 1982, 1989; Tarrier et al., 1988). Levels of EE have also been shown to change spontaneously even in the absence of family treatment. Brown et al. (1972) noted that approximately one third of initially high-EE relatives became low EE over a 9-month period, primarily because of decreases in levels of criticism. Similarly, Dulz and Hand (1986) reported a 50% change in EE status over the same period.[9] The finding by Leff et al. (1990) that 79% of Indian relatives initially classified as high EE were rated as low EE at 1-year follow-up is particularly dramatic. Moreover, the fact that relatives do not typically change from low to high EE suggests that these results do not simply reflect regression to the mean. Changes from low to high levels of EE do occur, but only in around 6% to 10% of cases (see Dulz & Hand, 1986; see also Tarrier et al., 1988).[10]

Our own observations from the Harvard Family Study, an ongoing longitudinal 9-month follow-up investigation of schizophrenic patients and their families, further suggest that marked decreases in criticism occur within a few months of the patient's discharge from the hospital. Preliminary analysis of CFI data collected from 21 family members of schizophrenic patients reveal that, although at the time of the patient's hospitalization relatives made an average of 11.3 critical remarks in the CFI, this had decreased to 4.3 criticisms by the time of the 3-month follow-up. These changes occurred in the absence of any formal intervention and reflect a 47% decrease in EE levels over the 3-month period.

That most of the decrease in criticism occurs in the early months after the patient returns home is further suggested by the results of Tarrier et al. (1988). High-EE relatives whose patient family member was receiving routine treatment showed significant decreases in criticism between admission and a CFI conducted at 4.5 months postdischarge. No significant changes in criticism were detected between the 4.5- and the 9-month CFI assessment, however.

The available evidence is therefore consistent with the notion that relatives' criticism typically decreases after patients return home from the hospital and (presumably) enter a period of less severe symptomatology or remission. It is important to note, however, that although absolute levels of criticism appear to be influenced by changes in patients' level of functioning, the *tendency* for relatives to be critical appears to be traitlike.

---

[9] Some of the relatives in this study received family-based intervention. However, no treatment effect was demonstrated for changes in levels of EE.
[10] The only evidence that contradicts this general assertion comes from Leff et al. (1982). These investigators found no overall decrease in EE in the small number of high-EE relatives who formed the control group of their intervention study. Parker et al. (1988) have argued that the only reason Leff et al. (1982) obtained an effect of their intervention (and thus assigned a causal role to EE) was because of the (unusual) lack of change in levels of EE that characterized the control group.

Preliminary data from the Harvard Family Study reveal a .74 correlation between the frequency of critical comments made by relatives in the CFI conducted at the time of the patient's index hospitalization and the CFI conducted 3 months after the patient returned home. This suggests some impressive stability in the rank ordering of relatives with regard to the number of critical comments that they make. Relatives who make most critical comments at the time of the patient's hospitalization are also those who make the most critical comments in the later interview—even though the number of critical remarks that they make has dropped considerably overall.

The data are thus consistent with the notion that high levels of EE are both traitlike and statelike. Although relatives may retain a tendency to be more or less critical, how critical they are at a given point in time may depend to some extent on the patient's level of functioning. These findings do not necessarily preclude a unidirectional causal model of EE and relapse. The tendency to be high EE may reflect a stable underlying trait, and it may be this trait (rather than frequency of criticism) that is causally related to relapse. Alternatively, high-EE effects could have a long half-life even if levels of EE are not themselves stable. Nonetheless, the possibility remains that high levels of EE (at least in some cases) reflect a reaction to an exacerbation of the patient's illness. As such, these findings pose problems for the most simple version of Model 1.

PREDICTION 3: FAMILY MEMBER CRITICISM LEVELS ARE NOT SIGNIFICANTLY CORRELATED WITH THE SYMPTOM TYPES, LEVELS, OR FUNCTIONING DEFICITS OF PATIENTS

In many ways, this is the most crucial prediction under Model 1. Support for this model would require the demonstration that EE levels are not influenced by aspects of the patient's clinical condition. Under this model, it is also necessary that no symptom or functioning differences be demonstrated to exist between patients living with high- and low-EE family members.

In general, the literature provides little evidence for an association between relatives' EE and such patient characteristics as gender, age, level of education, and severity of illness (Bertrando et al., 1992; Karno et al., 1987; Miklowitz et al., 1983; Nuechterlein et al., 1986; Stirling et al., 1991; Vaughn et al. 1984). These same studies further suggest that factors such as medication, premorbid adjustment, acuteness of illness onset, age at onset of symptoms, number of previous hospitalizations, and duration of illness also typically fail to discriminate across patients in the high- and low-EE groups. However, scattered throughout the literature are a number of reports of significant correlations between high levels of EE (or frequency of criticism) and a greater degree of behavior disturbance (Brown et al., 1972; J. Mintz et al., 1987), work or occupational role impairment (Brown et al., 1972; Hogarty et al., 1988; Stirling et al., 1991; Vaughan et al., 1992), increased number of previous hospitalizations (Bertrando et al., 1982; Vaughan et al., 1992), longer duration of illness (MacMillan et al., 1986; Stirling et al., 1991), lower levels of social functioning (Barrowclough & Tarrier, 1990), higher levels of personal distress (Hogarty et al., 1988), increased irritability (Karno et al., 1987), more depression (Glynn et al., 1990; Strachan et al., 1986), more severe delusional thinking (Glynn et al., 1990), higher global ratings of anhedonia/asociality (Glynn et al., 1990), and more negative verbal and nonverbal interpersonal behavior (Hahlweg et al., 1989; Mueser et al., 1993) in schizophrenic patients living in high-EE households.

It must be emphasized that these patient correlates of high EE represent isolated findings in the literature. In some cases, a significant correlation in one study needs to be evaluated against nonsignificant correlations in other studies. Given the large number of studies examined and the number of statistical tests performed, it is quite possible that these findings reflect nothing more than random (sample-specific) fluctuations in the EE-wide data set.

What is interesting, however, is the overall pattern of the results. Regardless of the variables reported, it is noteworthy that the correlations are invariably in the direction of higher levels of EE being associated with patients doing worse.[11] This overall consistency suggests that we are not dealing with chance phenomena.

It is, of course, quite possible that some of the patient differences identified previously are the consequences rather than the causes of high EE. This would not challenge the postulates of Model 1. Patients with high-EE relatives may tend to break down earlier and so have earlier ages of first onset or longer durations of illness. At the present time, however, we are unable to rule out the possibility that the association between high EE and increased risk of relapse results from relatives' reactions to aspects of patient's pathology that are themselves (independently) predictive of poor outcome. This is consistent with Model 2, and it is to a more detailed consideration of this model that we now turn.

Model 2: Patient Symptoms and/or Role Functioning Impairments Engender Criticism From Key Relatives and Serve as Markers of Risk for Relapse

Model 2 (Figure 3.2) holds that the EE–relapse association arises because patient-related criticism in key relatives is engendered by identifiable patient factors, symptoms, and/or role functioning impairments. Initial

---

[11] The only exception in favor of patients living in high-EE households is a tendency reported by Hogarty et al. (1988) for such patients to be involved in more frequent free-time activities with others.

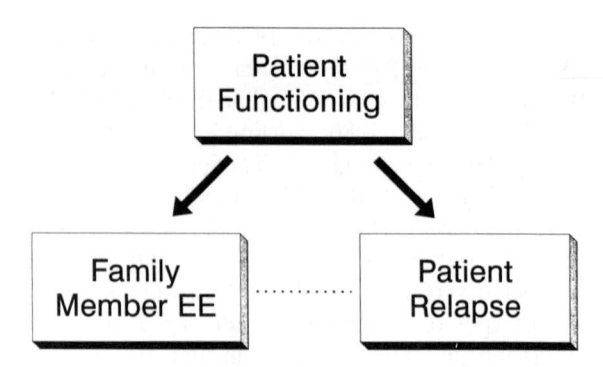

```
                    ┌─────────────┐
                    │   Patient   │
                    │ Functioning │
                    └─────────────┘
                      ↙         ↘
        ┌─────────────┐          ┌─────────────┐
        │   Family    │ ········ │   Patient   │
        │ Member EE   │          │   Relapse   │
        └─────────────┘          └─────────────┘
```

FIGURE 3.2. Model 2: Patient symptoms and/or role-functioning impairments engender criticism from key relatives and serve as markers of risk for relapse.

support for this model requires a demonstration that patient characteristics are significantly correlated with the criticism levels of key relatives (see preceding discussion).

Although no consistent findings have yet emerged, the findings reported previously are suggestive and provide preliminary evidence consistent with Model 2. However, the identification of characteristics of patients that correlate with high levels of relatives' criticism is only a first step. To provide support for Model 2 rather than for Model 1, these patient attributes also need to be revealed by subsequent regression analyses to be associated with poor outcome (i.e., relapse). If these patient characteristics explain a significant proportion of the outcome variance, *and* if the subsequent entry of variables reflecting criticism of key relatives does not result in a significant increase in explained variance, *and* if reversing the order of entry of the variables reveals *only* a significant effect for criticism, then the viability of Model 2 will have been demonstrated. If, on the other hand, *both* patient variables and criticism levels make unique and significant contributions to predictions of relapse, and especially if the entry of the interaction term further significantly increases the explained variance, then Model 3 (the interactional model) emerges as a more viable explanatory model of the EE–relapse link.

Although researchers have reported patient variables that are associated with high levels of EE, they do not routinely make such variables compete with EE for relapse variance. In some cases, however, potential "confounding variables" have been identified and statistically controlled in EE analyses. Brown and co-workers' (1972) study, for example, demonstrated that EE was still associated with relapse when such factors as work impairment and degree of behavior disturbance were considered. L.I. Mintz and colleagues' (1989) reanalysis of the Northwick Park data also reported a trend ($p = .07$, one-tailed) for EE to be significantly predictive of relapse even when the patient's duration of untreated illness was statistically controlled. Because duration of untreated illness was based on parental estimates and thus also likely to be confounded with EE in this study (see L.I. Mintz et al., 1989), it is possible that the Northwick Park researchers underestimated the relapse variance attributable to EE. In short, although the available literature provides no strong tests of the plausibility of Model 2, the few data points that do exist suggest that the EE–relapse relationship is not likely to result solely from family members reacting to relapse-related patient characteristics; EE still appears to make a contribution to the prediction of relapse even when patient characteristics are considered. This point is further echoed in data recently published by Nuechterlein and colleagues (1992). Taken together, the available evidence suggests that we may need to look beyond the simple main effects postulated by Models 1 and 2 in our efforts to understand the EE–relapse link.

Model 3: Patients' and Relatives' Characteristics Interact Negatively and in Ways That Influence the Likelihood of Relapse

Within Model 3, EE is still assigned a causal role (Figure 3.3). However, high levels of EE are also considered to reflect a response of some relatives to aspects of the patient's illness. The fact that EE levels are not stable over time is consistent with the notion that, at least in some relatives, high levels of EE may be a reaction to the clinical state of the patient. As indicated previously, empirical evidence suggests that relatives' criticism levels decrease after patients leave the hospital and (presumably) begin to show some clinical improvement. These findings suggest that the current clinical state of the patient (or correlates of the patient's clinical state) may be linked to levels of criticism in the family.

However, the interaction may well be more complex than this. Even during a patient's episode of illness requiring hospitalization, EE levels

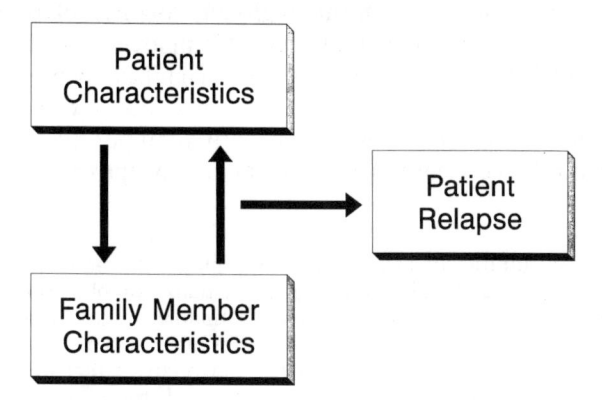

FIGURE 3.3. Model 3: Patients' and relatives' characteristics interact negatively and in ways that influence the likelihood of relapse.

differ across families. That these differences are unlikely to reflect simple clinical differences in the patients is further suggested by the fact that, even *within families*, levels of EE sometimes differ. Moreover, as Table 3.3 indicates, the distribution of high- and low-EE relatives shows clear cultural variation. Although in North American and European samples high-EE relatives are in the majority (a fact that itself is important for destigmatizing high EE), the reverse is true in Indian and Mexican-American samples. Although it is possible that there may be some culture-based differences in the way patients manifest symptoms, this explanation is unlikely to account for much (if any) of the variability in the distribution of EE. This is because, even across the different cultures, all the patients are carrying the same diagnosis and all are experiencing an index episode of illness at the time of the CFI assessment. This raises the important question of why all relatives, when exposed to a psychiatrically impaired family member, do not react in the same way.[12]

Elsewhere we have suggested (Hooley, 1987) that the answer to these questions may lie in an examination of how relatives attempt to understand the marked behavioral changes that typically accompany psychiatric impairment. If high-EE relatives view aspects of the patient's behavior as being (at least to some degree) under voluntary control, they may make efforts to change the behaviors that they perceive as undesirable. Furthermore, because in EE assessments, relatives are rated as critical because they make it clear that there are aspects of the patient's personality or behavior that they do not like, this idea at least has face validity. Criticism, by its very nature, implies that the relative would like the patient to be different. If relatives both desire, and consider patients' capable of change, it is plausible to suggest that this may be associated with efforts to modify those elements of patients' behavior that are considered aversive, hence the designation of high EE.

What characteristics of relatives might be associated with their striving to obtain change in their patient family member? Preliminary results from the Harvard Family Study reveal that high- and low-EE relatives differ in a number of theoretically coherent ways. Compared with their low-EE counterparts, high-EE family members have a significantly more internally based locus of control. Moreover, the number of critical comments made by relatives during the CFI is also significantly (and positively) correlated with how controllable relatives rate psychiatric symptoms (in others) as

---

[12] Although cross-cultural investigations are highly informative, we consider Jenkins's view that criticism reflects nothing more than "kin objections to cultural rule violations" (1991, p. 404) to be inherently unable to explain why, *within* a given culture, relatives differ in how critical they are. Again, we return to the need to examine the factors in both patients and relatives that might engender criticism, regardless of the cultural differences in the distribution of high and low levels of EE.

TABLE 3.3. The prevalence of High EE.

Study	Household-based classification % High EE	Individual-based classification % High EE	Cutoff*
Predominantly high EE			
Australia			
Parker et al. (1988)	74	71% mothers[a]	a
Vaughan et al. (1992)	54	58% fathers	
Mean	64		
Czechoslovakia			
Možný & Votýpková (1992)	55	N/A	b
Denmark			
Wig et al. (1987)	N/A	54	a
Germany			
Köttgen et al. (1984)[b]	56	N/A	b
Niedermeier et al. (1992)	57	49	a
Mean	57		
Italy			
Bertrando et al. (1992)	76	64	a
	58	52	b
Spain			
Arévalo & Vizcarro (1989)	58		
Montero et al. (1992)	50	50	c
Mean	54		
Switzerland			
Barrelet et al. (1990)	67	N/A	a
United Kingdom			
Brown et al. (1962)	52	N/A	e
Brown et al. (1972)	45	N/A	d
MacMillan et al. (1986)	53	57	a
Stirling et al. (1991)	48	45	a
Tarrier et al. (1988)	77	N/A	a
Vaughn & Leff (1976a)	57	N/A	a
Mean	55		
USA			
Moline et al. (1985)	71		a
Nuechterlein et al. (1992)	72	N/A	b
Vaughn et al. (1984)	67	N/A	a
Mean	70		
Predominantly low EE[c]			
USA: (Mexican American)			
Karno et al. (1987)	41	28	b
India			
Leff et al. (1987)	24	23[d]	a

* a (conventional cutoff): Criticism = 6 or more or EOI = 3 or more or Hostility = 1 or more; b: CC = 6 or more or EOI = 4 or 5 or H = 1 or more; c: CC = 4 or more; d: CC = 7 or more or EOI = 4 or 5 or H = 1 or more; e: cutoffs unknown.
[a] Reported in Parker & Johnston (1987).
[b] Reported in Dulz & Hand (1986).
[c] Ivanović & Vuletić (1989) report a 48% prevalence of high EE in Belgrade, in the former Yugoslavia. Because this is within the range of estimates found in predominantly high-EE cultures, this culture cannot be classified as predominantly high or low EE at the present time.
[d] Reported in Wig et al. (1987).

being. Taken together, these findings suggest that high levels of EE may be associated with an underlying belief about the patient's ability to deal with some of the behavioral consequences of their illness.

Recent findings from our ongoing investigations further indicate that high levels of EE in the relatives of both schizophrenic and depressed patients are associated with significantly more spontaneous causal attributions to factors perceived as controllable by the patient. Brewin, MacCarthy, Duda, and Vaughn (1991) report essentially similar results from a sample of relatives of schizophrenic patients. These results lend further support to this hypothesis. In a related vein, Leff and his colleagues (1990) report that hostility in Indian relatives is linked to the belief that the patient is responsible for the disturbed behavior that occurs during an episode of schizophrenia.

These differences in the belief systems of high- and low-EE relatives may explain why levels of expressed emotion vary cross-culturally. Although clearly only speculation at the present time, in the absence of clear differences in the patients with whom they have to cope, this idea may go some way toward explaining why relatives of schizophrenic patients in India are more likely to be low rather than high EE.

It is also worth noting that, at least in Western culture, an internally based locus of control is generally regarded as a good thing. It reflects an individual "can-do" approach to dealing with problems that is culturally valued. This is important, because the EE literature often creates the impression that high-EE relatives are doing something wrong—that they harm by not being more sympathetic and supportive toward a suffering family member. In fact, the reverse may be true. The underlying motivation of high-EE family members may be to help patients get better, or to deal with their problems more effectively. Unfortunately, in many cases, these efforts to help may not be well received or may not lead to the kind of results hoped for by the relative. Thus, over time, initially supportive relatives may become increasingly frustrated by the failure of their efforts to help, eventually becoming angry with the patient and blaming him or her for the continuation of problem behaviors.

Implicit in the preceding statement is the idea that high levels of EE reflect the product of a mutually interactive and developmental process. If this is the case, we might predict that EE would become a more reliable predictor of relapse as time progresses. Here it is interesting to note that several of the nonreplication studies involved only first- or recent-onset patients (e.g., MacMillan et al., 1986; Stirling et al., 1991) or contained a high proportion of these patients in the total sample (e.g., Köttgen et al., 1984). However, the fact that Nuechterlein et al. (1986, 1992) and Leff et al. (1987) also used recently diagnosed patients and still demonstrated an association between EE and relapse suggests that this is not the whole story. Expressed emotion is a phenomenon that is unlikely to yield itself to an easy answer. That it reflects an ongoing, complex, and likely developmental process, however, is perhaps the most tenable hypothesis at the present time.

Summary and Concluding Comments

An examination of the current EE literature suggests that, despite recent concern about nonreplications, EE is still a variable of importance to those interested in understanding psychiatric relapse. However, early enthusiasm and simplified interpretations of the EE construct should be tempered with caution. Although tentative evidence suggests a causal role for EE in patient relapse, other evidence suggests that it may to an important extent be a reaction to the functioning deficits of patients. Moreover, although specific patient characteristics that may elicit high EE have not yet been reliably identified, there is reason to believe that they will be characteristics that family members perceive as controllable by the patient (Hooley, 1987). Rather than neglecting to examine patient correlates of high and low levels of relative's EE (as is the case in many studies), we make a plea for this issue to be examined in a thorough and systematic manner.

The fact that high levels of EE are found in family members of patients suffering from a wide range of clinical conditions further suggests that, in investigations of this type, researchers would do well to look beyond specific aspects of psychiatric symptomatology and to focus instead on more-widely distributed characteristics. Personality factors are obvious candidates for this search, a conclusion supported by recent data collected on long-stay psychiatric inpatients and members of their treatment teams (Heinssen et al., 1994). These authors documented high levels of criticism in hospital staff members who were unrelated to patients and identified three particular characteristics of patients that were associated with staff member criticism and/or hostility. Specifically, these were uncooperativeness, hostility, and impulsivity. These three patient characteristics were also associated with patients' doing poorly (earning little money) in a token economy. It is therefore intriguing to speculate whether some of the patient characteristics previously associated with high levels of criticism in family members (e.g., poor work performance) might also reflect an underlying tendency in patients toward oppositional or uncooperative behavior. That these characteristics of patients might also be considered by treatment personnel (and family members) to reflect aspects of behavior that are potentially controllable by patients provides further (albeit tentative) support for the hypothesized link between high EE and attributions of controllability of behavior.

It is clear that continued replications of the EE–relapse link, and/or mere extensions of EE research to other psychiatric and medical populations will add little to our understanding of relapse in the absence of data relevant to understanding the underlying causal processes. The question of why the risk of relapse in depressed patients residing in high-EE home environments is so high (see Table 3.1) obviously also demands further study. The course of schizophrenia may be strongly determined by biological factors. Although these may also be important for the course of depression, compared with their schizophrenic counterparts, mood-disordered patients (especially unipolar depressives) may have a course of illness that is more susceptible to environmental influences. We should also bear in mind that the mechanism through which EE is associated with relapse is likely to be different across different diagnoses and perhaps also across different cultures. Although for schizophrenic patients, psychophysiological models of overstimulation and hyperautonomic arousal are being examined as possible pathways linking high family stress to relapse (Nuechterlein & Dawson, 1984; Sturgeon, Kuipers, Berkowitz, Turpin, & Leff, 1981; Tarrier, Barrowclough, Porceddu, & Watts, 1988; Valone, Goldstein, & Norton, 1984), the mechanism through which EE and relapse may be associated in depressed samples has received no attention to date. Quite possibly, however, the route to relapse is not via psychophysiological overstimulation caused by an inability to process complex social stimuli but via the consequences of criticism for an individual whose vulnerability may be more cognitive or interpersonal in nature. Clearly, issues such as these warrant attention from researchers in the coming years.

Finally, as others have also noted (e.g., Lefley, 1992), we believe that there is much to gain by examining EE-related phenomena in nonfamilial settings. Not all psychiatric patients live with their families, and EE researchers need to remain cognizant of this fact. We also need to bear in mind that relapse is not the only measure of negative outcome. The extensive focus on psychiatric relapse in the EE literature has left us with little knowledge of the association between EE and other aspects of patient functioning such as social adjustment, symptomatology, and overall levels of functioning between psychiatric episodes. Researchers should bear in mind that psychiatric relapse marks the end point in a process that almost certainly begins well before patients have decompensated sufficiently to warrant rehospitalization or to meet predetermined relapse criteria. The significance of this lies in the fact that understanding the EE–relapse link is inherently a developmental task (see Hooley & Richters, in press; see also Stirling et al., 1993) and will require studying the process of relapse from its earliest manifestations.

In conclusion, we call for systematic research that examines the nature of the EE–relapse relationship in a more theoretically driven manner. The processes that underlie the EE–relapse link are, in all probability, likely to be much more complex than once thought and will require sophisticated research designs to disentangle them. Expressed emotion is clearly indexing something interesting and important. With more focused and targeted research, we will perhaps finally come to learn why the EE construct has proved so difficult to understand.

References

Arévalo, J., & Vizcarro, C. (1989). "Emoción Expressada" ay curso de la esquizofrenia en una muestra Española. *Analisis y Modificación de Conducta, 15*(43), 3–25.

Barrelet, L., Ferrero, F., Szigethy, L., Giddey, C., & Pellizer, G. (1990). Expressed emotion and first-admission schizophrenia. Nine-month follow-up in a French cultural environment. *British Journal of Psychiatry, 156*, 357–362.

Barrowclough, C., & Tarrier, N. (1990). Social functioning in schizophrenic patients: I. The effects of expressed emotion and family intervention. *Social Psychiatry and Psychiatric Epidemiology, 25*, 125–129.

Bertrando, P., Beltz, J., Bressi, C., Clerci, M., Farma, T., Invernizzi, G., & Cazzullo, C.L. (1992). Expressed emotion and schizophrenia in Italy: A study of an urban population. *British Journal of Psychiatry, 161*, 223–229.

Bledin, K.D., MacCarthy, B., Kuipers, L., & Woods, R.T. (1990). Daughters of people with dementia: Expressed emotion, strain, and coping. *British Journal of Psychiatry, 157*, 221–227.

Brewin, C.R., MacCarthy, B., Duda, K., & Vaughn, C.E. (1991). Attribution and expressed emotion in the relatives of patients with schizophrenia. *Journal of Abnormal Psychology, 100*(3), 546–554.

Brown, G.W., Birley, J.L.T., & Wing, J.K. (1972). Influence of family life on the course of schizophrenic disorders: A replication. *British Journal of Psychiatry, 121*, 241–258.

Brown, G.W., Monck, E.M., Carstairs, G.M., & Wing, J.K. (1962). Influence of family life on the course of schizophrenic illness. *Journal of Preventive and Social Medicine, 16*, 55–68.

Buchkremer, G., Stricker, K., Holle, R., & Kuhs, H. (1991). The predictability of relapses in schizophrenic patients. *European Archives of Psychiatry and Clinical Neuroscience, 240*, 292–300.

Butzlaff, R., & Hooley, J.M. (1995). *A meta-analysis of the EE-relapse link.* Manuscript in preparation.

Cazzullo, C.L., Bressi, C., Bertrando, P., Clerici, M., & Maffei, C. (1989). Schizophrénie et expression émotionnelle familiale: Étude d'une population italienne. *L'Encéphale, 15*, 1–6.

Cicchetti, D. (1989). Developmental psychopathology: Past, present, and future. In D. Cicchetti (Ed.), *Rochester symposium on developmental psychopathology, Vol. 1: The emergence of a discipline*, (pp. 1–12). Hillsdale, NJ: Lawrence Erlbaum.

Dulz B., & Hand, I. (1986). Short-term relapse in young schizophrenics: Can it be predicted by family (CFI), patient, and treatment variables? An experimental study. In M.J. Goldstein, I. Hand, & K. Hahlweg (Eds.), *Treatment of schizophrenia: Family assessment and intervention*, (pp. 59–75) New York: Springer-Verlag.

Fischmann-Havstad, L., & Marston, A.R. (1984). Weight loss maintenance as an aspect of family emotion and process. *British Journal of Clinical Psychology, 23*, 265–271.

Glynn, S.M., Randolph, E.T., Eth, S., Paz, G.G., Leong, G.B., Shaner, A.L., & Strachan, A. (1990). Patient psychopathology and expressed emotion in schizophrenia. *British Journal of Psychiatry, 157*, 877–880.

Goldstein, M.J., Talovic, S.A., Nuechterlein, K.H., Fogelson, D.L., Subotnik, K.L., & Asarnow, R.F. (1992). Family interaction versus individual psychopathology. Do they indicate the same processes in the families of schizophrenics? *British Journal of Psychiatry, 161*(Suppl. 18), 97–102.

Guttierrez, E., Escudero, V., Valero, J.A., Vasquez, M.C., Castro, J.A., Alvarez, L.C., Baltar, M., Blanco, J., Gonzalez, I., & Gomez, I. (1988). Expresion de emociónes y curso de la esquizofrenia. II Expresion de emociónes y el curso de la esquizofrenia en pacientes en remision. *Analisis y Modificacion de Conducta, 15*, 275–316.

Hahlweg, K., Goldstein, M.J., Nuechterlein, K.H., Magaña, A., Mintz, J., Doane, J.A., & Snyder, K.S. (1989). Expressed emotion and patient–relative interaction in families of recent onset schizophrenics. *Journal of Consulting and Clinical Psychology, 57*, 11–18.

Heckelman, L.R., & Hooley, J.M. (March, 1993). *Maternal expressed emotion and child attachment*. Poster presentation at the meeting of the Society for Research in Child Development, New Orleans, LA.

Heinssen, R.K., Hooley, J.M., & Minarik, M.E., Israel, S.B., & Fenton, W. (1994). *Expressed emotion in psychiatric hospital staff: It's not all relative*. Unpublished manuscript.

Hogarty, G.E. (1985). Expressed emotion and schizophrenic relapse: Implications from the Pittsburgh study. In M. Alpert (Ed.), *Controversies in schizophrenia*, (pp. 354–363). New York: Guilford Press.

Hogarty, G.E., Anderson, C.M., Reiss, D.J., Kornblith, S.J., Greenwald, D.P., Javna, C.D., & Madonia, M.J. (1986). Family psychoeducation, social skills training, and maintenance chemotherapy in the aftercare treatment of schizophrenia. *Archives of General Psychiatry, 43*, 633–642.

Hogarty, G.E., McEvoy, J.P., Munetz, M., Dibarry, A.L., Bartone, P., Cather, R., Cooley, S.J., Ulrich, R.F., Carter, M., & Madonia, M.J. (1988). Dose of fluphenazine, familial expressed emotion, and outcome in schizophrenia: Results of a two-year controlled study. *Archives of General Psychiatry, 45*, 797–805.

Hooley, J.M. (1985). Expressed emotion: A review of the critical literature. *Clinical Psychology Review, 5*, 119–139.

Hooley, J.M. (1986). Expressed emotion and depression: Interactions between patients and high versus low expressed emotion spouses. *Journal of Abnormal Psychology, 95(3)*, 237–246.

Hooley, J.M. (1987). The nature and origins of expressed emotion. In M.J. Goldstein & K. Hahlweg (Eds.), *Understanding major mental disorder: The contributions of family interaction research*, (pp. 176–194). New York: Family Process Press.

Hooley, J.M. (1990). Expressed emotion and depression. In G.I. Keitner (Ed.), *Depression and families: Impact and treatment*, (pp. 57–83). Washington, DC: American Psychiatric Press.

Hooley, J.M., & Hahlweg, K. (1986). Interaction patterns of depressed patients and their spouses: Comparing high and low EE dyads. In M.J. Goldstein, I. Hand, & K. Hahlweg (Eds.), *Treatment of schizophrenia: Family assessment and intervention*. Berlin: Springer-Verlag.

Hooley, J.M., Orley, J., & Teasdale, J.D. (1986). Levels of expressed emotion and relapse in depressed patients. *British Journal of Psychiatry, 148,* 642–647.
Hooley, J.M., & Richters, J.E. (in press). Expressed emotion: A developmental perspective. In D. Cicchetti & S.L. Toth (Eds.), *Rochester symposium on developmental psychopathology: Vol. VI. Emotion, cognition, and representation.* Rochester, NY: University of Rochester Press.
Iacono, W.G., & Beiser, M. (1992). Where are the women in first episode studies of schizophrenia? *Schizophrenia Bulletin, 18(3),* 471–480.
Ivanović, M., & Vuletić, Z. (1989). *Expressed emotion in families of patients with frequent types of schizophrenia and influence on the course of illness.* Unpublished manuscript.
Jenkins, J.H. (1991). Anthropology, expressed emotion and schizophrenia. *Ethos, 19(4),* 387–431.
Karno, M., Jenkins, J.H., de la Selva, A., Santana, F., Telles, C., Lopez, S., & Mintz, J. (1987). Expressed emotion and schizophrenic outcome among Mexican-American families. *The Journal of Nervous and Mental Disease, 175(3),* 143–151.
Kavanagh, D.J. (1992). Recent developments in expressed emotion and schizophrenia. *British Journal of Psychiatry, 160,* 601–620.
Koenigsberg, H.W., & Handley, R. (1986). Expressed emotion: From predictive index to clinical construct. *American Journal of Psychiatry, 143(11),* 1361–1373.
Koenigsberg, H.W., Klausner, E., Pellino, D., Rosnick, P., & Campbell, R. (1993). Expressed emotion and glucose control in insulin-dependent *diabetes mellitus. American Journal of Psychiatry, 150,* 114–115.
Köttgen, C., Sönischen, I., Mollenhauer, K., & Jurth, R. (1984). Group therapy with the families of schizophrenic patients: Results of the Hamburg Camberwell-Family Interview Study III. *International Journal of Family Psychiatry, 5,* 83–94.
Kuipers, L. (1979). Expressed emotion: A review. *British Journal of Social and Clinical Psychology, 18,* 237–243.
Kuipers, L., Sturgeon, D., Berkowitz, R., & Leff, J. (1983). Characteristics of expressed emotion: Its relationship to speech and looking in schizophrenic patients and their relatives. *British Journal of Clinical Psychology, 22,* 257–264.
Lam, D.H. (1991). Psychosocial family intervention in schizophrenia: A review of empirical studies. *Psychological Medicine, 21,* 423–441.
Leff, J., Berkowitz, R., Shavit, N., Strachan, A., Glass, I., & Vaughn, C. (1989). A trial of family therapy v. a relatives' group for schizophrenia. *British Journal of Psychiatry, 154,* 58–66.
Leff, J., Kuipers, L., Berkowitz, R., Eberlein-Fries, R., & Sturgeon, D. (1982). A controlled trial of social intervention in the families of schizophrenic patients. *British Journal of Psychiatry, 141,* 121–134.
Leff, J., & Vaughn, C. (1985). *Expressed emotion in families.* New York: Guilford Press.
Leff, J., & Vaughn, C. (1986). First episodes of schizophrenia. [Letter to the editor] *British Journal of Psychiatry, 148,* 215–216.
Leff, J., Wig, N.N., Bedi, H., Menon, D.K., Kuipers, L., Korten, A., Ernberg, G., Day, R., Sartorius, N., & Jablensky, A. (1990). Relatives' expressed emotion and the course of schizophrenia in Chandigarh: A two-year follow-up of a first-contact sample. *British Journal of Psychiatry, 156,* 351–356.

Leff, J., Wig, N.N., Ghosh, A., Bedi, H., Menon, D.K., Kuipers, L., Korten, A., Ernberg, G., Day, R., Sartorius, N., & Jablensky, A. (1987). Influence of relatives' expressed emotion in the course of schizophrenia in Chandigarh. *British Journal of Psychiatry, 151*, 166–173.

Lefley, H.P. (1992). Expressed emotion: Conceptual, clinical, and social policy issues. *Hospital and Community Psychiatry, 43(6)*, 591–598.

Lewis, S. (1992). Sex and schizophrenia: Vive la différence. *British Journal of Psychiatry, 161*, 445–450.

MacMillan, J.F., Crow, T.J., Johnson, A.L., & Johnstone, E.C. (1987). Expressed emotion and relapse in first episodes of schizophrenia. *British Journal of Psychiatry, 151*, 320–323.

MacMillan, J.F., Gold, A., Crow, T.J., Johnson, A.L., & Johnstone, E.C. (1986). Expressed emotion and relapse. *British Journal of Psychiatry, 148*, 133–148.

McCreadie, R.G., & Phillips, K. (1988). The Nithsdale schizophrenia survey. 7. Does relative high expressed emotion predict relapse? *British Journal of Psychiatry, 152*, 477–481.

Meehl, P.M. (1978). Theoretical risks and tabular asterisks: Sir Karl, Sir Ronald, and the slow progress of soft psychology. *Journal of Consulting and Clinical Psychology, 46*, 806–834.

Miklowitz, D.J. (November, 1992). *Family risk indicators in bipolar disorder*. Paper presented at the 6th annual meeting of the Society for Research in Psychopathology, Palm Springs, CA.

Miklowitz, D.J., Goldstein, M.J., & Falloon, I.R.H. (1983). Premorbid and symptomatic characteristics of schizophrenics from families with high and low levels of expressed emotion. *Journal of Abnormal Psychology, 92(3)*, 359–367.

Miklowitz, D.J., Goldstein, M.J., Falloon, I.R.H., & Doane, J.A. (1984). Interactional correlates of expressed emotion in the families of schizophrenics. *British Journal of Psychiatry, 144*, 482–487.

Miklowitz, D.J., Goldstein, M.J., Nuechterlein, K.H., Snyder, K.S., & Doane, J.A. (1986). Expressed emotion, affective style, lithium compliance, and relapse in recent onset mania. *Psychopharmacology Bulletin, 22(3)*, 628–632.

Miklowitz, D.J., Goldstein, M.J., Nuechterlein, K.H., Snyder, K.S., & Mintz, J. (1988). Family factors and the course of bipolar affective disorder. *Archives of General Psychiatry, 45*, 225–231.

Mintz, J., Mintz, L., & Goldstein, M. (1987). Expressed emotion and relapse in first episodes of schizophrenia: A rejoinder to MacMillan et al. (1986). *British Journal of Psychiatry, 151*, 314–320.

Mintz, L.I., Nuechterlein, K.H., Goldstein, M.J., Mintz, J., & Snyder, K.S. (1989). The initial onset of schizophrenia and family expressed emotion: Some methodological considerations. *British Journal of Psychiatry, 154*, 212–217.

Moline, R.A., Singh, S., Morris, A., & Meltzer, H.Y. (1985). Family expressed emotion and relapse in schizophrenia in 24 urban American patients. *American Journal of Psychiatry, 142(9)*, 1078–1081.

Montero, I., Gómez-Beneyto, M., Ruiz, I., Puche, E., & Adam, A. (1992). The influence of family expressed emotion on the course of schizophrenia in a sample of Spanish patients: A two-year follow-up study. *British Journal of Psychiatry, 161*, 217–222.

Moore, E., Ball, R.A., & Kuipers, L. (1992). Expressed emotion in staff working with the long-term adult mentally ill. *British Journal of Psychiatry, 161,* 802–808.

Možný, P., & Votýpková, P. (1992). Expressed emotion, relapse rate and utilization of psychiatric inpatient care in schizophrenia. *Social Psychiatry and Psychiatric Epidemiology, 27,* 174–179.

Mueser, K.T., Bellack, A.S., Wade, J.H., Sayers, S.L., Tierney, A., & Haas, G. (1993). Expressed emotion, social skill, and response to negative affect in schizophenia. *Journal of Abnormal Psychology, 102(3),* 339–351.

Niedermeier, T., Watzl, H., & Cohen, R. (1992). Prediction of relapse of schizophrenic patients: Camberwell Family Interview versus content analysis of verbal behavior. *Psychiatry Research, 41,* 275–282.

Nuechterlein, K.H., & Dawson, M.E. (1984). A heuristic vulnerability/stress model of schizophrenic episodes. *Schizophrenia Bulletin, 10,* 300–312.

Nuechterlein, K.H., Snyder, K.S., Dawson, M.E., Rappe, S., Gitlin, M., & Folfelson, D. (1986). Expressed emotion, fixed-dose fluphenazine decanoate maintenance, and relapse in recent-onset schizophrenia. *Psychopharmacology Bulletin, 22(3),* 633–639.

Nuechterlein, K., Snyder, K.S., & Mintz, J. (1992). Paths to relapse: Possible transactional processes connecting patient illness onset, expressed emotion, and psychotic relapse. *British Journal of Psychiatry, 161*(Suppl. 18), 88–96.

Okasha, A., El Akabawi, A.S., Snyder, K.S., Wilson, A.K., Youssef, I., & El Dawla, A.S. (1994). Expressed emotion, perceived criticism, and relapse in depression: A replication in an Egyptian community. *American Journal of Psychiatry, 151, 7,* 1001–1005.

Parker, G., & Hadzi-Pavlovic, D. (1990). Expressed emotion as a predictor of schizophrenic relapse: An analysis of aggregated data. *Psychological Medicine, 20,* 961–965.

Parker, G., & Johnston, P. (1987). Parenting and schizophrenia: An Australian study of expressed emotion. *Australian and New Zealand Journal of Psychiatry, 21,* 60–66.

Parker, G., Johnston, P., & Hayward, L. (1988). Parental "expressed emotion" as a predictor of schizophrenic relapse. *Archives of General Psychiatry, 45,* 806–813.

Peter, H., & Hand, I. (1988). Patterns of patient-spouse interviews in agraphobics: Assessment by Camberwell Family Interview (CFI) and impact on outcome of self-exposure treatment. In H.U. Wittchen and I. Hand (Eds.). *Panic & Phobia. II.* (pp. 240–251). Berlin: Springer-Verlag.

Rosenthal, R. (1991). *Meta-analytic procedures for social research.* Newbury Park, CA: Sage.

Stirling, J., Tantam, D., Thomas, P., Newby, D., Montague, L., Ring, N., & Rowe, S. (1991). Expressed emotion and early onset schizophrenia; a one year follow-up. *Psychological Medicine, 21,* 675–685.

Stirling, J., Tantam, D., Thomas, P., Newby, D., Montague, L., Ring, N., & Rowe, S. (1993). Expressed emotion and schizophrenia: The ontogeny of EE during an 18-month follow-up. *Psychological Medicine, 23,* 771–778.

Strachan, A.M., Leff, J.P., Goldstein, M.J., Doane, J.A., & Burtt, C. (1986). Emotional attitudes and direct communication in the families of schizophrenics: A cross-national replication. *British Journal of Psychiatry, 149,* 279–287.

Sturgeon, D., Kuipers, L., Berkowitz, R., Turpin, G., & Leff, J. (1981). Psychophysiological responses of schizophrenic patients to high and low expressed emotion relatives. *British Journal of Psychiatry, 138*, 40–45.

Szmuckler, G.I., Eisler, I., Russell, G.F.M., & Dare, C. (1985). Anorexia nervosa, parental 'expressed emotion' and dropping out of treatment. *British Journal of Psychiatry, 147*, 265–271.

Tarrier, N., Barrowclough, C., Porceddu, K., & Watts, S. (1988). The assessment of physiological reactivity to the expressed emotion of the relative of schizophrenic patients. *British Journal of Psychiatry, 152*, 618–624.

Tarrier, N., Barrowclough, C., Vaughn, C., Bamrah, J.S., Porceddu, K., Watts, S., & Freeman, H. (1988). The community management of schizophrenia: A controlled trial of a behavioural intervention with families to reduce relapse. *British Journal of Psychiatry, 153*, 532–542.

Tarrier, N., Barrowclough, C., Vaughn, C., Bamrah, J.S., Porceddu, K., Watts, S., & Freeman, H. (1989). Community management of schizophrenia: A two-year follow-up of a behavioural intervention with families. *British Journal of Psychiatry, 154*, 625–628.

Valone, K., Goldstein, M.J., & Norton, J.P. (1984). Parental expressed emotion and psychophysiological reactivity in an adolescent sample at risk for schizophrenia spectrum disorders. *Journal of Abnormal Psychology, 93*, 448–457.

Vaughan, K., Doyle, M., McConaghy, N., Blaszcynski, A., Fox, A., & Tarrier, N. (1992). The relationship between relatives' Expressed Emotion and schizophrenic relapse: an Australian replication. *Social Psychiatry and Psychiatric Epidemiology, 27*, 10–15.

Vaughn, C.E. (1986). Comment on Chapter 5. In M.J. Goldstein, I. Hand, & K. Hahlweg (Eds.), *Treatment of schizophrenia: Family assessment and intervention*, (pp. 76–77). Berlin: Springer-Verlag.

Vaughn, C., & Leff, J. (1976a). The influence of family and social factors on the course of psychiatric illness. *British Journal of Psychiatry, 129*, 125–137.

Vaughn, C., & Leff, J. (1976b). The measurement of expressed emotion in the families of psychiatric patients. *British Journal of Social and Clinical Psychology, 15*, 157–165.

Vaughn, C., Snyder, K.S., Jones, S., Freeman, W.B., & Falloon, I.R.H. (1984). Family factors in schizophrenic relapse: Replication in California of British research on expressed emotion. *Archives of General Psychiatry, 41*, 1169–1177.

Vitaliano, P.P., Becker, J., Russo, J., Magaña-Amato, A., & Maiuro, R.D. (1988). Expressed emotion in spouse caregivers of patients with Alzheimer's disease. *The Journal of Applied Social Sciences, 13*(1), 215–249.

Wig, N.N., Menon, D.K., Bedi, H., Leff, J., Kuipers, L., Ghosh, A., Day, R., Kosten, A., Ernberg, G., Sartorius, N., & Jablensky, A. (1986). Distribution of expressed emotion components among relatives of schizophrenic patients in Aarhus and Chandigarh. *British Journal of Psychiatry, 151*, 160–165.

Wing, J.K., Cooper, J.E., & Sartorius, N. (1974). *Measurement and classification of psychiatric symptoms*. London: Cambridge University Press.

Zubin, J., & Spring, B. (1977). Vulnerability—A new view of schizophrenia. *Journal of Abnormal Psychology, 86*, 103–126.

4
Modal Developmental Aspects of Schizophrenia Across the Life Span

ELAINE WALKER, DANA DAVIS, JAY WEINSTEIN, TAMMY SAVOIE, KATHLEEN GRIMES, AND KYM BAUM

One of the most well-established features of the schizophrenic syndrome is its modal age at onset; namely, late adolescence/early adulthood (Loranger, 1984). Because of the apparent salience of this developmental period for the emergence of psychotic symptoms, researchers in the field have focused a great deal of attention on the psychosocial and biological events that characterize it. A widely held assumption has been that certain unique aspects of this period must be playing a role in triggering the expression of psychotic symptoms. Most recently, with the increasing emphasis on biological origins of psychopathology, several writers have pointed out the potential triggering role of postpubescent maturational events in the central nervous system (Benes, 1989, 1991; Feinberg, 1982–1983; Weinberger, 1987).

Although there is no doubt that late adolescence/early adulthood must be viewed as a critical period in the ontogenesis of schizophrenia, it has become increasingly apparent that both the premorbid and postmorbid periods are characterized by significant behavioral changes that may hold clues to etiology. Specifically, evidence of significant behavioral dysfunction in preschizophrenic children has gradually accumulated, and recent studies on the long-term course of schizophrenia have revealed changes in the behavioral expression of the disorder. In this chapter, we explore the developmental aspects of schizophrenia across the life span, with a particular focus on our ongoing research on childhood precursors. The findings from research covering various periods of the life span are integrated in an attempt to identify modal periods of developmental change in the expression of behavioral dysfunction. We then consider the potential implications of the modal developmental aspects of the illness for conceptualizing the underlying neurodevelopmental process. Specifically, we propose that the expression of congenital, subcortical brain abnormality is moderated by maturational changes in the central nervous system.

The Origins of Schizophrenia: Prenatal, Birth and Neonatal Factors

It has become apparent to all in the field that a comprehensive picture of the ontogenesis of schizophrenia must encompass the beginning of life. The findings from behavioral genetics research clearly indicate that at least a subgroup of schizophrenic patients enters life with a genotype that produces a constitutional vulnerability to schizophrenia (Gottesman, 1991). In addition, the results of research on obstetrical complications suggest that early biological insults may interact with the genetic predisposition in producing constitutional vulnerability or may be sufficient to produce vulnerability in the absence of a genetic predisposition (McNeil, 1987; McNeil & Kaij, 1978; Walker & Emory, 1981).

The literature on the genetic determinants of schizophrenia has been reviewed in detail elsewhere (e.g., Gottesman, 1991) and is not discussed here. In this section, we focus on the research literature that deals with obstetrical factors in schizophrenia, particularly the evidence that developmental deviations originate in the prenatal period.

General Obstetrical Factors

Although many questions regarding the role of obstetrical factors in the etiology of schizophrenia remain to be answered, the association of obstetrical complications (OCs) with schizophrenia is fairly well established. In a recent review of the literature, McNeil (1987) points out that this relationship has been demonstrated by systematic investigation of samples of childhood schizophrenic patients, adult schizophrenic patients, schizophrenic patients adopted early in life, high-risk offspring who later develop schizophrenia, and monozygotic twin pairs discordant for schizophrenia. Further, positive evidence of elevated OCs has come from schizophrenic samples that have been drawn from widely separated birth cohorts and different countries, and who have experienced a variety of prenatal and obstetrical care routines. Various forms of oxygen deprivation appear to be a common denominator among several of the specific OCs that occur with excess in schizophrenic samples. Complications that repeatedly show up in the literature as being over-represented include toxemia, bleeding during pregnancy, threatened spontaneous abortion, and asphyxia (McNeil & Kaij, 1987; McNeil, 1987).

Subsequent to McNeil's review, several reports have confirmed the association between OCs and schizophrenia and also indicate that OCs may be more common in patients without a family history of the disorder. For example, Wilcox and Nasrallah (1987) examined the birth records of patients diagnosed with schizophrenia or schizophreniform disorder. They found that 21% of schizophrenics had a perinatal insult (prolonged labor, instrument delivery, or cyanosis after birth) compared to only 13.5% of schizophreniform patients. They also investigated possible genetic influences and found that perinatal complications occurred at a higher rate in patients without a family history of the disorder.

In another study, Lewis and Murray (1987) compared rates of OCs in schizophrenic, anorexic, and bipolar disorder patients. They found that OCs were significantly more common in schizophrenic patients; in particular, hemorrhage and preeclampsia in pregnancy and instrumental delivery were more common in the schizophrenic group. Also, consistent with Wilcox and Nasrallah (1987), they found that only 6% of schizophrenics with a positive family history of mental illness had perinatal complications, whereas 24% of patients without a family history had experienced perinatal complications. Similarly, Schwarzkopf and colleagues (Schwarzkopf, Nasrallah, Olson, Coffman, & Mclaughlin, 1988) reported that prolonged labor and instrumental delivery were significantly more common among schizophrenic patients than among schizoaffective and bipolar patients. And, confirming previous reports, patients with a positive family history of schizophrenia had fewer OCs than those without a family history of the disorder. Taken together, the findings from these studies suggest that OCs and genetic factors make independent contributions to vulnerability for schizophrenia.

Distinguishing Among Subtypes of Obstetrical Complications

Despite the strength of the evidence linking OCs with schizophrenia, one problem that has characterized many of the studies in this area is the failure to distinguish among the subtypes of complications; particularly, prenatal, birth and neonatal complications. Within OCs, prenatal complications (PCs) cover the period from conception to the onset of labor, birth complications (BCs) include the period during labor and delivery, and neonatal complications (NCs) cover the period from the birth of the child through 2 to 4 weeks post partum. Prematurity and some other signs of prenatal developmental deviation (e.g., small for gestational age) have been inconsistently classified as PCs, BCs, or NCs across studies.

This distinction among subtypes of OCs is important in interpreting their etiological significance. For example, the neurological consequences of hypoxic damage sustained during delivery in full-term neonates (i.e., BCs) appear to differ from those sustained prenatally, especially damage before 35 weeks' gestation (Towbin, 1986). Further, PCs, BCs, and NCs may be causally interrelated. Prenatal complications, for example, may induce BCs and NCs. Also, NCs, more so than PCs and BCs, may reflect preexisting abnormalities in the fetus, rather than exogenous insults. Thus, differentiating the *effects* of complications from the *induction* of complications by constitutional vulnerability requires a focus on the specific nature and timing of OCs.

Another problem in this literature concerns the reliability of data on complications. Although it has been shown that retrospective maternal reports agree with medical records (O'Callaghan, Sham, Takei, Glover, & Murray, 1991), both may be subject to substantial errors of omission. This is especially true of the prenatal period, when clinically relevant problems, such as maternal viral infections or exposure to teratagens, may be undetected or unreported. McNeil (1987) suggests that the neurological structures/systems relevant to schizophrenia may be sensitive to subclinical levels of hypoxia, or that subjects with a genetic predisposition to schizophrenia may be hypersensitive to even mild levels of hypoxia.

Prenatal Factors

Although elevated rates of PCs, BCs, and NCs have been found in numerous studies of schizophrenia, some of the most recent research findings suggest that the prenatal period may be especially significant. Several lines of evidence implicate disturbances of fetal neural development in the ontogenesis of schizophrenia. These include findings from studies of viral epidemics, dermatoglyphics, and neuropathology.

Viral epidemic studies provide evidence linking prenatal complications during the second trimester with schizophrenia. Mednick, Machon, Huttunen, and Bonnett (1988) found that Helsinki residents whose second trimester of gestation overlapped with a severe viral epidemic evidenced an increased rate of hospital diagnoses of schizophrenia. Further, psychiatric patients were more likely to be diagnosed with schizophrenia if their second trimester of fetal life overlapped with the 1957 epidemic. O'Callaghan and colleagues (O'Callaghan et al., 1991) examined the rate of schizophrenic births 5 months after the peak infection period for the A2 influenza epidemic that occurred in 1957 in England and Wales. The number of births of individuals who later developed schizophrenia was 88% higher than the average number of such births in the corresponding periods of the 2 previous and the 2 subsequent years.

Although these investigations provide impressive evidence to suggest that viral infection during the second trimester may be a predisposing factor in schizophrenia, there is an obvious discrepancy between the incidence of schizophrenia in the general population, that is, 1% to 2%, and the fact that viral infections are ubiquitous. Machon and colleagues (Machon, Mednick, & Schulsinger, 1987) suggest that this may indicate that viral infections only predispose genetically vulnerable individuals to develop schizophrenia. Alternatively, the window of prenatal vulnerability for damage to structures related to schizophrenia may be extremely small, that is, a matter of days. In addition, the drugs used to treat viral infections may be a critical factor, or the severity of the mother's clinical symptoms (e.g., fever, dehydration) may determine the effect of the viral infection on the developing fetus.

Related to this, it is important to note that maternal influenza during pregnancy has been linked to a variety of other neurodevelopmental abnormalities and cannot, therefore, be viewed as specific to the neuropathology of schizophrenia. Maternal influenza is associated with a general increase in congenital abnormalities of the central nervous system (CNS) (Coffey & Jessup, 1959), including anencephaly in female offspring (Leck, 1963). Thus, the outcome of maternal viral infection during pregnancy, with respect to the site and extent of neuropathology, may vary as a function of maternal, fetal, and temporal characteristics.

Another approach that researchers have used to determine the timing of prenatal injury is the study of subtle dysmorphological features (Newell-Morris, Fahrenbruch, & Sackett, 1989; Wakita, Narahara, & Kimoto, 1988). Fetal organogenesis occurs in the first trimester (Gluck, 1977; Hamilton, Boyd, & Mossman, 1972). The second trimester is the critical period of neuronal migration from the periventricular germinal matrix to the cortex (Freide, 1989; Jakob & Beckman, 1986; Rakic, 1988). During this time, the distal upper limbs develop simultaneously with neural migration to the cortex. The ectodermal cells of the fetal upper limb migrate to form the hand skin during the fourth and fifth months of gestation.

In twin pairs, prenatal insults do not always affect both members to the same extent. It has been found that intrapair discrepancies in dysmorphological signs between monozygotic co-twins can serve as temporal markers of intrauterine insults that affected one co-twin more than the other (Schaumann & Alter, 1976). Given these findings, Bracha and colleagues (Bracha, Torrey, Bigelow, Lohr, & Linington, 1991) postulated that aberrations during the second trimester could be determined by examining putative markers of prenatal dysmorphogenesis of the hand. Differences in hand morphology between the two members of discordant monozygotic twin pairs may be a temporal marker of ischemic and other nongenetic insults that affected one fetus more than the other during the early part of the second trimester. Bracha et al. (1991) studied the hand morphology of 24 monozygotic twin pairs discordant for schizophrenia or delusional disorder. Compared with well co-twins, the affected co-twins had significantly higher total scores for fourth- and fifth-month dysmorphological signs. Results indicated that hand maldevelopment scores in the symptomatic co-twin group were four times higher than in the well co-twin group. These dysmorphological signs may be the result of random and mostly unrecognized insults that take place during the second trimester (Cummins & Mildo, 1943; Schaumann & Alter, 1976). Bracha et al. point out that insults in the second trimester of pregnancy that could result in hand anomalies include ischemia, anoxia, anemia, toxic exposure, twin transfusion syndrome, and various infectious insults.

In interpreting the significance of second-trimester insults for the locus of neuropathology in schizophrenia, the sequence of maturational processes in fetal development must be taken into consideration. As previously mentioned, in the fetus and premature newborn infant, brain damage occurs primarily in subcortical regions, affecting germinal tissue, periventricular white matter, and neighboring basal ganglia (Towbin, 1986). In contrast, full-term infants subjected to perinatal insult typically manifest damage to neocortical regions. The specificity in location of fetal–neonatal acute cerebral hypoxic damage—deep in the premature infant and cortical in the infant at term—is determined by three biologic factors that are related to gestational age: (a) the presence or absence of germinal matrix tissue, (b) the momentum of local organogenesis, and (c) the degree of development of the local vascular elements (Towbin, 1986).

The germinal matrix contributes to the formation of subcortical structures, including the basal ganglia, hippocampus, and thalamus during the second trimester. Hypoxic damage is therefore more likely to occur in the deep cerebral structures before 35 weeks' gestation. Later, as the fetus becomes mature, with the basal ganglia and other deep structures now well formed and the germinal matrix depleted, developmental activity at the core of the cerebrum declines, and the momentum of organogenesis shifts to the cortex. Thus, because the germinal tissue becomes attenuated near term, it is no longer the primary target tissue for hypoxic damage. Instead, for full-term infants, the primary site of hypoxic damage shifts to the neocortex, as is also the case with adults (Towbin, 1986).

Evidence for temporal differences between cortical and subcortical regions in sensitivity to insult is provided by a variety of neuropathological studies, including autopsies of fetuses exposed to prenatal stress (Towbin, 1986). For example, Gossey, Golaire, and Larroche (1982) present a case of a mother who attempted suicide at 30 weeks of pregnancy and was delivered at 36 weeks. The newborn died, and autopsy revealed cystic areas and calcification in the basal ganglia, together with widespread necrosis in white matter.

Neuropathological studies of the brains of schizophrenic patients have yielded evidence of cytoarchitectural deviations consistent with a disturbance in neural development in the second trimester (Lyon, Barr, Cannon, Mednick, & Shore, 1989). For example, Kovelman and Scheibel (1984) studied postmortem brain tissue and observed disarray in the orientation of the pyramidal cells of the hippocampus in schizophrenics—an aberration most likely due to disruptions in brain development early in the second trimester. Similarly, Jakob and Beckman (1986) found cytoarchitectonic deviations of the entorhinal region, which they ascribe to genetic or environmental disturbances in fetal brain development occurring in the second trimester. In addition to these reports, several other studies of schizophrenia have revealed subcortical structural abnormalities (Heckers, Heinsen, Heinsen, & Beckmann, 1991; Jernigan et al., 1991; Bogerts et al., 1990; Suddath, Christison, Torrey, Casanova, & Weinberger, 1990) and metabolic abnormalities (Abdel-Dayem et al., 1990; Gur, Resnick, Alavi, et al., 1987; Gur, Resnick, Gur, et al., 1987; Resnick, Gur, Alavi, Gur, & Reivich, 1988).

To date, the research findings do not provide a basis for any strong conclusions about unique functional or structural subcortical abnormalities in schizophrenia. However, trends in the data suggest relatively heightened subcortical metabolic activity in schizophrenia, with elevated dopamine (DA) concentrations (Toru et al., 1988; Reynolds, 1983) and D2 receptors in the basal ganglia (Cross, Crow, & Owen, 1981; Crow, Johnstone, Longden, & Owen, 1978; Hess, Bracha, Kleinman, & Creese, 1987; Seeman et al., 1987; Toru et al., 1988; Wong et al., 1986). Limbic structures, in contrast, tend to be reduced in volume (Altshuler, Conrad, Kovelman, & Scheibel, 1987; Bogerts et al., 1990; Falkai & Bogerts, 1986; Jeste & Lohr, 1989; Suddath et al., 1990) and metabolic activation (Tamminga et al., 1992). The findings on striatal abnormalities are consistent with the reports of movement abnormalities in preschizophrenic children and nonmedicated schizophrenic patients that are discussed later in this chapter.

Summary

In summary, the findings from studies of viral epidemics, dermatoglyphics, and neuropathology indicate that vulnerability to schizophrenia is congenital and that, at least for some patients, it may reflect an aberration in prenatal neurodevelopment during the second trimester. Because subcortical brain regions are highly vulnerable to insult during this period, they constitute important sites for further exploration of the neuropathology of schizophrenia. Moreover, the site of dysfunction, namely, subcortical versus cortical, may have implications for developmental changes in behavior. In the following sections, we discuss the developmental aspects of behavioral abnormalities in schizophrenia.

Childhood Development

Bleuler (1950) argues that character anomalies exist in the majority of individuals who eventually develop schizophrenia and suggests that these anomalies revolve around the tendency to seclusion, withdrawal, and irritability. Findings from research on the behavior of preschizophrenic children generally support Bleuler's assumption of premorbid behavioral abnormality, although the research has not found a singular characteristic pattern of preschizophrenic behavior. Given that schizophrenia is heterogeneous with respect to clinical symptoms, it is not surprising that the premorbid behavioral course also varies.

In this section, we discuss some of the research findings on the development of preschizophrenic children. The discussion highlights recent findings from our archival-observational studies of the childhood home movies of schizophrenic patients, as well as studies using parent and teacher reports. Both reveal behavioral abnormalities in childhood that vary in nature with age.

Observational Studies

The archival-observational method has several advantages relative to other strategies that have been employed in the study of preschizophrenic development. First, childhood behavior can be directly and systematically observed in individuals whose adult psychiatric outcome is known. Thus, unlike prospective, "high-risk" studies of the biological offspring of schizophrenic patients, this method does not require following the subjects for extended periods of time to determine psychiatric outcome. Second, given that films spanning from infancy through adolescence are available for many of the subjects, behavior can be observed over the entire course of childhood development.

In one of the first studies from this project (Walker, Grimes, Davis, & Smith, 1993), a systematic emotion coding system (Izard, Dougherty, & Hembree, 1983) was used to test the hypothesis that preschizophrenic children differ from their healthy siblings in facial expressions of emotion. Specifically, it was predicted that they would show less positive and more negative emotion than their siblings. The subjects were 30 schizophrenic patients and 30 of their healthy siblings. The modal period of illness onset for the patients were late adolescence/early adulthood; none of the patients was diagnosed with a psychotic disorder before 16 years of age. The nearest-in-age, same-sex healthy sibling was chosen as the comparison subject, with the exception of sibships in which there was no same-sex sibling, in which case the nearest-in-age opposite-sex sibling was selected.

Analysis of the facial expression data showed that preschizophrenics could be distinguished from sibling controls and that the diagnostic group differences varied as a function of age and sex. The data on expression of positive (joy) and negative (a composite of anger, sadness, disgust, pain, contempt and fear) emotion are presented in Figures 4.1 and 4.2. When compared to same-sex controls, female preschizophrenics were characterized by a significant reduction in facial expressions of joy beginning in infancy (<1 year) and continuing through adolescence. Although male preschizophrenics showed fewer positive expressions of emotion than same-sex controls in infancy, they showed subsequent age-related increases in joy expressions leading to a trend toward more displays of joy than same-sex controls in adolescence. As predicted, the diagnostic groups also differed in the rates of expression of negative emotion. Male pre-

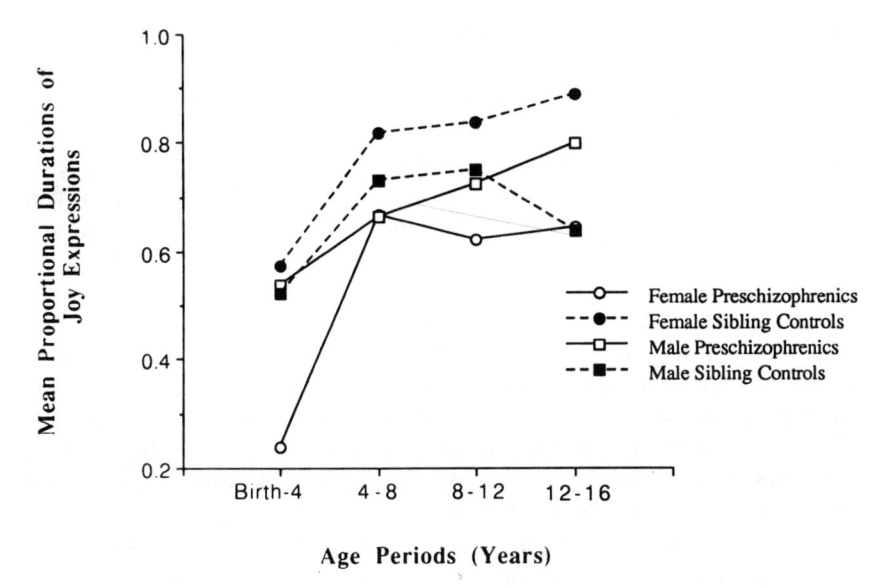

FIGURE 4.1. Mean proportions of joy expressions by age, sex, and diagnostic group.

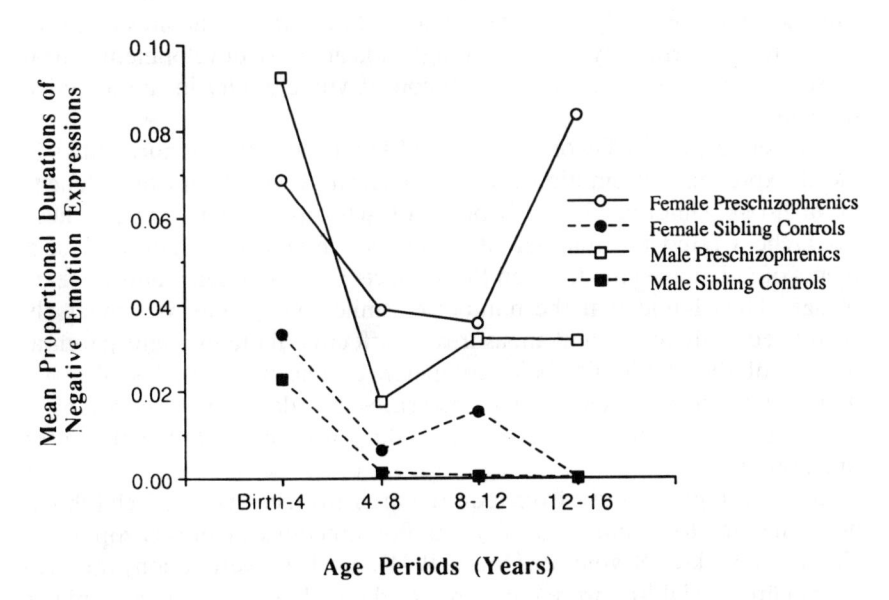

FIGURE 4.2. Mean proportions of negative facial expressions by age, sex, and diagnostic group.

schizophrenics showed significantly more facial expressions of negative emotion in infancy and late childhood relative to same-sex siblings. Female preschizophrenics showed significantly more negative emotion than their sibling controls in late childhood/adolescence. Comparisons of the sexes within diagnostic group revealed that preschizophrenic girls showed less joy than preschizophrenic boys in early childhood. In contrast, no significant sex differences were found for healthy siblings, although there was a trend toward greater positive emotion (i.e., joy) in girls.

The results of this study indicate that abnormalities in the expression of emotion predate the onset of schizophrenic symptoms by many years. Thus, the affective abnormalities that are often part of the schizophrenic syndrome may be subclinically manifested in childhood behavior. Further, the significant sex differences in facial expressions suggest that the behavioral expression of vulnerability is moderated by biological sex. Other research on the behavioral precursors of schizophrenia, discussed later, also points to the moderating influence of sex. Finally, it is of interest to note that the diagnostic group differences (i.e., effect sizes) in facial expressions tended to be most pronounced in the birth to 4-year and adolescent (12 to 16 years) age periods. The magnitude of the group differences tended to be smaller in the intervening developmental period. These developmental trends are illustrated in Figures 4.1 and 4.2. Presuming that the early affective abnormalities are a manifestation of congenital vulnerability, these findings indicate that developmental processes serve to moderate the expression of vulnerability in the affective domain.

A recent report by Dworkin et al. (1991) indicates that abnormalities in facial expression of emotion are also associated with behavioral deficits in high-risk children (i.e., offspring of schizophrenic parents). These researchers rated the number of smiles per minute from 3-minute segments of videotapes made when the subjects were between 9 and 12 years of age. They found that the number of smiles per minute was inversely correlated with concurrent measures of affective flattening and parental reports of the children's social competence. The number of smiles per minute was also correlated with measures of adolescent functioning 5 years later; a higher frequency of smiling was associated with better functioning.

In a subsequent study from our research project, we used childhood home movies to examine the neuromotor functions of preschizophrenic children (Walker, Savoie, & Davis, 1994). In this investigation, the preschizophrenic children were compared to their siblings, as well as subjects with adult-onset affective disorder, healthy siblings of the affective patients, and subjects from families with no mental illness in first-degree relatives. Two neuromotor rating subscales were used; the Neuromotor Abnormality (NA) Scale rated the presence/absence of neurological soft signs and movement abnormalities, and the Motor Skills (MS) Scale rated the quality of the subjects' motor skills. Group comparisons revealed that the preschizophrenics showed significantly more neuromotor abnormalities than their healthy siblings, the siblings of patients with affective disorder, and subjects from families with no mental illness. The preschizophrenics also showed significantly poorer motor skills than their healthy siblings and preaffective disorder subjects. (Comparisons of the preschizophrenics with the other two groups yielded marginally significant differences.) The neuromotor deficits of the preschizophrenics were most pronounced in the first 2 years of life, and their neuromotor abnormalities tended to predominate on the left side of the body. Analysis of individual items from the NA scale revealed that preschizophrenics showed more abnormal hand postures, associated reactions, and choreoathetoid movements of the upper limbs. In contrast to the data on facial expressions, no evidence of sex differences in the neuromotor development of preschizophrenic subjects was found.

The abnormalities of hand posture that were observed in the preschizophrenic subjects are similar to those that have been noted to occur at an elevated rate throughout development in children with known or suspected CNS damage (Knobloch & Pasamanick, 1974; Saint-Anne Dargassies, 1982; Stengel, Attermeier, Bly, & Heriza, 1984). Abnormalities of hand posture are viewed by some as a mild manifestation of athetotic movement (Knobloch & Pasamanick, 1974). This conceptualization is consistent with our findings. Examination of the NA data for the preschizophrenic children suggests a continuum of severity of neuromotor dysfunction that involves abnormal hand posture and associated reactions in its mildest form and choreoathetoid movements in the most extreme form. Of 30 schizophrenic subjects, 19 showed abnormal hand posture, 6 showed associated reactions, and 7 showed choreoathetoid movements. Of the 7 subjects with choreoathetoid movements, 4 also showed associated reactions and abnormal hand posture. Abnormal tonicity tended to co-occur with abnormalities of posture and movement.

It is important to point out that all of the signs we observed sometimes occur in children with no suspected neurological impairment. For example, a substantial number of normal neonates (Weggmann, Brown, Fulford, & Minns, 1987) and a small proportion (about 5%) of older children (Kennard, 1960) show subtle choreoathetoid movements when subjected to systematic neurological examination. However, the rates of postural and movement abnormalities decrease dramatically with age (Knobloch & Pasamanick, 1974; Saint-Anne Dargassies, 1982; Stengel et al., 1984); the persistence of fisting, finger hyperextension, and subtle choreoathetoid movements past the first few months of life is generally viewed as pathognomonic. Although the precise neural mechanism responsible for the decrease in movement abnormalities with age is not known, it is generally assumed that cortical maturation plays a role and that cortical activity gradually serves to reduce the influence of subcortical regions (Chugani,

Mazziotta, & Phelps, 1987; Knobloch & Pasamanick, 1974; Saint-Anne Dargassies, 1982; Stengel et al., 1984).

Although we are aware of no previous reports of movement abnormalities in preschizophrenic children, the general pattern of findings from our study of neuromotor precursors is consistent with previous reports from high-risk research. In a seminal prospective study, Fish, Marcus, Hans, Auerbach, and Perdue (1992) followed the development of 12 infants born to schizophrenic mothers. One of these subjects developed schizophrenia in adulthood, and 6 showed schizotypal or paranoid personality disorder. All 7 of these ill subjects showed what Fish has described as "pandysmaturation," including transient retardation in motor development. Further, consistent with the findings from our research, these developmental delays were most pronounced in the first 2 years of life.

Fish's high-risk project is the only one that was initiated when the subjects were infants and extended through the adult risk period for the onset of schizophrenia. However, several other high-risk projects have compared the motor development of high-risk children with that of control subjects whose parents have no mental illness. The results indicate that motor deficits are characteristic of high-risk subjects throughout childhood (for reviews, see Fish et al., 1992; Hans & Marcus, 1991), although they are most pronounced during the first 2 years of life.

Parent and Teacher Reports

The findings we have discussed up to this point have been based on the application of systematic assessment procedures and, therefore, do not address the question of whether preschizophrenic children manifest behavioral abnormalities that are apparent to others in their social environments. However, the findings from several high-risk and follow-back studies confirm that at least some preschizophrenic children manifest signs of behavioral abnormality in the home and in school.

The first large-scale study of children at high-risk for schizophrenia (i.e., offspring of schizophrenic parents) was initiated in Copenhagen by Sarnoff Mednick and Fini Schulsinger in 1962 (Mednick & Schulsinger, 1965). These investigators and their colleagues have followed the index and control subjects from childhood into the modal period of risk for schizophrenia onset. In relating adult psychiatric status to childhood characteristics, several items from school and parental reports were found to differentiate between the preschizophrenic children and those who remained healthy (Parnas et al., 1982). The following items from the school report were significantly more characteristic of preschizophrenics: (a) rejection by others/difficulty in making friends; (b) being uneasy about criticism; (c) exhibiting difficulty in affective control by becoming easily upset; (d) persisting in excitement; (e) disturbing the class by unusual behavior; and (f) presenting disciplinary problems (Parnas et al.,

1982). Parental interviews indicated that the schizophrenic and the borderline schizophrenic subjects were significantly more passive as babies and had poorer concentration during play when compared with the high-risk children who remained healthy as adults (Parnas et al., 1982). In a subsequent study from the Copenhagen project, gender differences were found in the teacher reports of behavior (John, Mednick, & Schulsinger, 1982). The male preschizophrenics tended to behave inappropriately, have discipline problems, and be characterized as anxious, lonely, and restrained. The female preschizophrenics were characterized as anhedonic, withdrawn, disengaged, and isolated.

Another longitudinal investigation of high-risk children is the New York High-Risk Project (Erlenmeyer-Kimling et al., 1984; N.F. Watt, Grubb, & Erlenmeyer-Kimling, 1984). Unlike the Danish study, the New York sample is only presently entering the risk period for schizophrenia onset; thus, most of the results from this group have not been associated with adult outcome. However, comparisons of the high-risk and control subjects have yielded findings that converge with those of the Danish project. For example, N.F. Watt et al. (1984) compared the school behavior of the high-risk and control subjects when they averaged 15 years of age and found that children of schizophrenic parents were rated as having less harmonious peer relations and as being more unpleasant, unpopular, negativistic, nervous, and maladjusted than the control children. The high-risk children also demonstrated significantly less scholastic motivation.

Retrospective and follow-back studies of schizophrenic adults have also revealed childhood behavioral abnormalities, as well as evidence of sex differences. Barthell and Holmes (1968) used data gathered from high school yearbooks of young men who developed schizophrenia in adulthood. They found that the preschizophrenic group participated in significantly fewer social activities than the normal control group. The number of social activities in which neurotics participated fell between the preschizophrenics and normal controls but did not significantly differ from either. It is interesting to note that the groups did not differ on the level of participation in service, performance, or sports groups. The researchers conclude that the preschizophrenic adolescents were more socially isolated than normal adolescents.

N.F. Watt (1978) examined the comments annually written by homeroom teachers from kindergarten through grade 12 for adult-onset schizophrenic patients and controls who were matched for sex, race, age, and parental social class. For the period between kindergarten and grade 6, there were no apparent differences between preschizophrenic boys and normal control boys. During the same period, however, the preschizophrenic girls were described as more emotionally unstable, introverted, and passive than the control girls. In the period between grades 7 and 12, the preschizophrenic boys were characterized as more emotionally unstable and more disagreeable than the normal control boys. During the same period, the preschizophrenic girls were still described as more emotionally unstable and introverted than their normal controls but were no longer considered more passive. Within-diagnostic group comparisons revealed that the preschizophrenic girls were more introverted than all other groups of children, particularly during adolescence. In contrast, the preschizophrenic boys were *never* more introverted than any other groups. Instead, the preschizophrenic boys were more disagreeable than all other groups during the adolescent years.

The research on intellectual performance has also revealed deficits in preschizophrenic children. A meta-analysis of the literature by Aylward, Walker, and Bettes (1984) indicated that preschizophrenic children score below their siblings and peers on standardized measures of cognitive ability. The cognitive performance deficits of preschizophrenic children are apparent early in elementary shcool and extend through the prodromal period. Further, as is the case with behavioral characteristics, there are sex differences in premorbid cognitive functioning, with boys showing greater deficits than girls.

Summary

The literature on childhood precursors of schizophrenia reveals deficits in neuromotor, cognitive, and social behavior. However, the deficits vary as a function of the sex and age of the child. Neuromotor deficits are apparent throughout childhood but are most pronounced in the first 2 years of life. No evidence from our study, or from other investigations of which we are aware, is suggestive of sex differences in the neuromotor functioning of preschizophrenic children. Cognitive deficits are apparent throughout childhood and appear to be more characteristic of preschizophrenic boys than girls. The data on facial expressions indicate that affective abnormalities are present, in childhood for both sexes but that preschizophrenic girls show greater affective abnormalities earlier in life and manifest less positive emotion than boys. These findings parallel those from high-risk and follow-back research on social behavior, which indicates that preschizophrenic girls differ from same-sex controls earlier and tend to show more dysphoria and social withdrawal than preschizophrenic boys. In contrast, the "modal" behavior pattern for preschizophrenic boys appears to involve externalized behavior problems accompanied by social detachment from peers. For both sexes, however, our data on facial expressions suggest that affective abnormalities are most pronounced early and late in childhood, with the latter period presumably reflecting the prodromal phase of the illness.

The Relation Between Clinical Symptoms and Premorbid Functioning

A central question raised by the findings on precursors of schizophrenia is whether childhood abnormalities are linked with clinical features of the illness in adulthood. If childhood behavioral characteristics are associated with the nature or severity of adult symptoms, then this would suggest that schizophrenia is subserved by a neurodevelopmental process that is moderated by individual differences across the life span.

Most of the literature relating premorbid functioning to postonset symptoms has used global measures of premorbid functioning during adolescence and early adulthood, rather than on specific behaviors or patterns of behavior in childhood. With respect to clinical symptoms, the focus has been on the positive and negative subtypes, both of which are observed in the majority of first-episode patients (Walker & Lewine, 1988). One result consistently reported is that ratings of negative symptoms (e.g., blunted affect, anhedonia, withdrawal), but not positive symptoms (e.g., hallucinations, delusions, positive thought disorder), are significantly correlated with poor premorbid functioning (Walker & Lewine, 1988). Specifically, negative symptoms have been found to be associated with fewer years of education (Andreasen & Olsen, 1982; Kay, Fiszbein, & Opler, 1987; Lindenmeyer, Kay, & Opler, 1984; O'Gureje, 1989; Opler, Kay, Rosado, & Lindenmeyer, 1984; Pogue-Guile & Harrow, 1984, 1985), poor social relationships (Andreasen & Olsen, 1982; Pogue-Guile & Harrow, 1984, 1985), and poor work adjustment (Andreasen & Olsen, 1982; Lindenmeyer, Kay & Opler, 1984). Moreover, it has been suggested that these findings are more characteristic of male than female patients (Pogue-Guile & Harrow, 1984, 1985) and especially males with a family history of psychotic disorder but not affective illness (Kay, Fiszbein, & Opler, 1987). This gender distinction is consistent with the observation that premorbid intellectual deficits are greater in boys than girls (Aylward, Walker, & Bettes, 1984).

In one of the few studies to examine positive and negative symptoms in high-risk children, Dworkin and colleagues (1990) used data from the New York High-Risk Project to compare offspring of depressed and schizophrenic parents with children of normal parents. Ratings of positive and negative symptoms were made by raters who viewed videotapes of the subjects as they underwent a structured interview. Social competence was also rated using subscales from the Premorbid Adjustment Scale (Cannon-Spoor, Potkin, & Wyatt, 1982). These researchers found that poorer social competence significantly distinguished adolescents at high risk for schizophrenia from children at high risk for depression and from control children. Furthermore, poorer social competence was significantly associated with higher scores for affective flattening and poverty of speech (negative symptoms) in all three groups. However, ratings of negative symptoms did not distinguish between the three groups. These results should be interpreted cautiously. They do not indicate that negative symptom precursors are nonexistent during the premorbid course but rather that adult-type schizophrenic symptomatology (on which the symptom rating scales were based) may not be predated by identical symptomatology in adolescence. The validity of this assumption cannot be evaluated, because these subjects have not yet been followed through the adult risk period for schizophrenia onset.

To date, only one prospective study has assessed adult psychiatric outcome and related clinical symptomatology to childhood behavioral characteristics. Cannon, Mednick, and Parnas (1990) investigated the antecedents of predominantly negative and predominantly positive symptom schizophrenia in a reanalysis of the Copenhagen high-risk data. The investigators used items from the questionnaires answered by the subjects' teachers to create scales of adolescent behavior disturbance. The five items chosen to represent "positive-type" characteristics were (a) is easily excited and irritated with little apparent reason; (b) when excited or emotional, persists in reaction and continues to be high-strung and distractible; (c) often disturbs class with completely inappropriate behavior; (d) is extremely violent and aggressive, frequently creating conflicts with classmates; and, (e) presents disciplinary problem for teacher. The seven items chosen to represent "negative-type" characteristics were (a) rarely takes part in spontaneous activities despite being asked to join; (b) seldom laughs or smiles when with other children, keeps serious "I don't care expression"; (c) has difficulty making friends, seems lonely and rejected by others; (d) seems content with isolation; (e) is very shy, reserved and silent; (f) does not react when praised or encouraged by teacher; and, (g) behavior is characteristically passive. Cronbach's alphas for the positive and negative groups were .76 and .84, respectively. The 10-year follow-up diagnostic evaluation was reassessed, using positive and negative symptom-rating scales.

Cannon et al. (1990) found that adult schizophrenics with predominantly negative symptoms had significantly more negative-type premorbid school disturbances than schizophrenics with predominantly positive symptoms. Conversely, schizophrenics with predominantly positive symptoms as adults were found to exhibit significantly more positive-type premorbid school behavior disturbances than schizophrenics with predominantly negative symptoms.

An earlier follow-up study by Roff, Knight, and Wertheim (1976) also revealed an association between childhood behavior and adult clinical outcome. These researchers conducted a follow-up assessment of men who had been at a child guidance clinic during their late childhood/early adolescent years. Thought disorder (a positive symptom) was rarely present in the childhood records; however, when it was present, it was not associated with poor adult outcome. On the other hand, disordered social relations and apathy (a negative symptom) in childhood were significantly associated with poor adult outcome. This lends additional support to the idea that positive and negative symptoms in adulthood may be differentially associated with features of the course of the illness.

Summary

In summary, negative symptoms have been found to be associated with poorer premorbid adjustment as evidenced by less education, and poorer social and occupational functioning, particularly in male schizophrenics. Positive symptoms appear to be less predictive of premorbid functioning. These results support the notions (a) that there is some continuity between premorbid behaviors and adult functioning and symptoms and (b) that there may be gender differences in the relation between premorbid functioning and clinical features of the illness.

Adult Development and Aging

Although often viewed as chronic and intractable, the recent longitudinal data suggest that schizophrenia is a developmentally diverse illness (Cohen, 1990; Miller & Cohen, 1987). A variety of studies reveal traceable life-course fluctuations in functioning, as well as some beneficial effects of the aging process. Changes in symptoms (Marneros, Deister, & Rohde, 1992), cognitive abilities (Heaton & Drexler, 1987), and motor signs (Casey & Hansen, 1984) have been reported.

As Cohen points out, "the notion that schizophrenia pursued an unfavorable course was prompted by the convictions of the key formulators of the disorder" (1990, p. 790). Likewise, Harding (1991) suggests that several assumptions of clinical psychiatry perpetuate the pessimistic views of aging and schizophrenia; perhaps the most damning of which is the assumption that patients who show signs of significant improvement were not schizophrenic in the first place. This has clear implications for the investigation and conceptualization of developmental changes.

Clinical Outcome

Although the long-term outcome of schizophrenia is worse than other major mental illnesses (McGlashan, 1984a, 1984b; Tsuang & Winokur, 1975), most patients do *not* show an extended deteriorating course (McGlashan, 1988). The estimated rates of recovery or significant improvement found in the major long-term outcome studies of schizophrenia have varied (e.g., 46%, Tsuang, Woolson, & Fleming, 1979; 49%, Ciompi, 1980; 53%, Bleuler, 1978; 57%, Huber, Gross, & Schuttler,

1979; 68%, Harding, Brooks, Ashikaga, Strauss, & Breir, 1987a, 1987b). However, given the diversity in the samples, it might be argued that the rates are impressively consistent.

McGlashan (1988) points out that the initial onset of schizophrenia is followed by a period of deterioration, but the decline in functioning eventually plateaus. After a review of North American long-term follow-up studies of schizophrenia, McGlashan concludes that the plateau typically occurs after about 5 to 10 years of illness. Therefore, he argues that follow-up into old age may not be necessary to determine the patient's overall outcome. However, other reports indicate that significant improvements occur well beyond the first decade of illness (Bridge, Cannon, & Wyatt, 1978; Harding, 1991).

Numerous correlates of favorable outcome have been identified. These include the following:

Family history of affective disorder (Stevens, 1978; Stevens, Astrup, & Mangrum, 1966; Vaillant, 1964a, 1964b)
Absence of a family history of schizophrenia (Fenton & McGlashan, 1987; McGlashan, 1986; Stevens, 1978; Stevens, & Astrup, 1963; Stevens et al., 1966)
Better premorbid social functioning (Carpenter, Strauss, Pulver, & Wolznier, 1991; Ciompi, 1980)
Being married (Ciompi, 1980; Stevens, 1978; Stevens et al., 1966)
Better premorbid work functioning (Ciompi, 1980; Jonsson & Nyman, 1991; Stevens, 1978; Stevens et al., 1966)
Higher premorbid IQ (Stevens, 1978; Stevens et al., 1966)
Acute onset (Ciompi, 1980; Jonsson & Nyman, 1991; Stevens, 1978; Stevens et al., 1966; Vaillant, 1964a, 1964b)
The presence of precipitating factors at onset (Jonsson & Nyman, 1991; Ojordsmoen, 1991; Stevens, 1978; Stevens et al., 1966; Vaillant, 1964a, 1964b)
Having depressive features (Fenton & McGlashan, 1987; Jonsson & Nyman, 1991; Kay & Murrill, 1990; McGlashan, 1986; Stevens, 1978; Stevens et al., 1966; Vaillant, 1964a, 1964b)
Shorter duration of prior hospitalizations (Carpenter et al., 1991; Engelhardt, Rosen, Feldman, Engelhardt, & Cohen, 1982)

Although the literature is generally consistent regarding the predictive power of the preceding characteristics, there is some controversy as to whether symptom patterns at onset are predictive of outcome. For example, some studies have found that negative symptoms at onset (Scottish Schizophrenia Research Group, 1988) or social withdrawal and inactivity (typically classified as negative symptoms) (Johnstone, Crow, Johnson, & MacMillan, 1986; Johnstone, MacMillan, Frith, Benn, & Crow, 1990) predict poor outcome. Kay and Murrill (1990), however, in a study of 58 chronic schizophrenic inpatients, found that positive symptoms predicted poorer outcome. Specifically, the presence of thought disorder was associated with poor outcome in terms of residual psychopathology, sustained hospitalization, and overall poorer social functioning. Similarly, Wilcox (1990) used psychiatric symptoms to predict relapse over a 24-month period and found that thought disorder at presentation was a significant predictor of relapse.

Much of the literature on outcome in schizophrenia does not distinguish between first admission and consecutive admission patients. Ram, Bromet, Eaton, Pato, and Schwartz (1992) reviewed the outcome research on first-admission patients and compared the findings with those from studies of consecutive admission samples. Although most of the first-admission studies covered only about 2 years, many of the characteristics predicting good versus poor outcome were similar to the long-term outcome studies. There were some differences, however, between the findings of first-admission and consecutive admission studies. Studies of consecutive admissions have shown earlier age at onset to be a consistent predictor of poor outcome (Ram et al., 1992) whereas first-admission studies have not (Bland & Orn, 1978; Rupp & Fletcher, 1940). Also, studies of consecutive admissions have found expressed emotion (EE) (i.e., hostility and high levels of criticism and overinvolvement) in family members to be a predictor of poor outcome (Brown, Birley, & Wing, 1972; Vaughn & Leff, 1976). Studies involving only first-admission patients, however, have failed to find a significant effect of relatives' EE on probability of relapse (MacMillan, Gold, Crow, Johnson, & Johnstone, 1986; Waring et al., 1988). These divergent findings suggest that sampling procedure (i.e., first vs. consecutive admissions) has important implications for study outcome. A subgroup of first-admission patients do not have any subsequent hospitalizations. In contrast, other patients have a high rate of rehospitalizations: the higher the rate, the more likely they are to be ascertained in studies based on consecutive admissions samples. Thus, samples of consecutively admitted patients may contain a higher proportion of severely ill, chronic patients than do first-admission samples.

Gender Differences in Outcome

One demographic variable that appears to account for a significant proportion of the variance in outcome is gender. In a 5-year follow-up study of 121 schizophrenics, D.C. Watt, Katz, and Shepherd (1983) found that for both the first-admission patients and for the group as a whole (first and subsequent admissions), outcome was significantly worse for men than for women. More women than men had only one episode without further impairment. In contrast, men were more likely to show increasing impairment with each episode and to be rehospitalized during the follow-up period. Consistent with these findings, Salokangas (1983) found that outcome, 8 years after first hospitalization, was poorer for men than women, especially in relation to interpersonal interaction and social competence. Men were not only admitted more frequently to the hospital but were also there for longer periods of time.

Various psychosocial and biological explanations for this gender difference have been offered (Goldstein & Kreisman, 1988; Lewine, 1991). For example, in the psychosocial realm, female schizophrenics are much more likely to marry than are male schizophrenics, and it has been suggested that this may put females at an advantage with respect to social support (Reich & Thompson, 1985; Seeman & Hauser, 1984; Test, Burke, & Wallisch, 1990; Wattie & Kedward, 1985). Also, high levels of EE have been found to be primarily a phenomenon of families of male patients (Vaughn, Snyder, Jones, Freeman, & Falloon, 1984), perhaps because family expectations for occupational functioning are higher for males. So it may be that the social environments of female patients are less stressful, and this may contribute to their better overall functioning and outcome.

With regard to the biological realm, some have proposed that estrogen plays a protective role and serves to delay symptom onset and ameliorate the clinical course in female patients (Lewine, 1991; Seeman, 1983). Although most studies note the overall superior functioning of women at follow-up (Bardenstein & McGlashan, 1990), some have noted changes for the worse around menopause (Harding et al., 1987b; Ojordsmoen, 1991). Further, although women at younger ages require lower maintenance doses of neuroleptics than men, they require higher doses as a group than men after age 40 (Seeman, 1983). These findings give credence to the possibility that estrogen accounts for the overall better prognosis among schizophrenic women. One likely mechanism is the effect of sex hormones on neurotransmitter systems, particularly dopamine (Lewine, 1991; Seeman, 1983).

Longitudinal Course of Clinical Symptoms

As previously mentioned, some evidence indicates that ratings of negative symptoms are more strongly associated with indices of premorbid functioning than are positive symptom ratings. Studies of longitudinal changes in symptoms indicate that the positive–negative symptom distinction (Andreasen et al., 1990) may also be important in understanding the clinical course of the schizophrenic syndrome. Marneros et al. (1992) evaluated subsequent episodes of illness in 148 schizophrenic patients over an average of 23 years (range 10 to 50 years). They found that positive symptoms decreased over time, whereas negative symptoms increased. Also, most patients (76%) showed a variable symptom profile over time; very few patients remained primarily positive or negative during the entire course of their illness.

Other studies have also found that positive symptoms occur at a lower rate later in the life course (Ciompi, 1980; Winokur, Pfolh, & Tsuang,

1987). For example, Winokur et al. (1987) found that, on initial evaluation, 71% of schizophrenics manifested persecutory delusions and 53% reported auditory hallucinations. At 40-year follow-up, only 6% had persecutory delusions and 18% had auditory hallucinations. Negative symptoms, however, appeared to be somewhat more resilient, because 80% of their sample presented with such symptoms, with little change and some exacerbation over time.

Because the majority of schizophrenic patients show both positive and negative symptoms (Guelfi, Faustman, & Csernansky, 1989; Walker & Lewine, 1988), it is reasonable to assume that the two types of symptoms are part of the same syndrome. However, the apparent differences in their temporal course, as well as in their relation with premorbid functioning, suggest that their expression is moderated by different developmental processes.

Several investigators have attempted to define schizophrenic subtypes based on global aspects of the longitudinal course (Carpenter & Kirkpatrick, 1988; Ciompi, 1980). For example, Ciompi (1980) discusses eight subtypes, defined according to type of onset, course, and end state. The onset epoch includes features of the premorbid and early postmorbid period and is classified as either acute or chronic. Course type is described by Ciompi as simple or undulating, and the end state, which refers to the clinical status in the latest stages of life, is classified as either "recovery or mild" or "moderate or severe."

Harding (1988) compared the illness courses of schizophrenic patients from the Lausanne investigations (Ciompi & Muller, 1976), the Burgholzli Hospital Study (Bleuler, 1972) and the Vermont Longitudinal Research Project (Harding et al., 1987a, 1987b). Although Harding found similarity in the overall outcome for these three samples of patients, she found differences in the way the end states were reached (i.e., course trajectories). The most common subtype in the Lausanne (25.4%) and Burgholzli (55% to 75%) studies was characterized by an acute onset, undulating course, and recovery/mild end state, but this subtype represents only 7% of the Vermont sample. In contrast, the most common subtype (38%) in the Vermont study involved chronic onset, undulating course, and recovery/mild end state, whereas this subtype represented only 9.6% in the Lausanne study and none of the subjects in the Burgholzli sample. Harding (1988) concludes that the discrepancies among the findings underscore the importance of heterogeneity among samples that is due to differences in sampling procedures.

Cognitive Changes

The impairment of schizophrenic patients in cognitive functioning has been well established (Heaton & Drexler, 1987; Wyatt & Saiz, 1991); however, changes in cognitive functioning over time have been less well understood. Furthermore, no specific set of cognitive deficits has been shown to uniquely characterize schizophrenic patients. Aggregate statistics on groups of schizophrenic patients suggest ubiquitous neuropsychological impairments (Walker, Lucas, & Lewine, 1992).

Heaton and Drexlers' (1987) review of 100 cross-sectional and 14 longitudinal studies led them to conclude that many schizophrenics show improvements in cognitive performance over time, and those who do tend to have more a favorable course and outcome. The authors note that patients' IQ scores tend *not* to be inversely correlated with age, suggesting that there is no progressive course of cognitive decline. Further, although longitudinal studies indicate that premorbid IQ scores are 5 to 10 points higher than at illness onset, IQ scores obtained after onset suggest that some patients show substantial improvements following their first episode. The authors conclude that "the cognitive impairment associated with the onset of schizophrenia is at least partially reversible in many subjects, but in some others it is probably progressive" (Heaton & Drexler, 1987, p. 160).

Neuroleptic Treatment

In a recent review of the literature on medication effects, Wyatt (1991) concluded that the majority of patients do show some improvement when treated with neuroleptics, and that neuroleptics improve the subsequent course of the illness, especially when given in the early phases of the disorder. For example, Rappaport, Hopkins, Hall, Belleza, and Silverman (1978) reported a study in which male schizophrenic patients were randomly assigned to either a placebo or neuroleptic treatment group. Not surprisingly, the neuroleptic group was significantly more improved on discharge. The two original groups were then divided based on their medication status at 3-year follow-up. The results from these group comparisons indicated that the group of patients who went unmedicated during hospitalization became more difficult to treat and had become poor responders when neuroleptics were subsequently introduced. Also, when the patients with good premorbid histories were compared across all four groups, it was found that the nonmedicated group did the worst of the good premorbid patients in the follow-up period. In sum, it appears that patients are more likely to have a favorable outcome, both at discharge and at follow-up, if neuroleptic medications are used early in the illness. Consistent with this, several studies have shown that patients have a better overall outcome if the duration of their initial episode (Rupp & Fletcher, 1940; Rabiner, Wegner, & Kane, 1986) and their pretreatment symptomatic period (Anzai et al., 1988; Crow, MacMillan, Johnson, & Johnstone, 1986) are shorter.

The apparent effect of neuroleptics on the natural course of schizophrenia raises challenging questions. It may be that the effect is essentially a psychosocial one, in that the reduction in symptoms achieved with neuroleptics ehhances the patients long-term capacity to cope with stress and acquire social and occupational skills. Alternatively, psychotic episodes may have cumulative organic consequences, and the greater their duration, the greater the likelihood of long-term impairment (Wyatt & Saiz, 1991). Whatever the case, the data point to the important effects of early treatment on the natural course of schizophrenia.

Movement Abnormalities in Schizophrenic Patients

Several reports on motor function in adolescent and adult schizophrenic patients provide specific descriptions of movement abnormalities that are similar to those we have observed in the childhood home movies of preschizophrenic subjects. Because neuroleptics have a variety of motor side effects, only studies of never-medicated patients can shed light on the natural course of motor dysfunction in schizophrenia. The findings from these studies indicate that abnormalities of limb posture and movement are common and that they increase with age.

In an early report, Reiter (1926) described 10 adult patients with "dementia praecox" who showed substantial postural and movement abnormalities, most notably "choreatic-athetotic restlessness." Every patient demonstrated abnormalities in hand posture and movement. Further, an apparent exacerbation with age was noted, especially in the months preceding death. Some evidence for diagnostic specificity of movement abnormalities is provided by Guttman (1936), who compared the medical histories of schizophrenic and manic depressive patients and found that chorea was significantly more common in the schizophrenics.

Casey and Hansen (1984) reviewed 28 studies of movement abnormalities in schizophrenia published between 1959 and 1984. These reports suggested a significant rate of movement disorders, including athetoid movements, in nonmedicated patients. Again, evidence indicates that the abnormalities occurred at a higher rate in older patients, with more significant diagnostic group differences after 60 years of age. A recently reported longitudinal study of four elderly schizophrenic patients who had never been medicated confirmed the exacerbation of movement abnormalities with age (Waddington & Youssef, 1990). In sum, although spontaneous movement abnormalities increase with age in normal elderly populations, the increase appears to be more pronounced in schizophrenics.

The literature on neuroleptic-induced motor disturbance also reveals age-related changes (Jeste & Wyatt, 1987). The prevalence of tardive dyskinesia in schizophrenic patients being treated with neuroleptics increases dramatically with age. Further, the probability of a remission of tardive dyskinesia following neuroleptic withdrawal decreases with age. However, the movement abnormalities induced by neuroleptics are primarily in the face, especially the tongue, lips, and jaw. This orofacial dyskinesia appears to constitute a subsyndrome that is separate from involuntary limb movements, and the neuropathophysiology underlying the two subsyndromes may differ (Gureje, 1988; Kidger, Barnes, Trauer, & Taylor, 1980). Although studies of medicated schizophrenic patients reveal a significant age-related increase in both orofacial and limb movement abnormalities, other findings suggest that these motor disturbances are not solely neuroleptic induced (Khot & Wyatt, 1991). This may be especially true of involuntary limb movements, and neuroleptics may, in fact, serve to mask or reduce some movement abnormalities.

Four points about the findings on neuromotor functions in schizophrenic patients warrant emphasis. First, the presence of movement abnormalities in patients who have never received medication indicates that motor dysfunction is part of, or at least a correlate of, the schizophrenic syndrome. (In other words, it either originates from the same neural substrate or can be a consequence of the same etiologic agent.) Second, there appears to be some diagnostic specificity of neuromotor abnormalities, in that they occur at a higher rate in schizophrenia than in other psychiatric disorders (Kennard, 1960; Hertzig & Birch, 1966; Rochford, Detre, Tucker, & Harrow, 1970; Virtunski, Simpson, & Meltzer, 1989). This is consistent with our findings on neuromotor abnormalities in preschizophrenic versus preaffective-disorder children. Third, the type of movement abnormality (i.e., choreoathetoid movement) that has been detected in both preschizophrenic and schizophrenic subjects has been shown to be associated with abnormalities of the basal ganglia (Harper, 1991; Hayashi, Satoh, Sakamoto, & Morimatsu, 1991; Klawans, 1988; Klawans & Weiner, 1975; Narabayashi, 1989). Finally, although evidence shows neurological soft signs and motor dysfunction throughout the course of schizophrenia, movement and postural abnormalities become more pronounced later in the life course of patients. Combined with the data on preschizophrenic children, the findings suggest an exaggerated version of the U-shaped relation between movement abnormalities and age that characterizes nonschizophrenic subjects.

Summary

In commenting on the longitudinal course of schizophrenia, Ciompi states that, "schizophrenia is affected by the aging process which, as in normal people, is capable of generating not only additional difficulties, but also amelioration and tranquilization" (1980, p. 616). This comment aptly describes the research findings. Evidence indicates that a substantial number of patients show significant improvement, and the florid symptoms of psychosis appear to be ameliorated by the aging process in most patients. Further, the nature of the long-term outcome is, in part, predictable on the basis of the patient's premorbid functioning, suggesting some continuity across the life span. Cognitive functions do not show a trend toward deterioration with age, and for some patients, cognitive performance may actually improve. In contrast, neuromotor abnormalities worsen with age, and the increase in movement abnormalities in elderly schizophrenic patients appears to be greater than that observed in the normal aging process.

Life-Course Developmental Trends

Although a comprehensive discussion of the research literature on the life course of schizophrenia is beyond the scope of this chapter, we have attempted to highlight some salient findings that bear on the nature of premorbid and postmorbid development. The key findings are briefly summarized here, then we turn to a discussion of their theoretical implications.

Before summarizing the findings, it is important to note that one of the most obvious features of the schizophrenic syndrome is its diversity. Individual differences are apparent when the disorder is viewed cross-sectionally, as well as longitudinally. It may be that this simply relfects the interaction between a unitary neuropathological process and widely varying environmental factors. Alternatively, these individual differences may reflect basic differences in etiology. Whatever the case, it is clear that no one characteristic is common to all patients. Thus, the following discussion focuses on modal developmental trends and is not intended to imply that all patients follow the same trajectory.

Key Developmental Aspects

Several lines of evidence point to abnormalities in the prenatal development of schizophrenic patients. Taken together, the findings on viral infections and dysmorphological features suggest that the second trimester, a critical period for the development of subcortical regions of the brain, may be particularly important. Research on brain morphology confirms that schizophrenic patients show structural and metabolic abnormalities in these regions. Given the evidence of congenital brain abnormalities in schizophrenia, it is not surprising that preschizophrenic children show subclinical signs of behavioral dysfunction.

Findings from our observational research, as well as other studies, indicate that the age and sex of the subject predict the nature and extent of dysfunction. In the domain of neuromotor functions, the life-course trend appears to be one in which abnormalities are most pronounced in infancy and with advanced age. There is no evidence of sex differences in this trend. In the domain of affective expression, our findings also suggest that abnormalities are pronounced during infancy. Unlike the neuromotor deficits, however, abnormalities in facial expressions of emotion appear to increase in adolescence, probably reflecting the prodromal phase of the illness. Similarly, behavior problems become more apparent in adolescence, although they are also reported earlier in childhood, especially in girls.

As previously mentioned, it is well established that late adolescence/early adulthood is the period when psychotic symptoms are most likely to emerge. Most patients show a combination of both positive and negative symptoms in their first episode and throughout their illness course. However, the severity of negative symptoms is positively correlated with the degree of premorbid impairment, whereas positive symptoms show relatively little association with premorbid functioning. These findings, and other results linking premorbid behavior with outcome, suggest there is some longitudinal stability for individual differences in functioning and that negative symptoms represent a more stable feature of the neuropathological process. Further, with advanced age, positive symptoms show a greater decline than negative symptoms. Thus, the expression of positive symptoms appears to be influenced by maturational changes to a greater extent than negative symptoms.

Theoretical Implications

Several conclusions are suggested by the findings we have discussed here. First, the neuropathology that constitutes the diathesis for schizophrenia appears to originate in the prenatal genesis of the CNS. Second, the behavioral expression of the diathesis is altered by the maturation of the CNS across the life span. Third, in addition to normative developmental processes, individual differences influence the course of premorbid and postmorbid development. Fourth, subcortical regions are a likely site for the neuropathology. This last assumption is suggested not only by the physical evidence of brain abnormalities in schizophrenia, but also by the developmental pattern of behavioral dysfunction observed in schizophrenia.

At the behavioral level, one of the strongest indicators of subcortical abnormalities is the presence of more pronounced neuromotor and affective deficits in the first few years of life than in the subsequent childhood years. This developmental pattern suggests that the dysfunctional region of the brain is having a relatively greater impact on behavior during these early years. Evidence from numerous sources, including studies of developmental changes in myelination (Benes, 1991; Konner, 1991), brain regional metabolism (Chugani, 1992; Chugani & Phelps, 1986), and cellular and synaptic density (Huttenlocher, 1990) indicate that, relative to cortical regions, subcortical regions are more mature and active during the infancy period. Gradually, cortical regions mature, and there is a reduction in the influence of subcortical regions on behavior.

Compared with other species, the development of the human CNS is protracted. Myelination, for example, is not completed until late adolescence/early adulthood (Konner, 1991). In addition, this developmental period is characterized by changes in neurotransmitter activity that are, at least partially, a consequence of hormonal changes (McGeer & McGeer, 1981; Ulloa-Aguirre et al., 1991). The maturational events that occur in the CNS during this period are assumed to account for the ascendance of complex cognitive abilities (Goldman-Rakic, 1987; Konner, 1991). As several authors have suggested (Benes, 1991; Feinberg, 1982–1983; Saugstad, 1989; Weinberger, 1987), one or more of these CNS maturational events may also be responsible for triggering the expression of psychotic symptoms in vulnerable individuals.

Significant changes in brain structure, biochemistry, and regional metabolism also occur with advanced age (Obrist, 1980; Winblad, Hardy, Backman, & Nilsson, 1985). These are manifested in changes in motor and cognitive functions. It is of particular interest to note that brain deterioration is greatest in nonmotor cortical regions, with relative sparing of subcortical areas. This may account for the re-emergence of some primitive motor reflexes in normal elderly subjects, as well as the apparent exacerbation of movement abnormalities and reduction in florid symptoms in elderly schizophrenic patients.

The research findings we have reviewed here also indicate that individual differences and sex-linked biological factors moderate the behavioral expression of vulnerability to schizophrenia. Our findings of sex differences in the very early emotional expressions of preschizophrenic children, but not normal controls, suggest that the sex differences that have been documented in schizophrenia across the life span are not entirely attributable to differential socialization experiences. Because knowledge of normal sex differences in neurodevelopment is extremely limited, no empirical data base is available for speculating on the neural mechanisms that might be responsible for sex differences in the developmental aspects of schizophrenia. However, it is clear from research on a variety of other mental disorders that sex plays a significant role in modifying their behavioral expression.

Conclusion

In summary, research on premorbid and postmorbid behavior in schizophrenia reveals developmental changes in behavior that may hold important clues to etiology. By mapping these behavioral changes onto maturational processes in the CNS, it may be possible to speculate, with greater precision, on the neuropathology of schizophrenia. Research on normal human neurodevelopment is burgeoning, largely because of the availability of new technologies for imaging the human brain in vivo. As the findings from this research accumulate, our opportunities for understanding the origins of schizophrenia will be greatly enhanced. In the interim, psychopathology researchers must continue in their efforts to document the longitudinal course of schizophrenia.

Acknowledgments. This work was supported by a research grant (MH46496) and a Research Scientist Development Award (MH00876) to the first author from the National Institute of Mental Health.

References

Abdel-Dayem, H.M., El-Hilu, S., Sehweil, A., Higazi, B., Jahan, S., Salhat, M., & Al-Mohannadi, S. (1990). Cerebral perfusion changes in schizophrenic patients using Tc-99m Hexamethylpropylene Amineoxime (HMPAO). *Clinical Nuclear Medicine, 15,* 468–472.

Altshuler, L.L., Conrad, A., Kovelman, J.A., & Scheibel, A. (1987). Hippocampal pyramidal cell orientation in schizophrenia. *Archives of General Psychiatry, 44,* 1094–1098.

Andreasen, N.C., Ehrardt, J.C., Swayze, V.W., Alliger, R.J., Yuh, W.T.C., Cohen, G., & Zeibell, S. (1990). Magnetic resonance imaging of the brain in schizophrenia. *Archives of General Psychiatry, 47,* 35–44.

Andreasen, N.C., & Olsen, S. (1982). Negative versus positive schizophrenia. *Archives of General Psychology, 39,* 789–794.

Anzai, N., Okazaki, Y., Miyauchi, M., Harada, S.I., Kanou, Y., Sasaki, T., Kumagai, N., Shikiba, N., Iwanami, A., Iadi, S., Hiramatsu, K.I., Niwa, S.I., & Ohta, M. (1988). Early neuroleptic medication within one year after onset can reduce risk of later relapses in schizophrenic patients. *Annual Report Pharmacopsychiatric Research Foundation, 19,* 258–265.

Aylward, E., Walker, E., & Bettes B. (1984). Intelligence in schizophrenia: Meta-analysis of the research. *Schizophrenia Bulletin, 10*(3), 430–459.

Bardenstein, K.K., & McGlashan, T.H. (1990). Gender differences in affective, schizoaffective and schizophrenic disorders. A review. *Schizophrenia Research, 3,* 159–172.

Barthell, C.N., & Holmes, D.S. (1968). High school yearbooks: A nonreactive measure of social isolation in graduates who later became schizophrenic. *Journal of Abnormal Psychology, 73*(4), 313–316.

Benes, F.M. (1989). Myelination of cortical hippocampal relays during late adolescence. *Schizophrenia Bulletin, 15,* 585–594.

Benes, F.M. (1991). Toward a neurodevelopmental understanding of schizophrenia and other psychiatric disorders. In D. Cicchetti & S.L. Toth (Eds.), *Rochester Symposium on Developmental Psychopathology, 3,* 161–184.

Bland, H.C., & Orn, H. (1978). 14-year outcome in early schizophrenia. *Acta Psychiatrica Scandinavia, 58,* 327–338.

Bleuler, E. (1950). *Dementia praecox or the group of schizophrenias.* (J. Zinkin, Trans.). New York: International University Press. (Original work published, 1911).

Bleuler, M. (1972). Die Schizophrenen Geistestorungen im Lichte langjahrigern Kranken—und Familiengeschichten. Stuttgart: George Thieme.

Bleuler, M. (1978). *The Schizophrenic Disorders; The Long Term Patient and Family Studies*. (S.M. Clemens, Trans.). New Haven: Yale University Press. (Original work published 1972).
Bogerts, B., Ashtari, M., Degreff, G., Alvir, J.M., Bilder, R.M., & Lieberman, J.A. (1990). Reduced temporal limbic structure volumes on magnetic resonance images in first episode schizophrenia. *Psychiatry Research, 35*, 1–13.
Bracha, H.S., Torrey, E.F., Bigelow, L.B., Lohr, J.B., & Linington, B.B. (1991). Subtle signs of prenatal maldevelopment of the hand ectoderm in schizophrenia: A preliminary monozygotic twin study. *Biological Psychiatry, 30*, 719–725.
Bridge, T.P., Cannon, H.E., & Wyatt, R.J. (1978). Burned-out schizophrenia: Evidence for age effects on schizophrenic symptomatology. *Journal of Gerontology, 33*, 312–318.
Brown, G.W., Birley, J.L.T., & Wing, J.K. (1972). Influence of family life on the course of schizophrenic disorders: A replication. *British Journal of Psychiatry, 121*, 241–258.
Cannon, T.D., Mednick, S.A., & Parnas, J. (1990). Antecedents of predominantly negative- and predominantly positive-symptom schizophrenia in a high-risk population. *Archives of General Psychiatry, 47*, 622–632.
Cannon-Spoor, E., Potkin, S.G., & Wyatt, R.J. (1982). Measurement of premorbid adjustment in chronic schizophrenia. *Schizophrenia Bulletin, 8*, 470–484.
Carpenter, W.T., & Kirkpatrick, B. (1988). The heterogeneity of the long-term course of schizophrenia. *Schizophrenia Bulletin, 14*, 645–652.
Carpenter, W.T., Strauss, J.S., Pulver, A.E., & Wolznier, P.S. (1991). The prediction of outcome in schizophrenia: IV. Eleven-year follow-up of the washington IPSS cohort. *Journal of Nervous and Mental Disease, 179*(9), 517–525.
Casey, D.E., & Hansen, T.E. (1984). Spontaneous dyskinesias. In D.V. Jeste & R.J. Wyatt (Eds.), *Neuropsychiatric movement disorders* (pp. 68–95). Washington, DC: American Psychiatric Press.
Chugani, H.T. (1992). Development of regional brain glucose metabolism in relation to behavior and plasticity. In G. Dawson & K. Fischer (Eds.), *Human behavior and the developing brain* (pp. 120–135). New York: Guilford Press.
Chugani, H.T., Mazziotta, J.C., & Phelps, M.E. (1987). Positron emission tomography study of human brain functional development. *Annals of Neurology, 22*, 487–497.
Chugani, H.T., & Phelps, M.E. (1986). Maturational changes in cerebral function in infants determined by [1]8FDG positron emission tomography. *Science, 231*, 840–843.
Ciompi, L. (1980). Catamnestic long-term study on the course of life and aging of schizophrenics. *Schizophrenia Bulletin, 6*, 606–618.
Ciompi, L., & Muller, C. (1976). *Lebensweg und Alter Schizophrenen eine Katamnestic Longzeitstudie bis ins Senium*. Berlin: Springer-Verlag.
Coffey, Y.P., & Jessup, W. (1959). Maternal influenza and congenital deformities: A prospective study. *Lancet, 2*, 935–938.
Cohen, C.I. (1990). Outcome of schizophrenia into later life: An overview. *The Gerontologist, 30*(6), 790–797.

Cross, A.J., Crow, T.J., & Owen, F. (1981). 3H-Flupenthixol binding in postmortem brains of schizophrenics: Evidence for a selective increase in dopamine D2 receptors. *Psychopharmacology, 74*, 122–124.

Crow, T.J., Johnstone, E.C., Longden, A.J., & Owen, F. (1978). Dopaminergic mechanisms in schizophrenia: The antipsychotic effect and the disease process. *Life Sciences, 23*, 563–567.

Crow, T.J., MacMillan, J.F., Johnson, A.L., & Johnstone, E.C. (1986). A randomized controlled trial of prophylactic neuroleptic treatment. *British Journal of Psychiatry, 148*, 120–127.

Cummins, H., & Mildo, C. (1943). *Fingerprints, palms and soles: An introduction to dermatoglyphics.* New York: Dover.

Dworkin, R.H., Friedman, R., Kaplansky, L.M., Lewis, J.A., Rinaldi, A., Shilliday, G., Cornblatt, B.A., & Erlenmeyer-Kimling, L. (1991, December). *Childhood precursors of affective and social deficits in adolescents at risk for schizophrenia.* Paper presented at the meeting of the Society for Research in Psychopathology, Cambridge, MA.

Dworkin, R.H., Green, S.R., Small, N.E., Warner, M.L., Cornblatt, B.A., & Erlenmeyer-Kimling, L. (1990). Positive and negative symptoms and social competence in adolescents at risk for schizophrenia and affective disorder. *Clinical and Research Reports, 147*(9), 1234–1236.

Engelhardt, D.M., Rosen, B., Feldman, J., Engelhardt, J.A.Z., & Cohen, P.A. (1982). A 15-year follow-up of 646 schizophrenic outpatients. *Schizophrenia Bulletin, 8*, 493–503.

Erlenmeyer-Kimling, L., Marcuse, Y., Cornblatt, B., Friedman, D., Rainer, J.D., & Rutschmann, J. (1984). The New York High-Risk Project. In N.F. Watt, E.J. Anthony, L.C. Wynne, & J. Roff (Eds.), *Children at risk for schizophrenia: A longitudinal perspective* (pp. 169–189). New York: Cambridge University Press.

Falkai, P., & Bogerts, B. (1986). Cell loss in the hippocampus of schizophrenics. *European Archives of Psychiatry and Neurological Science, 236*, 154–161.

Feinberg, I. (1982–1983). Schizophrenia: Caused by a fault in programmed synaptic elimination during adolescence. *Journal of Psychiatric Research, 17*, 319–334.

Fenton, W.S., & McGlashan, T.H. (1987). Sustained remission in drug-free schizophrenics. *American Journal of Psychiatry, 144*, 1306–1309.

Fish, B., Marcus, J., Hans, S.L., Auerbach, J.G., & Perdue, S. (1992). Infants at risk for schizophrenia: Sequelae of a genetic neurointegrative defect: A review and replication analysis of pandysmaturation in the Jerusalem Infant Development Study. *Archives of General Psychiatry, 49*, 221–235.

Freide, R.L. (1989). *Developmental neuropathology.* Berlin: Springer-Verlag.

Gluck, L. (1977). *Intrauterine asphyxia in the developing fetal brain.* Chicago: Year Book.

Goldman-Rakic, P.S. (1987). Development of cortical circuitry and cognitive function. *Child Development, 58*, 601–622.

Goldstein, J.M., & Kreisman, D. (1988). Gender, family environment and schizophrenia. *Psychological Medicine, 18*, 861–872.

Gossey, S., Golaire, M.C., & Larroche, J.C. (1982). Cerebral, renal, and splenic lesions due to fetal anoxia and their relationship to malformations. *Developmental Medicine and Child Neurology, 24*, 510–518.

Gottesman, I.I. (1991). *Schizophrenia genesis*. New York: W.H. Freeman.
Gur, R.E., Resnick, S.M., Alavi, A., Gur, R.C., Caroff, S., Dann, R., Silver, F.L., Saykin, A.J., Chawluk, J.B., Kushner, M., & Reivich, M. (1987). Regional brain function in schizophrenia: A positron emission tomography study. *Archives of General Psychiatry, 44*, 119–125.
Gur, R.E., Resnick, S.M., Gur, R.C., Alavi, A., Caroff, S., Kushner, M., & Reivich, M. (1987). Regional brain function in schizophrenia: Repeated evaluation with positron emission tomography. *Archives of General Psychiatry, 44*, 126–129.
Gureje, O. (1988). Topographic subtypes of tardive dyskinesia in schizophrenic patients less than 60 years. *Journal of Neurology, Neurosurgery and Psychiatry, 51*, 1525–1530.
Guttman, E. (1936). On some constitutional aspects of chorea and on its sequela. *Journal of Neurology and Psychopathology, 17*, 16–26.
Hamilton, W.J., Boyd, J.D., & Mossman, H.W. (1972). *Human embryology: Prenatal development of form and function*. London: Heffer.
Hans, S.L., & Marcus, J. (1991). Neurobehavioral development of infants at risk for schizophrenia. In E. Walker (Ed.), *Schizophrenia: A life-course developmental perspective* (pp. 35–53). New York: Academic Press.
Harding, C.M. (1988). Course types in schizophrenia: An analysis of European and American studies. *Schizophrenia Bulletin, 14*, 633–643.
Harding, C.M. (1991). Aging and schizophrenia: Plasticity, reversibility, and/or compensation. In E. Walker (Ed.), *Schizophrenia: A life-course developmental perspective* (pp. 257–273). New York: Academic Press.
Harding, C.M., Brooks, G.W., Ashikaga, T., Strauss, J.S., & Breier, A. (1987a). The Vermont longitudinal study of persons with severe mental illness: I. Methodology, study sample, and overall status 32 years later. *American Journal of Psychiatry, 144*, 718–726.
Harding, C.M., Brooks, G.W., Ashikaga, T., Strauss, J.S., & Breier, A. (1987b). The Vermont longitudinal study of persons with severe mental illness: II. Long-term outcome for subjects who retrospectively met DSM-III criteria for schizophrenia. *American Journal of Psychiatry, 144*, 727–735.
Harper, P.S. (Ed.). (1991). *Huntington's disease*. London: W.B. Saunders.
Hayashi, M., Satoh, J., Sakamoto, K., & Morimatsu, Y. (1991). Clinical and neuropathological findings in severe athetoid cerebral palsy. *Brain and Development, 13*, 47–51.
Heaton, R.K., & Drexler, M. (1987). Clinical neuropsychological findings in schizophrenia and aging. In N.E. Miller & G.D. Cohen (Eds.), *Schizophrenia and aging*. New York: Guilford Press.
Heckers, S., Heinsen, H., Heinsen, Y., & Beckmann, H. (1991). Cortex, white matter, and basal ganglia in schizophrenia: A volumetric postmortem study. *Biological Psychiatry, 29*, 143–149.
Hertzig, M.E., & Birch, H.G. (1966). Neurologic organization in psychiatrically disturbed adolescent girls. *Archives of General Psychiatry, 15*, 590–598.
Hess, E.J., Bracha, H.S., Kleinman, J.E., & Creese, I. (1987). Dopamine receptor subtype imbalance in schizophrenia. *Life Science, 40*, 1487–1497.
Huber, G., Gross, G., & Schuttler, R. (1979). Schizophrenie. Verlaufs-und sozialpsychiatrische Langzeitunter-suchungen an den 1945 his 1959 in Bonn

hospitalisierten schizophrenen Kranken. *Monographien aus dem Gesamtgebiete der Psychiatrie.* Bd. 21. Berlin: Springer-Verlag.

Huttenlocher, P.R. (1990). Morphometric study of human cerebral cortex development. *Neuropsychologia, 28,* 517-527.

Izard, C.E., Dougherty, L.M., & Hembree, E.A. (1983). *A System for identifying affect expressions by holistic judgements (AFFEX).* Neward, DE: Instructional Resources Center, University of Delaware.

Jakob, H., & Beckman, H. (1986). Prenatal development disturbances in the limbic allocortex in schizophrenics. *Journal of Neural Transmission, 65,* 303-326.

Jernigan, T.L., Zisook, S., Heaton, R.K., Moranville, J.T., Hesselink, J.R., & Braff, D.L. (1991). Magnetic resonance imaging abnormalities in lenticular nuclei and cerebral cortex in schizophrenia. *Archives of General Psychiatry, 48,* 881-890.

Jeste, D.V., & Lohr, J.B. (1989). Hippocampal pathologic findings in schizophrenia. *Archives of General Psychiatry, 46,* 1019-1026.

Jeste, D.V., & Wyatt R.J. (1987). Aging and tardive dyskinesia. In N. Miller & G. Cohen (Eds.), *Schizophrenia and aging* (pp. 275-283). New York: Guilford Press.

John, R.S., Mednick, S.A., & Schulsinger, F. (1982). Teacher reports as a predictor of schizophrenia and borderline schizophrenia: A Bayesian decision analysis. *Journal of Abnormal Psychology, 91*(6), 399-413.

Johnstone, E.C., Crow, T.J., Johnson, A.L., & MacMillan, J.F. (1986). The Northwick Park Study of first episodes of schizophrenia: I. Presentation of the illness and problems relating to admission. *British Journal of Psychiatry, 148,* 115-120.

Johnstone, E.C., MacMillan, J.F., Frith, C.D., Benn, D.K., & Crow, T.J. (1990). Further investigation of predictors of outcome following first schizophrenic episodes. *British Journal of Psychiatry, 157,* 182-189.

Jonsson, H., & Nyman, A.K. (1991). Predicting long-term outcome in schizophrenia. *Acta Psychiatrica Scandinavia, 83,* 342-346.

Kay, S.R., Fiszbein, A., & Opler, A. (1987). The positive and negative syndrome scale (PANSS) for schizophrenia. *Schizophrenia Bulletin, 13*(2), 261-276.

Kay, S.R., & Murrill, L.M. (1990) Predicting outcome of schizophrenia: Significance of symptom profiles and outcome dimensions. *Comprehensive Psychiatry, 31,* 91-102.

Kennard, M.A. (1960). Value of equivocal signs in neurologic diagnosis. *Neurology, 10,* 753-764.

Khot, V., & Wyatt, R.J. (1991). Not all that moves is tardive dyskinesia. *American Journal of Psychiatry, 148,* 661-666.

Kidger, T., Barnes, T.R.E., Trauer, T., & Taylor, P.J. (1980). Subsyndromes of tardive dyskinesia. *Psychological Medicine, 10,* 513-520.

Klawans, H.L. (1988). The pathophysiology of drug-induced movement disorders. In J. Jankonic & E. Tolosa (Eds.), *Parkinson's disease and movement disorder* (pp. 315-324). Baltimore: Urban and Schagenberg.

Klawans, H.L., & Weiner, W.J. (1975). The pharmacology of choreatic movement disorders. In G.A. Kerkut & J.W. Philips (Eds.), *Progress in neurobiology* (pp. 1-32). Oxford, England: Pergamon Press.

Knobloch, H., & Pasamanick, B. (1974). *Developmental diagnosis.* New York: Harper & Row.

Konner, M. (1991). Universals of behavioral development in relation to brain myelination. In K. Gibson & A. Peterson (Eds.), *Brain maturation and cognitive development* (pp. 181–224). New York: Aldine De Gruyter.

Kovelman, J., & Scheibel, A. (1984). A neurohistological correlate of schizophrenia. *Biological Psychiatry, 19*, 1601–1621.

Leck, T. (1963). Incidence of malformations following influenza symptoms. *British Journal of Preventative and Social Medicine, 17*, 70–73.

Lewine, R.R.J. (1991). Ontogenetic implications of Sex Differences in Schizophrenia. In E. Walker (Ed.), *Schizophrenia: A life-course developmental perspective*. New York: Academic Press.

Lewis, & Murray. (1987). Obstetric Complications, neurodevelopmental deviance and risk of schizophrenia. *Journal of Psychiatric Research, 21*, 413–421.

Lindenmeyer, J., Kay, S.R., & Opler, L. (1984). Positive and negative subtypes in acute schizophrenia. *Comprehensive Paychiatry, 25*(4), 455–464.

Loranger, A. (1984). Sex differences in age at onset of schizophrenia. *Archives of General Psychiatry, 41*, 157–161.

Lyon, M., Barr, C.E., Cannon, T.D., Mednick, S.A., & Shore, D. (1989). Fetal neural development and schizophrenia. *Schizophrenia Bulletin, 15*, 149–160.

Machon, R.A., Mednick, S.A., & Schulsinger, F. (1987). Seasonality, birth complications, and schizophrenia in a high risk sample. *British Journal of Psychiatry, 151*, 122–124.

MacMillan, J.F., Gold, A., Crow, T.J., Johnson, A.L., & Johnstone, E.C. (1986). The Northwick Park Study of first episodes of schizophrenia: IV. Expressed emotion and relapse. *British Journal of Psychiatry, 148*, 133–143.

Marneros, A., Deister, A., & Rohde, A. (1992). Validity of the negative/positive dichotomy for schizophrenic disorders under long-term conditions. *Schizophrenia Research, 7*, 117–123.

McGeer, P.L., McGeer, E.G. (1981). Neurotransmitters in the aging brain. In A.M. Darrison & R.M. Thompson (Eds.), *The nolecular basis of neuropathology* (pp. 631–648). London: Edward Arnold Ltd.

McGlashan, T.H. (1984a). The Chestnut Lodge follow-up study: I. Follow-up methodology and study sample. *Archives of General Paychiatry, 41*, 573–585.

McGlashan, T.H. (1984b). The Chestnut Lodge follow-up study: II. Longterm outcome of schizophrenia and the affective disorders. *Archives of General Psychiatry, 41*, 586–601.

McGlashan, T.H. (1986). Predictors of shorter-, medium-, and longer-term outcome in schizophrenia. *American Journal of Psychiatry, 143*, 50–55.

McGlashan, T.H. (1988). A selective review of recent North American long-term followup studies of schizophrenia. *Schizophrenia Bulletin, 14*, 515–542.

McNeil, T.F. (1987). Perinatal influences in the development of schizophrenia. In H. Helmehen & F.A. Henn (Eds.), *Biological perspective of schizophrenia*. (pp. 125–138).

McNeil, T.F., & Kaij, L. (1978). Obstetric factors in the development of schizophrenia: Complications in the births of preschizophrenice and in reproduction by schizophrenic parents. In L.C. Wynne, R.L. Cromwell, & S. Matthysee (Eds.), *The nature of schizophrenia* (pp. 401–429), New York: John Wiley and Sons Limited.

Mednick, S., Machon, R.A., Huttunen, M.O., & Bonnett, D. (1988). Adult schizophrenia following prenatal exposure to an influenza epidemic. *Archives of General Psychiatry, 45,* 15–23.

Mednick, S.A., & Schulsinger, F. (1965). A longitudinal study of children with a high risk for schizophrenia: A preliminary report. In S. Vandenberg (Ed.), *Methods and goals in human behavior genetics* (pp. 255–296). New York: Academic Press.

Miller, N.E., & Cohen, G.D. (Eds.) (1987). *Schizophrenia and Aging.* New York: The Guilford Press.

Narabayashi, H. (1989). Striatal symptoms. *Stereotactic Functional Neurosurgery, 52,* 200–204.

Newell-Morris, L.L., Fahrenbruch, C.E., Sackett, G.P. (1989). Prenatal psychological stress, dermatoglyphic asymmetry and pregnancy outcome in the pigtailed macaque *(Macaca nemestrina). Biology of the Neonate, 56,* 61–75.

O'Callaghan, E., Sham, P., Takei, N., Glover, G., & Murray, R.M. (1991). Schizophrenia after prenatal exposure to 1957 A2 influenza epidemic. *Lancet, 337,* 1248–1250.

Obrist, W.D. (1980). Cerebral blood flow and EEG changes associated with aging and dementia. In E.W. Busse & D.G. Blazer (Eds.), *Handbook of geriatric psychiatry* (pp. 83–101). New York: Van Nostrand Reinhold.

O'Gureje, O. (1989). Correlates of positive and negative schizophrenic syndromes in Nigerian patients. *British Journal of Psychiatry, 155,* 628–632.

Ojordsmoen, S. (1991). Long-term clinical outcome of schizophrenia with special reference to gender differences. *Acta Psychiatrica Scandinavia, 83,* 307–313.

Opler, L.A., Kay, S.R., Rosado, V., & Lindenmeyer, J. (1984). Positive and negative syndromes in chronic schizophrenic inpatients. *Journal of Nervous and Mental Disease, 172*(6), 317–324.

Parnas, J., Schulsinger, F., Teasdale, T.W., Schulsinger, H., Feldman, P.M., & Mednick, S.A. (1982). Perinatal complications and clinical outcome with the schizophrenia spectrum. *British Journal of Psychiatry, 340,* 416–420.

Pogue-Guile, M.F., & Harrow, M. (1984). Negative and positive symptoms in schizophrenia and depression: A follow-up. *Schizophrenia Bulletin, 10*(3), 371–387.

Pogue-Guild, M.F., & Harrow, M. (1985). Negative symptoms in schizophrenia: Their longitudinal course and prognostic importance. *Schizophrenia Bulletin, 11*(3), 427–439.

Rabiner, C.J., Wegner, J.T., & Kane, J.M. (1986). Outcome study of first episode psychosis: I. Relapse rate after one year. *American Journal of Psychiatry, 143,* 1155–1158.

Rakic, P. (1988). Defects of neuronal migration and the pathogenesis of cortical malformations. *Progress in Brain Research, 73,* 15–35.

Ram, R., Bromet, E.J., Eaton, W., Pato, E.C., & Schwartz, J.E. (1992). The natural course of schizophrenia: A review of first-admission studies. *Schizophrenia Belletin, 18,* 185–207.

Rappaport, M., Hopkins, H.K., Hall, K., Belleza, T., & Silverman, J. (1978). Are there schizophrenics for whom drugs may be unnecessary of contraindicated? *International Pharmacopsychiatry, 13,* 100–111.

Reich, J., & Thompson, W.D. (1985). Marital status of schizophrenic and alcoholic patients. *The Journal of Nervous and Mental Disease, 173,* 499–502.

Reither, P.J. (1926). Extrapyramidal motor-disturbances in dementia praecox. *Acta Psychiatrica et Neurologica Scandinavien, 1*, 287–309.

Resnick, S.M., Gur, R.E., Alavi, A., Gur, R.C., & Reivich, M. (1988). Positron emission tomography and subcortical glucose metabolism in schizophrenia. *Psychiatry Research, 24*, 1–11.

Reynolds, G.P. (1983). Increased concentrations and lateral asymmetry of amygdala dopamine in schizophrenia. *Nature, 305*, 527–529.

Rochford, J.M., Detre, T., Tucker, G.J., & Harrow, M. (1970). Neuropsychological impairments in functional psychiatric diseases. *Archives of General Psychiatry, 22*, 114–119.

Roff, J.D., Knight, R., & Wertheim, E. (1976). A factor-analytic study of childhood symptoms antecedent to schizophrenia. *Journal of Abnormal Psychology, 85*(6), 543–549.

Rupp, C., & Fletcher, E.K. (1940). A five to ten year follow-up study of 641 schizophrenic cases. *American Journal of Psychiatry, 96*, 877–888.

Saint-Anne Dargassies, S. (1982). *The neuromotor and psychoaffective development of the infant.* Amsterdam, the Netherlands: Elsevier.

Salokangas, R.K.R. (1983). Prognostic implications of the sex of schizophrenic patients. *British Journal of Psychiatry, 142*, 145–151.

Saugstad L. (1989). Social class, marriage and fertility in schizophrenia. *Schizophrenia Bulletin, 15*, 9–44.

Schaumann, B., & Alter, A. (1976). *Dermatoglyphics in medical disorders.* New York: Springer-Verlag.

Schwarzkopf, S., Nasrallah, A., Olson, S.C., Coffman, J., & Mclaughlin, J. (1988). Perinatal complications and genetic loading in schizophrenia: Preliminary findings. *Psychiatry Research, 27*, 233–239.

Scottish Schizophrenia Research Group. (1988). The Scottish first episode schizophrenia study: V. One year follow-up. *British Journal of Psychiatry, 152*, 470–476.

Seeman, M.V. (1983). Interaction of sex, age, and neuroleptic dose. *Comprehensive Psychiatry, 24*, 125–128.

Seeman, M.V., & Hauser, P. (1984). Schizophrenia: The influence of gender on family environment. *International Journal of Family Psychiatry, 5*, 227–232.

Seeman, P., Bzowej, N.H., Guan, H.C., Bergeron, C., Reynolds, G.P., Bird, E.D., Riederer, P., Jellinger, K., & Tourtellotte, W.W. (1987). Human brain D1 and D2 dopamine receptors in schizophrenia, Alzheimer's, Parkinson's, and Huntington's diseases. *Neuropsychopharmacology, 1*, 5–15.

Stengel, T.J., Attermeier, S.M., Bly, L., & Heriza, C.B. (1984). Evaluation of sensorimotor dysfunction. In S.K. Campbell (Ed.), *Pediatric neurologic physical therapy* (pp. 13–81). New York: Churchill Livingstone.

Stevens, J.H. (1978). Long-term prognosis and followup in schizophrenia. *Schizophrenia Bulletin, 4*, 25–47.

Stevens, J.H., & Astrup, C. (1963). Prognosis in "process" and "nonprocess" schizophrenia. *American Journal of Psychiatry, 119*, 945–953.

Stevens, J.H., Astrup, C., & Mangrum, J.C. (1966). Prognostic factors in recovered and deteriorated schizophrenics. American Journal of Psychiatry, 122, 1116–1121.

Suddath, R.L., Christison, G.W., Torrey, E.F., Casanova, M.F., & Weinberger, D.R. (1990). Anatomical abnormalities in the brains of monozygotic twins discordant for schizophrenia. *New England Journal of Medicine, 332*, 789–794.

Tamminga, C.A., Thaker, G.K., Buchanan, R., Kirkpatrick, B., Alphs, L.D., Chase, T.N., & Carpenter, W.T. (1992). Limbic system abnormalities identified in schizophrenia using positron emission tomography with fluorodeoxyglucose and neocortical alterations with deficit syndrome. *Archives of General Psychiatry, 49*, 522–530.

Test, M.A., Burke, S.S., & Wallisch L.S. (1990). Gender differences of young adults with schizophrenic disorders in community care. *Schizophrenia Bulletin, 16*(2), 331–344.

Toru, M., Watanabe, S., Shibuya, H., Nishikawa, T., Noda, K., Mitsushio, H., Ichikawa, H., Kurumaji, A., Takaskhima, M., Mataga, N., & Ogawa, A. (1988). Neurotransmitters, receptors and neuropeptides in post-mortem brains of chronic schizophrenic patients. *Acta Psychiatrica Scandinavica, 78*, 121–137.

Towbin, A. (1986). Obstetric malpractice litigation: The pathologist's view. *American Journal of Obstetrics and Gynecology, 155*, 927–935.

Tsuang, M.T., & Winokur, G. (1975). The Iowa 500: Field work in a 35 year follow-up of depression, mania, and schizophrenia. *Canadian Psychiatric Journal, 20*, 389.

Tsuang, M.T., Woolson, R.F., & Fleming, J.A. (1979). Long-term outcome of major psychosis: I. Schizophrenia and affective disorders compared with psychiatrically symptom-free surgical conditions. *Archives of General Psychiatry, 39*, 1295–1301.

Ulloa-Aguirre, A., Tarraso, J., Mendez, J.P., Garza-Flores, J., Diaz-Sanchez, V., & Perez-Palacios, G. (1991). Changes in the responsiveness of prolactin secretion to depaminergic blockade and TRH stimulation throughout sexual maturation in men. *Psychoneuroendocrinology, 15*, 279–286.

Vaillant, G.E. (1964a). An historical review of the remitting schizophrenias. *Journal of Nervous and Mental Diseases, 138*, 48–56.

Vaillant, G.E. (1964b). Prospective prediction of schizophrenic remission. *Archives of General Psychiatry, 11*, 509–518.

Vaughn, C., & Leff, J. (1976). The influence of family and social functions on the course of psychiatric illness. *British Journal of Psychiatry, 129*, 125–137.

Vaughn, C.E., Snyder, K.S., Jones, S., Freeman, W.B., & Falloon, R.H. (1984). Family factors in schizophrenic relapse: Replication in California of British research on expressed emotion. *Archives of General Psychiatry, 41*, 1169–1177.

Virtunski, P.B., Simpson, D.M., & Meltzer, H.Y. (1989) Voluntary movement dysfunction in schizophrenia. *Biological Psychiatry, 25*, 529–539.

Waddington, J.L., & Youssef, H.A. (1990). The lifetime outcome and involuntary movements of schizophrenia never treated with neuroleptic drugs: Four rare cases in Ireland. *British Journal of Psychiatry, 156*, 106–108.

Wakita, Y., Narahara, K., & Kimoto, H. (1988). Multivariate analysis dermatoglyphics of severe mental retardates: An application of the constellation in graphical method for discriminate analysis. *Acta Med Okayama, 42*, 159–168.

Walker, E., & Emory, E. (1981). Infants at risk for psychopathology: Offspring of schizophrenic parents. *Child Development, 54*, 1269–1285.

Walker, E., Grimes, K., Davis, D., & Smith, A. (1993). Childhood precursors of schizophrenia: Facial expressions of emotion. *American Journal of Psychiatry, 150*, 1654–1660.

Walker, E., & Lewine, R. (1988). The positive/negative symptom distinction in schizophrenia: Validity and etiological relevance. *Schizophrenia Research, 1*, 315–328.

Walker, E., Lucas, M., & Lewine, R. (1992). Schizophrenic disorders. In A. Puente & R. McCaffrey (Eds.), *Handbook of neuropsychological assessment: A biopsychosocial perspective* (pp. 309–334). New York: Plenum Press.

Walker, E., Savoie, T., & Davis, D. (1994). Neuromotor precursors of schizophrenia. *Schizophrenia Bulletin, 20*, 441–452.

Waring, E.M., Lefese, D.H., Carver, C., Barnes, S., Fry, R., & Abraham, B. (1988). The course and outcome of early schizophrenia. *The Psychiatric Journal of the University of Ottawa, 13*, 194–197.

Watt, D.C., Katz, K., & Shepherd, M. (1983). The natural history of schizophrenia: A 5-year prospective followup of a representative sample of schizophrenics by means of a standardized clinical and social assessment. *Psychological Medicine, 13*, 663–670.

Watt, N.F. (1978). Patterns of childhood social development in adult schizophrenics. *Archives of General Psychiatry, 35*, 160–165.

Watt, N.F., Grubb, T.W., & Erlenmeyer-Kimling, L. (1984). Social, emotional, and intellectual behavior at school among children at high risk for schizophrenia. In N.F. Watt, J. Anthony, L.C. Wynne, & J.E. Rolf (Eds.), *Children at risk for schizophrenia* (pp. 212–226). New York: Cambridge University Press.

Wattie, B.J.S., & Kedward, H.B. (1985). Gender differences in living conditions found among male and female schizophrenic patients on a follow-up study. *International Journal of Social Psychiatry, 31*, 205–216.

Weggmann, T., Brown, J.K., Fulford, G.E., & Minns, R.A. (1987). A study of normal baby movements. *Child: Care, Health, and Development, 13*, 41–58.

Weinberger, D. (1987). Implications of normal brain development for the pathogenesis of schizophrenia. *Archives of General Psychiatry, 44*, 660–669.

Wilcox. J.A. (1990). Thought disorder and relapse in schizophrenia. *Pscyhopathology, 23*, 153–156.

Wilcox, J.A., & Nasrallah, H.A. (1987). Perinatal distress and prognosis of psychotic illness. *Neuropsychobiology, 17*, 173–175.

Winblad, B., Hardy, J., Backman, L., & Nilsson, L.G. (1985). Memory function and brain biochemistry in normal aging and in senile dementia. *Annals of the New York Academy of Science, 444*, 255–268.

Winokur, G., Pfohl, B., & Tsuang, M. (1987). A 40-year follow-up of hebephrenic-catatonic schizophrenia. In N.E. Miller & G.D. Cohen (Eds.), *Schizophrenia and aging*. New York: Guilford Press.

Wong, D.F., Wagner, H.N., Jr., Tune, L.E., Dannals, R.F., Pearlson, G.D., Links, J.M., Tamminga, C.A., Brosusolle, E.P., Ravert, H.T., Wilson, A.A., Toung, J.K.T., Malat, J., Williams, J.A., O'Tuama, L.A., Snyder, S.H., Kuhar, M.J., & Gjedde, A. (1986). Positron emission tomography reveals elevated D2 dopamine receptors in neuroleptic-naive schizophrenics. *Science, 234*, 1558–1563; correction (1987) *235*, 623.

Wyatt, N. (1991). Neuroleptics and the natural course of schizophrenia. *Schizophrenia Bulletin, 17*, 325–351.

5
Cognitive Psychophysiological Indicators of Vulnerability in Relatives of Schizophrenic Patients

STUART R. STEINHAUER and DAVID FRIEDMAN

Schizophrenia may be studied at a number of different levels. In this chapter, we are concerned with identifying underlying psychophysiological mechanisms associated with the phenomena of schizophrenia, and their relevance as markers for identifying individuals who are at risk for the development of schizophrenia.

Psychopathological disorders are identified in an individual at first because of the behavioral anomalies observed, including verbal reports given by the patient. In the case of schizophrenia and other psychoses, it is not only the self-reports of discomfort that are so characteristic, but also the unusual aspects of behavioral conduct and verbal expressions that are key to the inference of abnormalities in thought processes.

Although a search for correlations between behavioral and physiological activities in patients provides an obvious impetus for psychophysiological investigations, several further advantages are provided by the methods to be discussed. Many of the psychophysiological responses of interest represent stages of information processing activities within 1 or 2s after information has been processed, reflecting basic components of deviant information processing. Furthermore, psychophysiological activity has revealed differences in the use of information by schizophrenic patients compared to control subjects that could not be detected from overt behavior alone. As examples, different approaches to assessing the electrical activity of the brain have provided objective evidence that schizophrenic patients exhibit physiological deficiencies in selective attention (Baribeau-Braun, Picton, & Gosselin, 1983) or in ability to respond fully to meaningful stimuli (Levit, Sutton, & Zubin, 1973). Autonomic measures have indicated that different subgroups of schizophrenic patients can be identified: some patients fail to show normal reactions to orienting stimuli, whereas others who respond well initially fail to habituate to repeated stimuli. In contrast, normal subjects show initial responses that typically tend to habituate on repeated stimulation (see Zahn, Frith, & Steinhauer, 1991).

A number of different etiological factors, including genetic, other biological, and environmental sources, appear to contribute to the development of schizophrenia. It has been proposed that the enduring characteristic of schizophrenia can be conceptualized as vulnerability to the disorder, a result of the contributing etiologies (Zubin & Spring, 1977). Thus, those subjects with greater vulnerability are most likely to develop episodes of schizophrenia when life stressors occur. The vulnerability model also emphasized that indicators of vulnerability could be sought from among the deviant responses seen among patients. A critical issue is that individuals from an affected family might or might not express schizophrenia, but even among unaffected family members, there should be a greater proportion of those who would exhibit the same deviant indicators, such as reduced amplitudes of event-related brain potentials or abnormalities of smooth-pursuit eye movements. The presence of a deviant response in an unaffected family member (one who has never exhibited an episode of disorder) could be indicative of a subject who has not expressed the disorder, but is at risk, or it could indicate the familial transmission of the marker, even in those who would never be ill. A more extensive discussion of the meaning of deviant response measures in patients and their relatives in relation to vulnerability has been detailed (Zubin & Steinhauer, 1981), and other discussions of vulnerability in relation to laboratory measures have followed (e.g., Nuechterlein & Dawson, 1984; Zubin, Steinhauer, Day, & van Kammen, 1985).

Empirical studies of risk factors in schizophrenia have focused on two classes of populations. Given that the best single predictor for the development of schizophrenia to date is still consanguinity, the greatest proportion of studies has focused on individuals at risk because they are relatives of schizophrenic patients. Several extensive programs have examined the children of schizophrenic patients (the so-called "high-risk" paradigm) on a variety of psychophysiological measures, including the Danish adoption study (Itil, Hsu, Saletu, & Mednick, 1974; Mednick & Schulsinger, 1968) and the New York High Risk Study (Friedman, Cornblatt, Vaughan, & Erlenmeyer-Kimling, 1986, 1988; Friedman, Vaughan, & Erlenmeyer-Kimling, 1982). Several investigators have tested patients and their siblings (Saitoh, Niwa, Hiramatsu, Kameyama, Rymar, & Itoh, 1984; Steinhauer et al., 1991), and studies of other first-degree relatives have also been conducted (Holzman, Proctor, Levy, Yasillo, Meltzer, & Hurt, 1974; Siegel, Waldo, Mizner, Adler, & Freedman, 1984; Siever et al., 1989). A few studies have focused on the parents of schizophrenic patients (e.g., Catts, Ward, Armstrong, & McConaghy, 1986, Catts, Ward, Armstrong, Fox, & McConaghy, 1986) or on identical twin pairs concordant or discordant for schizophrenia (Holzman, Kringlen, Levy, Proctor, & Haberman, 1978).

A second approach for studying at-risk subjects has involved populations identified through behavioral and psychophysiological testing. This has included a limited number of population studies, such as the initial evaluation of children on the island of Mauritius (Venables, 1977). More typical is the assessment of college students on batteries of tests that are believed to assess some of the same characteristics reported for schizophrenic patients. For example, patients with high scores on measures of anhedonia and groups of students who also score high on the Chapman Anhedonia Scales (Chapman & Chapman, 1978; Chapman, Chapman, & Raulin, 1976) have been compared with normally scoring students. The premise of the latter studies is that if subjects are identified who respond more like schizophrenic patients on certain clinical measures, and also respond similarly to patients on laboratory assessments of psychophysiological activity, then there is stronger evidence of risk for future disorder among these populations (Miller, 1986; Simons & Miles, 1990). Studies of high-risk groups identified through such procedures are examined in detail by Fernandes and Miller (Chapter 2).

Before proceeding further, a general caveat concerning the meaning of risk markers is needed. Virtually all of the psychophysiological measures employed in these studies have been pursued because they were found to be deviant in the patient samples studied and have been interpreted as being associated with schizophrenic disorder. However, an apparent genetic marker may reflect one of several conditions. When the measure is present because it is related to the etiological manifestation of disorder, it indicates association with the disorder. That is, some measurable trait is influenced directly by the genetic makeup of the individual. Another aspect of genetic markers, however, is linkage, in which some trait occurs in connection with the disorder, because the chromosomal sites for both the disorder and the trait lie nearby along the same chromosome and are more likely to be transmitted together than other sites more distant on the same chromosome.

The difficulty encountered in the study of psychophysiology (or most behavioral or chemical markers, for that matter) is the fact that they have been identified because they are associated with the expression of the disorder. Thus, some laboratory measures reflect signs of the disorder itself. Alternatively, it is possible that the trait seen may represent a true genetic marker because it shows close linkage to a major chromosal site for a disorder yet is not likely to represent a level of dysfunction causally associated with the disorder. Consequently, any assumption that the laboratory measures examined are "pure" indicators of genetic risk must be regarded with skepticism.

Several psychophysiological systems related to primarily passive or simple behavioral requirements, but which may also serve as models for examining more complex psychophysiological functions, will be mentioned only briefly.

Studies of Adult Relatives of Schizophrenic Patients

Smooth-Pursuit Eye Movements

A straightforward task involves instructing a subject to gaze at a sinusoidally moving object, such as a pendulum, and recording the subject's eye movements. Holzman et al. (1974) observed that the smooth-pursuit eye movements (SPEM) seen in normal subjects had a high rate of impairment of various types among schizophrenic patients. In subsequent work, Holzman and co-workers showed that there was high concordance for such abnormalities in identical twin pairs (Holzman, Kringlen, Levy, & Haberman, 1980). Moreover, high rates of SPEM abnormalities were observed in relatives of schizophrenic patients, including nonschizophrenic relatives (Holzman et al., 1974). In addition, they noted that immediate relatives of the proband could exhibit this impairment even when the index patient did not show such an impairment. From these data, they have suggested that one possible model of transmission involves a latent trait, which may be expressed as either the SPEM impairment, schizophrenia, or both (Matthysse, Holzman, & Lange, 1986). A summary of SPEM findings is provided by Holzman (1991).

Pupil Diameter

Autonomic level and phasic changes in information can also be detected by measurement of pupillary diameter (Loewenfeld, 1993). Although deviations in the response to light have long been noted in schizophrenia (Zahn et al., 1991), of greater interest for the present discussion is the fact that dilation of the pupil tends to be proportional to the amount of information provided to the subject (Steinhauer & Hakerem, 1992). In an early evaluation of 12 schizophrenic inpatients and 12 of their brothers, subjects were asked to guess which of two possible auditory stimuli was likely to occur. Patients were observed to show decreased pupillary amplitudes compared to their brothers or control subjects ($n = 15$) (Steinhauer, Hakerem, & Spring, 1979). However, no differences were observed between controls and the siblings.

We present several findings taken from the more recent Pittsburgh Vulnerability Study (Steinhauer et al., 1991). In this research effort, male schizophrenic outpatients who were clinically stable, and who had at least one male sibling, were recruited for participation in a large battery of behavioral and neurophysiological tests. Several distinguishing aspects were observed in this sample. First, to minimize ascertainment bias, male siblings were randomly selected from the available pool of brothers for each patient. Second, a comprehensive diagnostic evaluation battery was used to assess both Axis I and Axis II personality disorders according to the *Diagnostic and Statistical Manual of Mental Disorders* (3rd ed., revised) (DSM-III-R) (American Psychiatric Association, 1987). Although controls were required to be free of both Axis I and Axis II disorders, it was possible to categorize each sibling as either (a) having a schizophrenia spectrum disorder (i.e., Schizotypal or Paranoid Personality Disorder), (b) meeting criteria for some other nonspectrum disorder, usually either substance abuse or history of depression, or (c) meeting criteria for no disease. Among the psychophysiological variables, pupillary diameter was recorded in 18 outpatients, 31 of their brothers, and 19 controls during tasks in which event probability was varied. Controls showed significantly larger dilations compared to patients. The brothers of the patients showed a mean dilation amplitude that was intermediate between the controls and patients but did not differ significantly from either the control or patient groups (Steinhauer, Condray, Saito, & Detka, 1992). In addition, it was observed that both the patients and their brothers exhibited significantly larger average diameters than did control subjects.

A model of pupillary function has been proposed in which separate contributions of sympathetic and parasympathetic activity have been hypothesized to contribute to the pupillary dilation response during cognitive tasks (Steinhauer & Hakerem, 1992). When the pupillary data for controls, patients, and their brothers were subjected to a Principal Components Analysis, three factors were extracted (Steinhauer, Condray et al., 1992). There were no differences among subject groups on two of these factors, which appear related to inhibitory activity of the parasympathetic pathways. However, analysis of the factor scores for the third factor, which was similar in time course and waveform shape to the sympathetic component of the model, suggested that this factor was responsible for differences among the patient, sibling, and control groups. Because sympathetic activity reflects final converging afferents from limbic and frontal cortical regions that are modulated through posterior hypothalamic centers, it suggests that functional physiological activity involving these regions may be disrupted in schizophrenic patients, as well as in some of their nonschizophrenic relatives. The possibility that sympathetically mediated dilation represents a familial indicator clearly requires further evaluation. In the discussion of event-related brain potential (ERP) activity recorded simultaneously in the same subjects, presented next, the possibility is entertained that psychophysiological signs of disruption may be greater among those relatives with milder symptomatology associated with the schizophrenia spectrum—that is, schizotypal personality disorder.

Cardiac Activity

A substantial literature has demonstrated higher tonic autonomic levels in schizophrenic patients, including higher average heart rates (Zahn et al., 1991). However, phasic aspects of cardiac activity recorded during cognitive performance have often been reported to be diminished in schizophrenic patients. For example, when information delivery is likely to resolve uncertainty at a particular time, heart rate tends to show an anticipatory deceleration, followed by cardiac acceleration after stimulus presentation. Schizophrenic inpatients exhibited reduction of both cardiac components, and higher heart rates, compared to controls (Steinhauer, Jennings, van Kammen, & Zubin, 1992). When similar tasks were carried out with clinically stable outpatients and their brothers in the Pittsburgh Vulnerability study, a tendency for greater anticipatory deceleration was observed under all conditions in both patients and a large proportion of their brothers, even when events were predictable and required no motor response (Steinhauer et al., 1990). The outpatients as well as their brothers could be characterized as hypervigilant—that is, overprepared for stimulus presentation as compared to controls. However, no evidence of higher heart rates among the brothers of patients was found, indicating that generalized cardiac arousal is not likely to be a familial factor involved in schizophrenia.

Event-Related Potentials

Part of the electroencephalographic activity that can be recorded at the scalp comprises responses evoked by auditory, visual, and tactile stimuli. The resulting ERP is seen as a series of voltage swings, called *waves*, or *deflections*. These are usually identified by their polarity (positive or negative) and their latency to peak voltage after the onset of the stimulus, as shown in Figure 5.1.

Thus, N100 refers to a scalp negativity at approximately 100 ms, and P300 refers to a positivity at approximately 300 ms after stimulus presentation.

One type of sensory evoked response deviation in schizophrenia has been termed *sensory gating* of information. When two clicks are presented over a short period (e.g., 500 ms), the P50 response (i.e., the scalp positivity at 50 ms) to the second click tends to be decreased in amplitude among normal subjects. Schizophrenic patients, however, show significantly less reduction of the second (test) response, which suggests an impairment in sensory gating (Adler et al., 1981; Freedman & Mirsky, 1991). Nonaffected adult relatives of schizophrenic patients also appear to show a high rate of this sensory gating impairment (Siegel et al., 1984), suggesting that it may also serve as a familial indicator of vulnerability.

However, a number of concerns may be raised regarding the robustness of the P50 findings. First, difficulties have been encountered in replicating the control/schizophrenic difference in P50 suppression to the second click (Deldin, Miller, & Gergen, 1991; Kathmann & Engel, 1990). Investigators have also expressed some concern that the reliability in evaluating P50 suppression using the ratio of its amplitudes in the first to

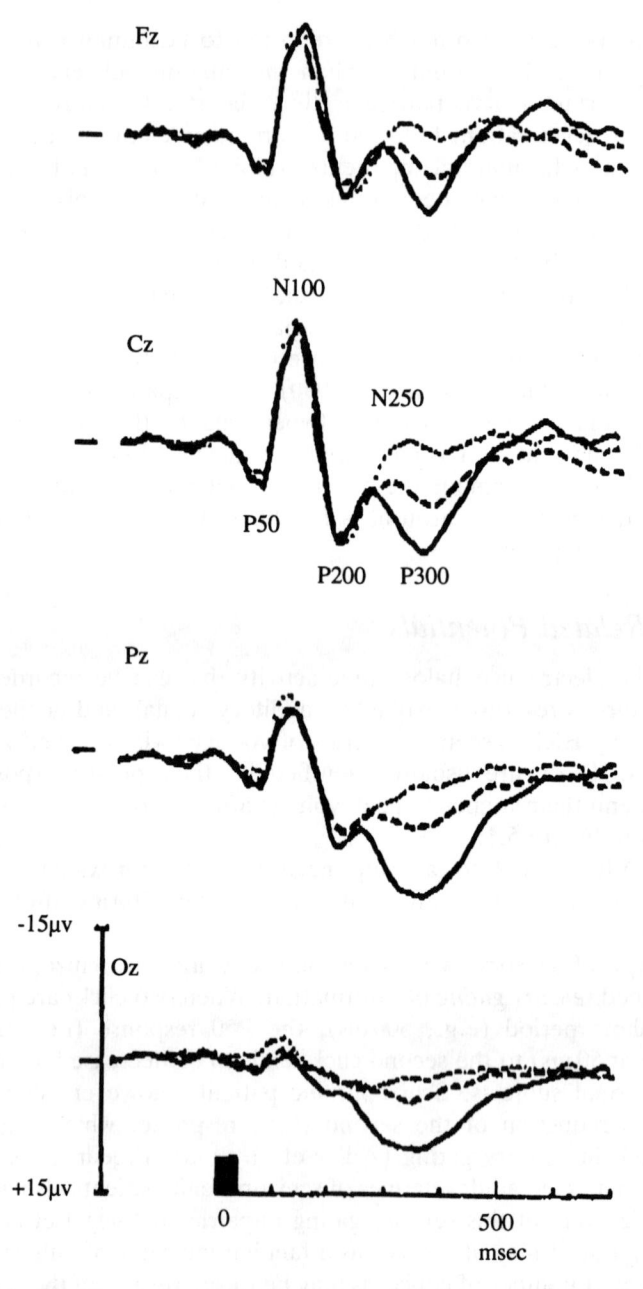

FIGURE 5.1. Grand mean event-related potentials from midline scalp locations obtained from 18 normal male volunteers. These averages were recorded in response to auditory stimuli occurring with varying probabilities. Positivity at the scalp is indicated by downward-going activity. The P300 component can be seen to be most positive for a rare event (event probability = .33, solid line), smaller for a more likely event (p = .67, dashed line), and least positive for a predictable event (p = 1.00, dotted line).

the second click is low (Boutros, Overall, & Zouridakis, 1991; Cardenas, Gerson, & Fein, 1993). Thus, the status of P50 as a familial indicator remains questionable.

In evaluating complex cognitive activities in schizophrenia, perhaps the most studied portion of the ERP involves those regions of the ERP that occur with a latency of 100 to 500 ms after a meaningful event. In particular, a great deal of attention has been directed toward the P300 component, a scalp positivity occurring with a peak latency between 300 and 1,000 ms, depending on task complexity.

P300 was first observed during prediction tasks (Sutton, Braren, Zubin, & John, 1965) and could be elicited in subjects even by the absence of a stimulus at a critically informative point in time (Sutton, Tueting, Zubin, & John, 1967). The fact that P300 and other long-latency components were found to be sensitive to a variety of task-related contingencies, including event probability, feedback, monetary value, and other aspects of salience, led to the designation of such activity as being event related—hence, the term *event-related potential*, or ERP, to distinguish such components from those that are modulated more by sensory aspects of the stimulus. In essence, it was this discovery of a brain wave component that was more sensitive to psychological conditions than to physical stimulus parameters that established the existence of the field of cognitive psychophysiology. Examples of cognitive ERP components in response to auditory stimuli may be seen in Figure 5.1. P300 is the large, downward-going (i.e., positive in the figure) wave, which occurs at approximately 300 ms. The solid line represents responses to a stimulus event that occurred with a probability of .33, the dashed line represents an event having a probability of .67, and the dotted line represents an event with a probability of 1.00 (i.e., an event that could always be predicted with certainty). In these data, P300 amplitude increases as the probability associated with different events decreases. In contrast, note that for P50, N100, and P200 there is essentially no variation in amplitude in relation to event probability.

In this review, the term *P300* is used to refer to the classical P3 (or P3b; Squires, Squires, & Hillyard, 1975) component (for a review, see Johnson, 1988) that is usually elicited with a parietal maximum scalp distribution (in a midline montage of scalp electrodes). It is typically of large amplitude when elicited by task-relevant events. This classical P300 component stands in contrast to other P3 components that have more anteriorly oriented scalp distributions, such as P3a (Squires et al., 1975), or the "novelty P3" (Courchesne, 1983). The novelty P3 component is produced by unexpected "novel" stimuli, in contrast to the P300 component, which is elicited by target events, to which the subject is instructed to respond in some fashion. The P3a component is best elicited when a series of frequent and infrequent deviant (or rare) stimuli are played in the background (e.g., when subjects are instructed to read a book and ignore the background stimuli). The P3a is largest when elicited by the rare background events. In contrast to the classical P300, the P3a occurs quite early (between 250 and 280 ms) and has a frontocentral scalp distribution. The classical P300 component can have a latency to peak of anywhere from 300 to 1,000 ms, depending on the requirements of the task.

Since the initial studies of schizophrenic patients (Levit et al., 1973; Roth & Cannon, 1972), there have been numerous replications of decreased P300 amplitudes in schizophrenic patients across a variety of paradigms (for reviews, see Ford, Pfefferbaum, & Roth, 1992; Friedman, 1991; Friedman & Squires-Wheeler, 1994; Pritchard, 1986). Moreover, the shape of the ERP, especially P300, has been shown to be more similar for genetically related individuals within normal families, with greatest similarity for identical twins (Bock, 1976; Polich & Burns, 1987; Rogers & Deary, 1991; Steinhauer, Hill, & Zubin, 1987; Surwillo, 1980).

The fact that P300 showed some genetic variance increased the likelihood that it might provide a vulnerability marker related to risk for schizophrenia and provided a strong thrust to research on high-risk populations (Friedman, 1990). The data obtained from adult relatives of probands, and from children of parents with a diagnosis of schizophrenia, are examined separately.

Event-Related Potentials in Adult Relatives of Schizophrenic Patients

Discrepant findings have resulted from the examination of ERPs in adult relatives of schizophrenic patients. Saitoh et al. (1984) reported ERP deficits in relatives of patients on a selective attention syllable-discrimination task. In this task, separate sequences of stimuli are presented to each ear, and subjects are asked to detect infrequent targets occurring in one or the other ear. The ear to which the subject's attention is directed is called the *attended channel*, and the other ear is labeled the *unattended channel*. Using a recording obtained at the midline vertex location (Cz; left and right temporal leads were also recorded), they observed that both siblings and controls showed enhancement of negativity in the attended channel (most likely, processing negativity; Näätänen, 1990), indicating an effect of selective attention. Siblings, however, showed decreased negative amplitudes overall and decreased amplitudes for the Late Positive Complex (which includes P300) in the attended target condition as compared to controls. The use of Cz, however, is not optimal for observing the Late Positive Complex, because the type of target task used usually produces a P300 that is largest at the midline parietal site (Pz). One further source of variability in this study concerns the comparison of patient and sibling data: Of the 20 siblings studied, 8 of their probands were not included in the patient group.

Catts et al. (Catts, Ward, Armstrong, & McConaghy, 1986; Catts, Ward, Armstrong, Fox et al., 1986) have examined ERP components recorded from parents of schizophrenic patients. They observed no differences between parents of schizophrenic patients and parents of controls in the contingent negative variation, a negative-going slow potential that precedes a forewarned stimulus (Catts, Ward, Armstrong, Fox, et al., 1986). They also reported no differences between parents of schizophrenic patients and parents of controls for a simple auditory evoked response recorded at vertex (Catts, Ward, Armstrong, & McConaghy, 1986). However, when the 10 patients were divided into high or low scorers on an Object Sorting Test (in which thought disorder among patients is associated with higher scores), the parents of the patients with high scores also had high scores, as well as lower amplitudes for the P200 component and late positive complex of the ERP. This suggests the likelihood of a familial indicator for thought disorder.

A recent study by Kidogami, Yoneda, Asaba, and Sakai (1992) employed an auditory oddball task, in which the subject was required to press a button after rare tones, which occurred on 20% of all trials. As is typical, P300 amplitude was largest for all groups over Pz. For 20 first-degree relatives of schizophrenic patients, P300 amplitudes were smaller than those recorded from 26 control subjects but did not differ significantly from 34 schizophrenic patients. However, there is a problem in accepting the conclusions of Kidogami et al. that the decreased P300 amplitudes were related to a familial factor. The authors assumed that because there was little effect of age on P300 among the controls, who had a mean age of 43.5 years, it was not necessary to account for the effects of age among either the patients (mean age 51.8) or the group of relatives of patients (mean age 61.8), which included 17 parents of patients. Although it is true that age effects tend to be somewhat minimal in adults until the period of approximately 45 to 50 years, at that point well-known increases in latency and decreases in amplitude of P300 occur. Consequently, it is impossible to determine whether the amplitude reductions reported for the group of relatives, who were approximately 18 years older than controls, are due to advancing age or to a familial vulnerability to psychopathology.

Event-related potentials were also measured in the Pittsburgh Vulnerability study, introduced previously, in which brothers of patients were diagnosed according to the presence of either schizophrenia spectrum disorders, other psychiatric disorders, or no psychiatric disorder. Using a modified auditory oddball procedure, ERPs were recorded during both a counting task ("count rare tones") and a choice reaction task ("press a separate button to rare vs. frequent tones") (Steinhauer et al., 1991).

What was most evident was that the outpatient group showed diminished ERP amplitudes, especially for P300, but the brothers did not appear to differ from controls. Next, the initial sample was broken down by the brothers' diagnoses. Those brothers meeting criteria for Schizotypal Personality Disorder (SPD) appeared to exhibit smaller P300 amplitudes (like the patients) and longer P300 latencies (Steinhauer et al., 1991). Among the group of brothers, regardless of diagnosis, there was a trend for increasing latency to be associated with a greater number of schizotypal symptoms. However, when a group of 10 siblings meeting criteria for SPD was compared with the controls and a group of 15 different SPD subjects for whom there was no family history for schizophrenia, it was not possible to demonstrate any differences in latency or amplitude among groups (Condray & Steinhauer, 1992). Consequently, current analyses are examining the relationships between severity of symptomatology (even for personality disorders in the schizophrenia spectrum) and psychophysiological parameters. For example, an examination of ERPs in an identical twin pair concordant for an overall diagnosis of schizophrenia indicated that the twin with the more severe clinical course, that is, with greater numbers of rehospitalizations, was also distinguished by a longer latency P300 than his co-twin (Condray, Steinhauer, van Kammen, & Zubin, 1992).

Blackwood, St. Clair, Muir, & Duffy (1991) tested a large sample of 96 schizophrenic patients, 196 relatives of schizophrenic patients drawn from 20 families (including 45 diagnosed with schizophrenia), and 212 controls. They recorded auditory ERPs at vertex (Cz), and for many of the subjects, SPEMs were also obtained. P300 latency was longest for schizophrenic patients (338 ms) and schizophrenic relatives (347 ms), shorter for nonschizophrenic relatives (328 ms), and shortest for controls (301 ms). P300 amplitude did not differ for controls and nonschizophrenic relatives of patients but was decreased for patients and family members meeting criteria for schizophrenia. A discriminant function indicated that P300 latency and P300 amplitude were the two strongest variables in discriminating among subject groups. While noting the prevalence of both P300 and SPEM abnormalities in the affected families, the authors also noted the absence of any abnormal measure from the members of three affected families.

One of the key difficulties in interpreting the strength of the Blackwood et al. (1991) and Saitoh et al. (1984) reports is the failure to record from more than a single midline electrode, which in both cases was also not the best single location to examine. Because Pz (midline parietal) is the site where the classical P300 (P3b) is maximal, and Cz (midline vertex) is the site where P3a is maximal, it is not clear from these studies that what has been measured is the classic P300 or P3a.

Psychophysiological Studies of Offspring of Schizophrenic Patients

Electrodermal Assessments

Probably the most widely studied autonomic response system in patients with psychopathology, especially schizophrenia, is activity of the electrodermal system, known as skin conductance or galvanic skin response. Measurements are obtained from electrodes placed on the fingers or hands in which small constant voltages are passed between the electrodes. Recordings may be made with the subject at rest, while being presented with strong orienting stimuli (e.g., loud tones), or during performance of a task. Because the recovery time of skin conductance responses is relatively slow, long interstimulus intervals (10 to 60 s) are typically used. Specific measurements that have been used to compare groups include differences in overall skin conductance level, in spontaneous fluctuations unrelated to stimulus presentation, and in the tendency to show lack of phasic responses to stimuli (i.e., nonresponding), habituation of responses, or lack of habituation following stimulus presentation (for review, see Zahn et al., 1991). The implications of deviant patterns of skin conductance responding have been of major interest in the examination of potential vulnerability markers (Nuechterlein & Dawson, 1984; Öhman, 1981; Olbrich, 1989).

Among high-risk studies, a report that generated much initial interest emanated from the Danish high-risk study. Mednick and Schulsinger (1968), using a classical conditioning procedure, reported that time to half-recovery of the phasic aspect of the galvanic skin conductance response was faster in those who were "sick" compared to those who were "well." This classification was made 5 years after the electrodermal assessments had been performed. This study provided the only initial evidence suggesting that high-risk subjects might be characterized by hyperarousal and was the first high-risk report of the possibility that a psychophysiological indicator could serve as a vulnerability marker.

Subsequent studies, however, have not been able to replicate this finding from the Danish cohort. Salzman and Klein (1978) indicated several differences between 10 children of schizophrenic mothers compared to 29 control children during several conditions or their study, but no differences between groups in measures comparable to those used by Mednick and Schulsinger (1968): trials to habituation, response latency, or spontaneous responses. The high-risk group was characterized by lower skin conductance levels than controls only during conditioning, though these subjects had higher heart rates during test and baseline conditions.

Janes, Hesselbrock, and Stern (1978) compared larger groups of children of schizophrenic ($n = 54$), manic-depressive ($n = 31$), or normal ($n = 76$) parents on electrodermal measures during a conditioning experiment. While a movement artifact factor was found to be highest among children of manic-depressives, they observed no differences among groups on electrodermal measures of conditioning or habituation.

In a unique approach, Israeli children at high or low risk for schizophrenia were also compared across two different environments: those raised in cities and those raised in the communal atmosphere of the Kibbutz (Kugelmass, Marcus, & Schmueli, 1985). Using a variety of measures that included several directly comparable to the Danish study, these workers were unable to demonstrate any replication of the Mednick and Schulsinger (1968) report and even found one completely opposite effect: The high-risk group showed a trend for slower skin conductance recovery than control subjects.

A similar study carried out by Van Dyke, Rosenthal, and Rasmussen (1974) was notable in its emphasis on adult offspring of schizophrenic patients, many of whom were diagnosed with schizophrenia spectrum disorders. However, they also failed to find any association of schizophrenia with skin conductance level or recovery, although they did observe a trend for greater responding in the adult offspring before unconditioned aversive stimuli, which had been reported by Mednick and Schulsinger (1968).

The New York High Risk Project is a unique long-term follow-up study in which two cohorts of children have been studied longitudinally (see Erlenmeyer-Kimling & Cornblatt, 1987, for an overview). Each cohort has three groups of subjects: children of parents with a diagnosis of schizophrenia, children of parents with a diagnosis of affective disorder (the so-called "psychiatric control" group), and children of psychiatrically normal parents. The study of the initial sample, Sample A (a subsequent sample, Sample B, was recruited several years later), was begun in 1971. Electrodermal assessments using galvanic skin conductance were performed only during the first round of testing the initial group of children (Sample A). The procedure used was similar, but not identical, to that employed by Mednick and Schulsinger (1968). Preliminary results from the New York High Risk Project were negative (Erlenmyer-Kimling, 1975). Later, more complete analyses, after the patients' parents had been rediagnosed using more up-to-date Research Diagnostic Criteria, were also negative (Erlenmeyer-Kimling, Friedman, Cornblatt, & Jacobsen, 1985). Time to half-recovery did not differentiate any of the three parental diagnosis groups. Moreover, similar to the electrophysiological reports from the New York project, discussed next (Friedman et al., 1986, 1988), several ancillary analyses of the electrodermal data failed to reveal relationships between the skin conductance measures and severity of parental illness, or global rating of behavioral adjustment in adolescence.

Taken together, subsequent studies have not been able to uphold the original contention of Mednick and Schulsinger (1968) that recovery of the skin conductance response could be an accurate predictor of risk for schizophrenia.

Event-Related Potentials in Children of Schizophrenic Parents

Sensory evoked responses have been studied in children at risk for schizophrenia. Itil, Hsu, Saletu, and Mednick (1974) and Saletu, Saletu, Marasa, Mednick, and Schulsinger (1975) suggested that portions of the auditory evoked response in the range of the second positive component

(P200) occurred earlier in children at risk for schizophrenia, although no differences in amplitude were observed. These and an additional report of differences in sensory evoked responses in a small sample of high-risk children (Herman, Mirsky, Ricks, & Gallant, 1977) have not yielded fruitful results.

In addition to the classical P300, two other ERP components can be elicited during variations of the "oddball" experiment. The mismatch negativity is an ERP component elicited by infrequent oddball or deviant stimuli. It can be clearly seen by subtracting the ERP elicited by the frequent or standard stimuli from the ERP elicited by the infrequent oddball or deviant stimuli delivered during versions of the oddball paradigm, including the selective attention paradigm. It is typically recorded to stimuli in an "ignore" condition. For example, subjects might be asked to pay attention to stimuli presented during a primary task. At the same time, frequent and infrequent auditory stimuli, which subjects are instructed to ignore, are presented in the background. The mismatch negativity is thought to be an index of automatic or preattentive processing, because it is typically elicited by stimuli that subjects are instructed to ignore (Shelley, Ward, Catts, Michie, Andrews, & McConaghy, 1991).

The "processing negativity" briefly alluded to earlier (see Näätänen, 1990, for a review) is elicited during selective attention paradigms and has been reported to be reduced in schizophrenic patients (e.g., Baribeau-Braun et al., 1983; Michie, Fox, Ward, Catts, & McCongaghy, 1990; Ward, Catts, Fox, Michie, & McConaghy, 1991). In the selective attention paradigm, this electrical activity is larger to stimuli in the "attended" compared to the "unattended" channel (see preceding description in "ERPs in Adult Relatives of Schizophrenic Patients"). An operational measure of this activity is obtained by subtracting the ERP elicited by the frequently occurring standard stimuli when they were unattended from the ERP elicited by these same stimuli when they were attended. This is labeled Nd (or negative difference waveform). The processing negativity is thought to reflect the selection of stimuli within the attended channel for further processing and is, therefore, assumed to reflect controlled attentional and cognitive mechanisms.

Schreiber and colleagues have examined a group of 24 children of schizophrenic parents, with a matched control for each child. Using an auditory oddball task, they reported prolonged latencies for the N250 and P300 components of the high-risk children (Schreiber et al., 1991). The authors noted that the lack of amplitude differences between groups might be attributable to the high variability of these measures. The same children were later tested during an auditory selective attention task (Schreiber, Stolz-Born, Kornhuber, & Born, 1992). In contrast to their earlier study, P300 amplitude was reduced among the high-risk children, but P300 latency was not affected. Difference waveforms were calculated to compare Nd, and mismatch negativity. Mismatch negativity did not significantly differ between groups, suggesting that preattentive processing was intact in the high-risk subjects. However, frontal Nd was reduced among the high-risk children, suggesting that controlled selective attention mechanisms might be deficient in these children at risk for schizophrenia. The lack of a psychiatric control group, however, did not allow these investigators to conclude the reduction of Nd was diagnostically specific to schizophrenia. Schreiber et al. (1991, 1992) noted that differences in behavioral performance among their groups of children might be reflected in the ERP data.

As part of the New York High Risk Project, in which children of schizophrenic, affective, or control parents were studied beginning in 1971, ERP assessments were added in 1975, when the third round of testing these children was begun. At that time, they ranged in age from 11 to 19. Electrodermal measures, however, were recorded during Round 1 of testing, when the children were between 7 and 12 years of age, and these findings have been described earlier (Erlenmeyer-Kimling et al., 1985). The children of mentally ill parents were ascertained through the admission in 1971 or 1972 of either one or both of the parents to one of several state psychiatric facilities in the New York Metropolitan area, and only those parents for whom there was full diagnostic agreement for the presence of schizophrenia (by two board-certified psychiatrists, using DSM-II (American Psychiatric Association, 1968) criteria) were included in the study. The parents of these children were later rediagnosed using the Schedule for Affective Disorders and Schizophrenia—Lifetime Version (SADS-L; Spitzer et al., 1977) in conjunction with the Research Diagnostic Criteria (RDC; Spitzer et al., 1975). In the data summarized here, the children are grouped according to the diagnoses of their parents using the latter criteria. The normal control group was obtained with the cooperation of two large school systems within the New York Metropolitan area. Families were excluded from the normal control group if either parent was found to have had psychiatric treatment.

The ERPs were recorded during tasks on which adult schizophrenic patients had been reported to differ from normal controls, either on the basis of their ERP or behavioral responses. The tasks eventually used were a three-stimulus version of the oddball paradigm and two versions of the continuous performance test. The former had yielded abnormal ERP responses in schizophrenic patients (e.g., Roth & Cannon, 1972; Roth, Horvath, Pfefferbaum, & Kopell, 1980), while the latter had produced poorer performance indices in adult schizophrenic patients (Orzack & Kornetsky, 1966), as well as relatively poorer accuracy in children of schizophrenic parents when tested via a noncomputerized playing card version of the continuous performance test (CPT) in the New York High Risk Project (the complete details were published by Cornblatt and Erlenmeyer-Kimling in 1985 but were reported in preliminary form by Erlenmeyer-Kimling in 1975).

Using auditory oddball and visual CPT paradigms recorded at four midline locations ensured that the P300 component and a later Slow

Wave component would be recorded, both of which had been implicated in the kind of cognitive information processing that had been hypothesized to be deficient in adult schizophrenic patients and in children at risk for schizophrenia. Moreover, the use of these tasks was motivated by the strategy that, if some ERP components were "trait" markers (enduring characteristics of the individual as compared with "state" markers), then using a task that had produced a deficit in the adult schizophrenic should also lead to a similar deficit in the child at genetic risk.

For the auditory task, preliminary analyses using just high-risk and normal control subjects had been published by Friedman et al. (1982). Analyses of the complete auditory data recorded from the three groups of children after the patients' parents had been rediagnosed using RDC revealed no significant group differences for P300 or Slow Wave (either frontal negative or parietal positive components) (Friedman et al., 1988). There was also no relationship between any of the identified ERP components and a global rating of behavioral adjustment in adolescence (Cornblatt & Erlenmeyer-Kimling, 1985) or between measures of attentional function and ERP amplitude. Because only a small percentage of individuals at risk are expected to develop schizophrenia, Friedman et al. (1988) performed ERP amplitude distributional analyses to determine if there were more outlying subjects in the group at high risk for schizophrenia compared to the other two groups. However, no such pattern was evident. Similar conclusions were reached using the ERP data recorded during the two versions of the CPT (Friedman et al., 1986).

One other finding arose from the children's data in the New York High Risk Project. Friedman, Brown, Vaughan, Cornblatt, and Erlenmeyer-Kimling (1984) noted that in normal adolescence, a prominent negativity over anterior regions of the brain in the 200- to 400-ms range tends to decline. Friedman et al. (1988) observed that the children of schizophrenic patients showed enhanced anterior negativity. A parallel finding has been observed for children of families at risk for the development of alcoholism in two cohorts (Hill, Steinhauer, Park, & Zubin, 1990; Steinhauer & Hill, 1993). In the latter study, greater negativity was most marked among older (14 to 18 years) children from families with a history of alcoholism. These findings suggest that failure for the enhanced anterior negativity to decrease in older subjects may reflect a maturational lag among children from families at risk for psychopathology or substance abuse but may not be specific to children of schizophrenic patients.

Squires-Wheeler, Friedman, Skodol, and Erlenmeyer-Kimling (1993) reexamined the ERP data that had been recorded years earlier in light of recent diagnoses of the adolescents (now young adults). These subjects (mean age, 24 years) were evaluated for DSM-III-R Schizophrenic and Schizoaffective Disorder and for DSM-III-R Axis II schizophrenia-related traits and disorders including schizotypal, schizoid, and paranoid features. Axis II assessments were also made at this time by means of blind, direct interviews using the Personality Disorder Examination (PDE; Loranger,

1985). The PDE is a standardized, semistructured clinical interview for eliciting information relevant to personality disorders defined in the DSM-III-R.

For this reevaluation of the data, it was opportune to test the hypothesis advocated by Duncan and her colleagues (Duncan et al., 1987) that P300-amplitude decrements in the auditory, but not the visual, modality might predict subsequent schizophrenic-related outcomes in subjects from the high risk for schizophrenia group (i.e., these would be "trait" markers). However, this hypothesis was not supported; there was no relationship between P300 amplitude and the presence of Axis I schizophrenic or schizoaffective disorder or Axis II schizophrenia-related traits and disorders. An unanticipated (but statistically reliable) result linking P300 reduction in both modalities with a global functioning outcome based on the PDE (i.e., poorer behavioral outcome associated with P300-amplitude reductions) was observed for offspring from both psychiatric parental groups, as well as for offspring of psychiatrically normal parents (although this latter result was not as statistically robust). Limitations of the study include the small sample size and low prevalence of Axis I disorders, reducing the power to detect a significant anticipated association between P300 decrements in adolescence and subsequent schizophrenic-related traits and disorder in young adulthood. Nevertheless, these data are consistent with the results of a large number of clinical studies of P300 component, in demonstrating P300 decrements in individuals characterized by deficits in behavioral dysfunction. Their importance lies in the fact that the reduced P300 amplitudes were detected long before the overt behavioral symptoms were identified. They are also consistent with the notion emerging from data on adult relatives of schizophrenic patients that the long-latency components may be predictive of future levels of psychopathological symptomatology in individuals, rather than only reflective of familial association with schizophrenia.

Conclusions

From the studies reviewed in this chapter, a number of strengths have emerged regarding psychophysiological assessment of schizophrenic patients and their relatives during tasks involving cognitive processing. One of the major advantages of this approach has been the examination of central mechanisms of information processing that cannot be obtained from behavioral observation alone. The temporal resolution of these techniques, in some cases permitting examination of changes in the range of milliseconds, is a critical adjunct to recent technological developments in brain imaging such as positron emission tomograhy and functional magnetic resonance imaging that provide excellent spatial resolution but limited temporal resolution.

In addition, psychophysiological measures appear to reflect various aspects of familial association with schizophrenia. Evidence has been published that indicates that markers of vulnerability, for example, eye-movement dysfunctions, occur in family members even though they may not be observed in all probands. Other measures, including the pupillary response and components of the ERP (e.g., the P300 component) appear to be associated with either the presence of psychopathology or with clusters of symptoms. Our review suggests less confidence in those studies that have proposed that P300 is a "pure" vulnerability marker, that is, that it can serve as a marker for a specific psychopathological syndrome (e.g., schizophrenia). The study of children of schizophrenic parents, however, indicates that psychophysiological assessment may provide predictor variables for the development of global personality dysfunction. Whether this will eventually reflect a general psychopathology factor is a subject for future research. Overall, these findings, though modest, imply that continued psychophysiological assessment of relatives of schizophrenic patients may increase our understanding of both the physiological and cognitive mechanisms underlying schizophrenia. It is an important goal of future research to determine whether many of these psychophysiological indices will also contribute to clinical prediction for those at risk for the development of schizophrenia as well as other psychopathological syndromes.

References

Adler, L.E., Pachtman, E., Franks, R.D., Pecevich, M., Waldo, M.C., & Freedman, R. (1982). Neurophysiological evidence for a defect in neuronal mechanisms involved in sensory gating in schizophrenia. *Biological Psychiatry, 17*, 639–654.

American Psychiatric Association (1968). *Diagnostic and statistical manual of mental disorders* (2nd ed.). Washington, DC.

American Psychiatric Association (1987). *Diagnostic and statistical manual of mental disorders* (3rd ed., revised). Washington, DC.

Baribeau-Braun, J., Picton, T.W., & Gosselin J.-Y. (1983). Schizophrenia: A neurophysiological evaluation of abnormal information processing. *Science, 219*, 874–876.

Blackwood, D., St. Clair, D., Muir, W., & Duffy, J.C. (1991). Auditory P300 and eye tracking dysfunction in schizophrenic pedigrees. *Archives of General Psychiatry, 48*, 899–909.

Bock, F.A. (1976). *Pupillary dilation and vertex evoked potential similarity in monozygotic and dizygotic twins and siblings.* Doctoral dissertation, City University of New York.

Bourtros, N.N., Overall, J., & Zouridakis, G. (1991). Test–retest reliability of the P50 mid-latency evoked response. *Psychiatry Research, 39*, 181–192.

Cardenas, V.A., Gerson, J., & Fein, G. (1993). The reliability of P50 suppression as measured by the conditioning/testing ratio is vastly impoved by dipole modeling. *Biological Psychiatry, 33*, 335–344.

Catts, S.V., Ward, P.B., Armstrong, M.S., & McConaghy, N. (1986). Auditory evoked potentials and allusive thinking in the parents of schizophrenics: A preliminary report. In W.C. McCallum, R. Zappoli, & F. Denoth (Eds.),

Cerebral psychophysiology: Studies in event-related potentials. *Electroencephalography and Clinical Neurophysiology* (Suppl. 38), 443-445.

Catts, S.V., Ward, P.B., Armstrong, M.S., Fox, A.M., & McConaghy, N. (1986). CNV/PINV in psychiatric patients and their parents. In *Abstracts of the 8th International Conference on Event-Related Slow Potentials of the Brain (EPIC VIII)* Palo Alto, CA.

Chapman, L.J., & Chapman, J.P. (1978). Revised physical anhedonia scale. Unpublished test.

Chapman, L.J., Chapman, J.P., & Raulin, M.L. (1976). Scales for physical and social anhedonia. *Journal of Abnormal Psychology, 85*, 374-382.

Condray, R., & Steinhauer, S.R. (1992). Schizotypal personality disorder in individuals with and without schizophrenic relatives: Similarities and contrasts in neurocognitive and clinical functioning. *Schizophrenia Research, 7*, 33-41.

Condray, R., Steinhauer, S.R., van Kammen, D.P., & Zubin, J. (1992). Dissociation of neurocognitive deficits in a monozygotic twin pair concordant for schizophrenia. *The Journal of Neuropsychiatry and Clinical Neuroscience, 4*, 449-453.

Cornblatt, B.A., & Erlenmeyer-Kimling, L. (1985). Global attentional deviance as a marker for schizophrenia: Specificity and predictive validity. *Journal of Abnormal Psychology, 94*, 470-486.

Courchesne, E. (1983). Cognitive components of the event-related potential: Changes associated with development. In A.W.K. Gaillard & W. Ritter (Eds.), *Tutorials in event-related potential research: Endogenous components* (pp. 329-344). Amsterdam, The Netherlands: North Holland.

Deldin, P.J., Miller, G.A., & Gergen, J.A. (1991, December 6). *Partial evidence for P50 suppression in schizophrenia.* Paper presented at the Society for Research in Psychopathology, Cambridge, MA.

Duncan, C.C., Morihisa, J.M., Fawcett, R.W., & Kirch, D.G. (1987). P300 in schizophrenia: State or trait marker? *Psychopharmacology Bulletin, 23*, 497-501.

Erlenmeyer-Kimling, L. (1975). A prospective study of children at risk for schizophrenia: Methodological considerations and some preliminary findings. In R.D. Wirt, G. Winokur & M. Roff (Eds.), *Life history research in psychopathology* (pp. 23-46). Minneapolis: University of Minnesota Press.

Erlenmeyer-Kimling, L., & Friedman, D., Cornblatt, B., & Jacobsen, R. (1985). Electrodermal recovery data on children of schizophrenic parents. *Psychiatry Research, 14*, 149-161.

Erlenmeyer-Kimling, L., & Cornblatt, B.A. (1987). The New York High Risk Project: A followup report. *Schizophrenia Bulletin, 13*, 451-461.

Ford, J.M., Pfefferbaum, A., & Roth, W.T. (1992). P3 and schizophrenia. In D. Friedman & G. Bruder (Eds.), Psychophysiology and experimental psychopathology: A tribute to Samuel Sutton. *Annals of the New York Academy of Sciences, 658*, 146-162.

Freedman, R., & Mirsky, A.F. (1991). Event-related potentials: Exogenous components. In S.R. Steinhauer, J.H. Gruzelier & J. Zubin (Eds.), *Handbook of schizophrenia: Vol. 5. Neuropsychology, psychophysiology, and information processing* (pp. 71-90). Amsterdam, The Netherlands: Elsevier.

Friedman, D. (1990). Event-related potentials in populations at genetic risk: A methodological review. In J.W. Rohrbaugh, R. Parasuraman & R. Johnson, Jr. (Eds.), *Event-related potentials: Basic issues and applications* (pp. 310-332). New York: Oxford University Press.

Friedman, D. (1991). Endogenous scalp-recorded brain potentials in schizoprehnia: A methodological review. In S.R. Steinhauer, J.H. Gruzelier & J. Zubin (Eds.), *Handbook of schizophrenia: Vol. 5. Neuropsychology, psychophysiology, and information processing* (pp. 91–127). Amsterdam, The Netherlands: Elsevier.

Friedman, D., Brown, C., Vaughan, H.G., Jr., Cornblatt, B., & Erlenmeyer-Kimling, L. (1984). Cognitive brain potential components in adolescents. *Psychophysiology, 21*, 83–96.

Friedman, D., Cornblatt, B., Vaughan, H., Jr., & Erlenmeyer-Kimling, L. (1986). Event-related potentials in children at risk for schizophrenia during two versions of the continuous performance test. *Psychiatry Research, 18*, 167–177.

Friedman, D., Cornblatt, B., Vaughan, H.G., Jr., & Erlenmeyer-Kimling, L. (1988). Auditory event-related potentials in children at risk for schizophrenia: The complete initial sample. *Psychiatry Research, 26*, 203–221.

Friedman, D., & Squires-Wheeler, E. (1994). Event-related potentials (ERPs) as indicators of risk for schizophrenia. *Schizophrenia Bulletin, 20*, 63–74.

Friedman, D., Vaughan, H.G., Jr., & Erlenmeyer-Kimling, L. (1982). Cognitive related brain potentials in children at risk for schizophrenia: Preliminary findings and methodological considerations. *Schizophrenia Bulletin, 8*, 514–531.

Herman, J., Mirsky, A.F., Ricks, N.L., & Gallant, D. (1977). Behavioral and electrographic measures of attention in children at risk for schizophrenia. *Journal of Abnormal Psychology, 86*, 27–33.

Hill, S.Y., Steinhauer, S.R., Park, J., & Zubin, J. (1990). Event-related potential characteristics in children of alcoholics from high density families. *Alcoholism: Clinical and Experimental Research, 14*, 6–16.

Holzman, P.S. (1991). Eye movement dysfunctions in schizophrenia. In S.R. Steinhauer, J.H. Gruzelier & J. Zubin (Eds.), *Handbook of schizophrenia: Vol. 5. Neuropsychology, psychophysiology, and information processing* (pp. 129–145). Amsterdam, The Netherlands: Elsevier.

Holzman, P.S., Kringlen, E., Levy, D.L., & Haberman, S.J. (1980). Deviant eye tracking in twins discordant for psychosis: A replication. *Archives of General Psychiatry, 37*, 627–631.

Holzman, P.S., Kringlen, E., Levy, D.L., Proctor, L.R., & Haberman, S.J. (1978). Smooth pursuit eye movements in twins discordant for schizophrenia. *Journal of Psychiatric Research, 14*, 111–120.

Holzman, P.S., Proctor, L.R., Levy, D.L., Yasillo, N.J., Meltzer, H.Y., & Hurt, S.W. (1974). Eye-tracking dysfunctions in schizophrenic patients and their relatives. *Archives of General Psychiatry, 31*, 143–151.

Itil, T.M., Hsu, W., Saletu, B., & Mednick, S.A. (1974). Computer EEG and auditory evoked potential investigations in children at high risk for schizophrenia. *American Journal of Psychiatry, 131*, 892–900.

Janes, C.L., Hesselbrock, V., & Stern, J.A. (1978). Parental psychopathology, age, and race as related to electrodermal activity of children. *Psychophysiology, 15*, 24–34.

Johnson, R., Jr. (1988). The amplitude of the P300 component of the event-related potential: Review and synthesis. In P.K. Ackles, J.R. Jennings, & M.G.H. Coles. (Eds.), *Advances in psychophysiology* (Vol. 3, pp. 69–138). Greenwich, CT: JAI Press.

Kathmann, N., & Engel, R.R. (1990). Sensory gating in normals and schizophrenics: A failure to find strong P50 suppression in normals. *Biological Psychiatry, 27*, 1216.

Kidogami, Y., Yoneda, H., Asaba, H., & Sakai, T. (1992). P300 in first degree relatives of schizophrenics. *Schizophrenia Research, 6*, 9–13.

Kugelmass, S., Marcus, J., & Schmueli, J. (1985). Psychophysiological reactivity in high-risk children. *Schizophrenia Bulletin, 11*, 66–73.

Levit, R.A., Sutton, S., & Zubin, J. (1973). Evoked potential correlates of information processing in psychiatric patients. *Psychological Medicine, 3*, 487–494.

Loewenfeld, I.E. (1993). *The pupil: Anatomy, physiology, and clinical applications.* Detroit: Wayne State University Press.

Loranger, A.W. (1985). *Personality disorder examination (PDE).* Yonkers, NY: DV Communications.

Matthysse, S., Holzman, P.S., & Lange, K. (1986). The genetic transmission of schizophrenia: Application of Mendelian latent structure analysis to eye tracking dysfunctions in schizophrenia and affective disorder. *Journal of Psychiatric Research, 20*, 57–76.

Mednick, S.A., & Schulsinger, F. (1968). Some premorbid characteristics related to breakdown in children of schizophrenic mothers. *Journal of Psychiatric Research, 6*, 354–362.

Michie, P.T., Fox, A.M., Ward, P.B., Catts, S.V., & McConaghy, N. (1990). Event-related potential indices of selective attention and cortical lateralization in schizophrenia. *Psychophysiology, 27*, 209–227.

Miller, G.A. (1986). Information processing deficits in anhedonia and perceptual aberration: A psychophysiological analysis. *Biological Psychiatry, 21*, 100–115.

Näätänen, R. (1990). The role of attention in auditory information processing as revealed by event-related potentials and other measures of cognitive function. *Behavioral and Brain Sciences, 13*, 201–288.

Nuechterlein, K.H., & Dawson, M.E. (1984). A heuristic vulnerability/stress of model of schizophrenic episodes. *Schizophrenia Bulletin, 10*, 300–312.

Öhman, A. (1981). Electrodermal activity and vulnerability to schizophrenia: A review. *Biological Psychology, 12*, 87–145.

Olbrich, R. (1989). Electrodermal activity and its relevance to vulnerability research in schizophrenics. *British Journal of Psychiatry, 155*(suppl. 5), 40–45.

Orzack, M.H., & Kornetsky, C. (1966). Attention dysfunction in chronic schizophrenia. *Archives of General Psychiatry, 14*, 323–326.

Polich, J., & Burns, T. (1987). P300 from identical twins. *Neuropsychologia, 25*, 299–304.

Pritchard, W.S. (1986). Cognitive event-related potential correlates of schizophrenia. *Psychological Bulletin, 100*, 43–66.

Rogers, T.D., & Deary, I. (1991). The P300 component of the auditory event-related potential in monozygotic and dizygotic twins. *Acta Psychiatrica Scandinavica, 83*, 412–416.

Roth, W.T., & Cannon, E.H. (1972). Some features of the auditory evoked response in schizophrenics. *Archives of General Psychiatry, 27*, 466–471.

Roth, W.T., Horvath, T.B., Pfefferbaum, A., & Kopell, B.S. (1980). Event-related potentials in schizophrenics. *Electroencephalography and Clinical Neurophysiology, 48*, 217–239.

Saitoh, O., Niwa, S.I., Hiramatsu, K.I., Kameyama, T., Rymar, K., & Itoh, K. (1984). Abnormalities in late positive components of event-related potentials may reflect a genetic predisposition to schizophrenia. *Biological Psychiatry, 19,* 292–303.

Saletu, B., Saletu, M., Marasa, J., Mednick, S.A., & Schulsinger, F. (1975). Acoustic evoked potentials in offsprings of schizophrenic mothers ("high risk" children for schizophrenia). *Clinical Electroencephalography, 6,* 92–102.

Salzman, L.F., & Klein, R.H. (1978). Habituation and conditioning of electrodermal responses in high risk children. *Schizophrenia Bulletin, 4,* 210–222.

Schreiber, H., Stolz-Born, G., Kornhuber, H.H., & Born, J. (1992). Event-related potential correlates of impaired selective attention in children at high risk for schizophrenia. *Biological Psychiatry, 32,* 634–651.

Schreiber, H., Stolz-Born, G., Rothmeier, J., Kornhuber, A., Kornhuber, H.H., & Born, J. (1991). Endogenous event-related potentials and psychometric performance in children at risk for schizophrenia. *Biological Psychiatry, 30,* 177–189.

Shelley, A.M., Ward, P.B., Catts, S.V., Michie, P.T., Andrews, S., & McConaghy, N. (1991). Mismatch negativity: An index of a preattentive processing deficit in schizophrenia. *Biological Psychiatry, 30,* 1059–1062.

Siegel, C., Waldo, M., Mizner, G., Adler, L.E., & Freedman, R. (1984). Deficits in sensory gating in schizophrenic patients and their relatives. *Archives of General Psychiatry, 41,* 607–612.

Siever, L.J., Coursey, R.D., Alterman, I.S., Zahn, T., Brody, L. Bernad, P., Buchsbaum, M., Lake, C.R., & Murphy, D. (1989). Clinical, psychophysiological, and neurological characteristics of volunteers with impaired smooth pursuit eye movements. *Biological Psychiatry, 26,* 35–51.

Simons, R.F., & Miles, M.A. (1990). Nonfamilial strategies for the identification of subjects at risk for severe psychopathology: Issues of reliability in the assessment of event-related potential and other marker variables. In J.W. Rohrbaugh, R. Parasuraman, & R. Johnson, Jr. (Eds.), *Event-related potentials: Basic issues and applications* (pp. 343–363). New York: Oxford University Press.

Spitzer, R.L., & Endicott, J. (1977). *Schedule for Affective Disorders and Schizophrenia—Lifetime Version.* New York: New York State Psychiatric Institute, Biometrics Research Department.

Spitzer, R.L., Endicott, J., & Robins, E. (1975). *Research Diagnostic Criteria (RDC) for a selected group of functional disorders,* 2nd ed. New York: New York State Psychiatric Institute, Biometrics Research Department.

Squires, N.K., Squires, K.C., & Hillyard, S.A. (1975). Two varieties of long-latency positive waves evoked by unpredictable auditory stimuli in man. *Electroencephalography and Clinical Neurophysiology, 38,* 387–410.

Squires-Wheeler, E., Friedman, D., Skodol, A., & Erlenmeyer-Kimling, L. (1993). A longitudinal study relating P3 amplitude to schizophrenia spectrum disorders and to global personality functioning. *Biological Psychiatry, 33,* 774–785.

Steinhauer, S.R., Condray, R., Saito, H., & Detka, C. (1992). A deficient sympathetic component of pupillary dilation in schizophrenics reflects impaired cognitive processing. *Biological Psychiatry, 31,* 116A.

Steinhauer, S.R., & Hakerem, G. (1992). The pupillary response in cognitive psychophysiology and schizophrenia. In D. Friedman & G. Bruder (Eds.),

Psychophysiology and experimental psychopathology: A tribute to Samuel Sutton. *Annals of the New York Academy of Sciences, 658*, 182–204.
Steinhauer, S., Hakerem, G., & Spring, B. (1979). The pupillary response as a potential indicator of vulnerability to schizophrenia. *Psychopharmacology Bulletin, 15*, 44–45.
Steinhauer, S.R., & Hill, S.Y. (1993). Auditory event-related potentials in children at high risk for alcoholism. *Journal of Studies on Alcohol, 54*, 408–421.
Steinhauer, S.R., Hill, S.Y., & Zubin, J. (1987). Event-related potentials in alcoholics and their first-degree relatives. *Alcohol, 4*, 307–314.
Steinhauer, S.R., Jennings, J.R., van Kammen, D.P., & Zubin, J. (1992). Beat-by-beat cardiac responses in normals and schizophrenics to events varying in conditional probability. *Psychophysiology, 29*, 223–231.
Steinhauer, S.R., Zubin, J., Condray, R., Peters, J.L., Jennings, J.R., & van Kammen, D.P. (1990). Cardiac responsivity and information processing in outpatient schizophrenics and their siblings. *Biological Psychiatry, 27*, 115A.
Steinhauer, S.R., Zubin, J., Condray, R., Shaw, D.B., Peters, J.L., & van Kammen, D.P. (1991). Electrophysiological and behavioral signs of attentional disturbance in schizophrenics and their siblings. In C.A. Tamminga & S.C. Schulz (Eds.), *Schizophrenia research. Advances in neuropsychiatry and psychopharmacology* (Vol. 1, pp. 169–178). New York: Raven Press.
Surwillo, W.W. (1980). Cortical evoked potentials in monozygotic twins and unrelated subjects: Comparisons of exogenous and endogenous components. *Behavior Genetics, 10*, 201–209.
Sutton, S., Braren, M., Zubin, J., & John, E.R. (1965). Evoked-potential correlates of stimulus uncertainty. *Science, 150*, 1187–1188.
Sutton, S., Tueting, P., Zubin, J., & John, E.R. (1967). Information delivery and the sensory evoked potential. *Science, 155*, 1436–1439.
Van Dyke, J.L., Rosenthal, D., & Rasmussen, P.V. (1974). Electrodermal functioning in adopted-away offspring of schizophrenics. *Journal of Psychiatric Research, 10*, 199–215.
Venables, P.H. (1977). The electrodermal psychophysiology of schizophrenics and children at risk for schizophrenia: Controversies and developments. *Schizophrenia Bulletin, 3*, 28–47.
Ward, P.B., Catts, S.V., Fox, A.M., Michie, P.T., & McConaghy, N. (1991). Auditory selective attention and event-related potentials in schizophrenia. *British Journal of Psychisatry, 158*, 534–539.
Zahn, T.P., Frith, C.D., & Steinhauer, S.R. (1991). Autonomic functioning in schizophrenia: Electrodermal activity, heart rate, pupillography. In S.R. Steinhauer, J.H. Gruzelier, & J. Zubin (Eds.), *Handbook of schizophrenia: Vol. 5: Neuropsychology, psychophysiology, and information processing* (pp. 185–224). Amsterdam: Elsevier.
Zubin, J., & Spring, B. (1977). Vulnerablity: A new view of schizophrenia. *Journal of Abnormal Psychology, 86*, 103–126.
Zubin, J., & Steinhauer, S.R. (1981). How to break the logjam in schizophrenia: A look beyond genetics. *The Journal of Nervous and Mental Disease, 169*, 477–492.
Zubin, J., Steinhauer, S.R., Day, R., & van Kammen, D.P. (1985). Schizophrenia at the crossroads: A blueprint for the 80's. *Comprehensive Psychiatry, 26*, 217–240.

6
Toward a Definition of Schizophrenia

HOWARD BERENBAUM

The five chapters in this section provide detailed and thoughtful reviews of factors that may be indicative of risk to schizophrenia. One thing that none of them do, however, is provide a comprehensive definition of schizophrenia. This is not surprising, because with few exceptions (e.g., Meehl, 1990), it is uncommon for contemporary psychopathologists to define schizophrenia. I believe psychopathologists are reluctant to provide definitions of schizophrenia for several reasons. First, most psychopathologists share similar views of many, though not all, of the more salient aspects of schizophrenia. For example, most psychopathologists would probably agree that being out of touch with reality (as indicated by the presence of hallucinations and/or delusions) is an important aspect of schizophrenia and can be used relatively reliably, albeit imperfectly, for identifying schizophrenic individuals. Second, there does not yet exist a widely shared *comprehensive* definition of schizophrenia. Finally, it is not yet possible to provide a comprehensive definition of schizophrenia that could be supported by empirical evidence. Consequently, when comprehensive definitions of schizophrenia are provided, they are admittedly speculative (e.g., Meehl, 1993).

Although the authors of these chapters do not provide definitions of schizophrenia, I believe the sort of research and theorizing they present are what will someday lead to the development of a comprehensive definition of the disorder that can be supported by empirical evidence. In the meantime, these chapters provide a glimpse of what such a definition will probably look like. Before returning to a discussion of the definition of schizophrenia, I discuss several themes and issues that appeared in these chapters which, in my opinion, need to be considered when embarking on the task of defining schizophrenia.

Theory

In addition to describing theories of schizophrenia and its developmental course, several of these chapters explicitly discuss the role of theory. For example, after reviewing the empirical evidence concerning the Wisconsin scales, Edell, in Chapter 1, discusses the role that theory will need to play if additional progress is to be made. He states that "strikingly absent is any coherent or unifying theory that links the particular trait(s) studied to the later development of psychosis." He then adds that "Without such a theory, it is unclear how far the current work correlating trait scores to other variables will take us." The importance of developing theoretical models is not limited to work using the Wisconsin scales. Hooley et al. conclude Chapter 3 with "a call for systematic research that examines the nature of the EE [expressed emotion]–relapse relationship in a more theoretically driven manner." Whereas Edell and Hooley et al. point out the potential value of theoretical models for guiding and interpreting future research, Fernandes and Miller, in Chapter 2, point out that research that leads to "empirically derived, reliable predictors of psychosis" has the potential to "contribute to theoretical models." Clearly, our understanding of schizophrenia can benefit from additional theory and data, ideally with each informing the other.

Development

If schizophrenia were a phenomenon that is here one moment and gone the next, or that was unchanging in its presentation, it would not make sense to talk about a behavioral high-risk paradigm. It is precisely because of developmental changes that one can speak of behaviors that are indicative of risk to schizophrenia. Walker et al. devote Chapter 4 to the issue of developmental changes across the entire life span. They describe schizophrenia as "a developmentally diverse illness" and provide numerous illustrations of why this is the case. One of the more impressive aspects of the evidence they review is the range of phenomena that change over time. They discuss changes in almost anything one can think of, ranging from brain morphology to smiling to overt psychotic symptoms. In fact, the Walker et al. chapter leads me to wonder what, if anything, does *not* change?

Although Hooley et al. do not describe how schizophrenic individuals change over time, they do suggest that the association between EE and relapse may reflect a "developmental process" and that those who study the association may benefit by "applying the principles of developmental psychopathology." I believe the most important contribution made by Hooley et al. is their discussion of how the process of change is dynamic and interactive. Their discussion reveals that it is often not possible to fully understand one aspect of a schizophrenic individual's life (e.g., relapse) without considering other contemporaneous factors (e.g., level of familial EE). Their discussion also suggests that, to understand factors at one point in time (e.g., level of familial EE during a psychotic episode), one must consider factors at different points in time (e.g., the schizophrenic individual's level of premorbid functioning).

Cutoff Points

One of the issues discussed in several of these chapters concerns where to draw cutoff points when defining risk. In Chapter 3, Hooley et al. note that different cutoff points may be needed in different populations to most efficiently identify risk. They add that such a state of affairs is "neither surprising nor problematic." In fact, some psychopathologists have discussed the potential advantages of using different cutoff points in different settings (Finn, 1982; Meehl & Rosen, 1955; Widiger, Hurt, Frances, Clarkin, & Gilmore, 1984).

In Chapter 2, Fernandes and Miller point out that "because the Chapman scales assess a continuous distribution of putatively psychosis-related features within a population, categorically labeling an individual as an anhedonic or perceptual aberrator can be arbitrary in some cases." In a similar vein, Edell (Chapter 1) suggests that examining the full spectrum of scores may be a "more powerful strategy than is the traditional method of dichotomizing the sample into high- and low-risk groups." The question of whether risk to schizophrenia should be conceptualized and/or measured on a continuum rather than in a categorical manner is not limited to work using the Wisconsin scales.

A related issue not discussed by any of the authors concerns which end of the distribution to focus on. For example, research using the Wisconsin scales typically uses a cutoff point intended to select individuals presumed to be at greater than average risk, such as individuals who experience less pleasure than average. It is possible, however, that some variables are relevant to the likelihood of an individual developing schizophrenia because they are associated with reduced vulnerability rather than increased vulnerability. For example, some individuals may be hyperhedonic, a characteristic that may protect them against developing schizophrenia. It has been suggested that to best understand the developmental pathways to schizophrenia, one must study resilience and protective factors, rather than focusing exclusively on deficits and disadvantages (e.g., Garmezy, 1987; Rutter, 1985). An understanding of resilience and protective factors may be particularly important for the purpose of primary prevention (Cowen & Work, 1988).

Focus on vulnerability and high risk, rather than invulnerability and low risk, is not limited to research using the Wisconsin scales. Recent research employing the EE construct has also typically ignored the positive, even though "warmth" and "positive remarks" are two of the five original EE scales (Leff & Vaughn, 1985). It is conceivable that certain aspects of positive family functioning and communication may be associated with risk to schizophrenia or to the developmental course of the disorder. Examining invulnerability might also be useful in the sort of family research that Steinhauer and Friedman review in Chapter 5. For example, in addition to viewing a family history of schizophrenia as a marker of vulnerability, one can examine whether the absence of family history of any sort of psychiatric disturbance is associated with relative invulnerability. Such an approach could lead to the identification of specific features or processes in less vulnerable individuals or families that would not be identified using a research strategy focused on the high end of the vulnerability continuum.

Diversity and Dimensionality

As Walker et al. point out in Chapter 4, "one of the most obvious features of the schizophrenic syndrome is its diversity." In fact, the magnitude of the diversity becomes almost overwhelming when one considers all of the aspects of schizophrenia described across these five chapters. Hardly a domain of human functioning is not mentioned in at least one of the chapters.

The disturbances in cognition and emotion associated with schizophrenia receive a lot of attention in these chapters. The evidence reviewed indicates that these disturbances are not independent. For example, Fernandes and Miller (Chapter 2) review evidence suggesting that anhedonia, which can be considered a form of emotional disturbance, is associated with deviant information processing strategies. Similarly, both Fernandes and Miller, and Edell (Chapter 1) review evidence suggesting that perceptual aberration, which might be considered a form of cognitive disturbance, is associated with emotional processes and disturbances. Fernandes and Miller posit that perceptual aberrators differ from control subjects "in their handling of the emotional valence dimension." Edell points out that perceptual aberrators are more likely than control subjects to have had episodes of mood disorder. The possibility that there is an intricate relationship between emotion and information processing among individuals who are vulnerable to schizophrenia or who have already developed the disorder is consistent with the results of research examining the relationship between emotion and information processing among individuals with emotional disorders (Mathews & MacLeod, 1994). The results of previous research, however, suggest that the relationships between deficits in information processing and emotional functioning among schizophrenic individuals are unlikely to be simple (Berenbaum &

Oltmanns, 1992). Consequently, the development of sophisticated new models that provide unitary explanations of cognition and emotion (e.g., Barnard & Teasdale, 1991; Johnson & Multhaup, 1992) may prove to be quite helpful in advancing understanding of schizophrenia.

Because of the remarkable diversity of schizophrenia, it would be tempting to treat phenomena in different domains (e.g., behavior and neurochemistry) independently. Fortunately, the authors of these chapters resist this temptation. Walker et al. (Chapter 4) attempt to integrate data regarding such diverse phenomena as congenital subcortical abnormalities, motor disturbances, and facial expressions of emotion. Fernandes and Miller (Chapter 2) review research that examined the relationship between psychophysiological anomalies and behaviors measured by the Wisconsin scales. Edell (Chapter 1) discusses the familial resemblance of behaviors measured by the Wisconsin scales, with an eye toward understanding the role of genetic factors in risk to schizophrenia. Steinhauer and Friedman (Chapter 5) review research involving psychophysiological assessment of schizophrenic individuals and their relatives, research that they believe may "increase our understanding of both the physiological and cognitive mechanisms underlying schizophrenia." Hooley et al. (Chapter 3) attempt to explicate the relationship between phenomena at the group level (e.g., family dynamics) with phenomena at the individual level (e.g., a schizophrenic individual's psychotic relapse). In addition, they mention the possibility that the link between EE and relapse might be mediated by "overstimulation and hyperautonomic arousal."

It would be far simpler to understand schizophrenia if all of the features described in these chapters were strongly associated with each other, both cross-sectionally and longitudinally. Unfortunately, this is not the case. Psychopathologists have postulated that multiple processes underlie the different signs and symptoms of schizophrenia (e.g., Crow, 1985; Strauss, Carpenter, & Bartko, 1974). The evidence reviewed in these chapters is consistent with the association of multiple dimensions with schizophrenia. Walker et al. (Chapter 4) present evidence that various features of schizophrenia differ in their developmental patterns. They focus on the positive and negative symptom dimensions and posit that "their expression is moderated by different developmental processes." Edell (Chapter 1) points out that implicit in much of the work employing the Wisconsin scales "is the view of the schizotypy construct as multidimensional and multidetermined." The empirical work reviewed by Fernandes and Miller (Chapter 2) and Edell indicates that the different schizotypic features measured by the Wisconsin scales have different clinical, social, and psychophysiological correlates. Such findings are not surprising, given that the different Wisconsin scales are not all highly correlated with each other and given that evidence suggests that the different traits are familially transmitted independently of one another (Berenbaum & McGrew, 1993). The evidence reviewed by Hooley et al. (Chapter 3) suggests that even the responses of families to schizophrenic family members is multidimensional. They point out that "criticism and emotional involvement appear to have little in common" even though both have been found to be associated with higher rates of relapse. Further, they point out that these two facets of EE may differ in their prognostic significance at different points in time. Although the sort of diversity illustrated by these examples has the potential to bog down researchers attempting to understand schizophrenia, it can also provide fertile ground for the development and testing of new theoretical models.

Process and Output

Fernandes and Miller (Chapter 2) emphasize "the importance of analyzing deviant processes themselves, rather than merely the overt output of these processes... as a means of differentiating psychopathology subgroups." They provide several clear illustrations of the distinction between process and output. For example, they review evidence suggesting that anhedonic subjects may perform as well as control subjects on some information processing tasks, yet they appear to do so by employing different strategies. Another important point Fernandes and Miller raise is that sometimes what appears to be a deviant process or output may be the result of a strategy being used by the subject to compensate for a different process that is disturbed. The proclivity of individuals to develop compensatory strategies following neurological insult is well established (e.g., Clarke, Assal, & de Tribolet, 1993; Kershner & Micallef, 1992). Recognition of this proclivity is critical for developing models of developmental pathways and is not limited to the domain of brain functioning. Hooley et al. (Chapter 3) suggest that emotional overinvolvement may "represent an early and natural reaction of family members to the development of symptoms in patients," though the persistence of such patterns appears to be associated with poor outcome.

Sex and Gender

Sex differences in schizophrenia have received increasing attention over the past 10 or so years (e.g., Goldstein & Tsuang, 1990; Iacono & Beiser, 1992; Lewine, 1981). Walker et al. (Chapter 4) discuss sex differences extensively, pointing out that there are sex differences in both the childhood precursors and outcome of schizophrenia. In fact, such data lead them to suggest that "sex-linked biological factors moderate the behavioral expression of vulnerability to schizophrenia." Sex differences have also emerged in research using the Wisconsin scales. For example, Edell (Chapter 1) points out that there are sex differences in the ability of physical anhedonia to predict the later development of psychosis among the offspring of schizophrenic parents (Erlenmeyer-Kimling et al., 1993).

Unfortunately, it is not possible to determine whether the differences found in previous research between males and females are the result of differences in biological sex or socially constructed gender. For example, sex differences in schizophrenia could be a consequence of sex chromosomes playing a role in the genetic transmission of the disorder (e.g., DeLisi & Crow, 1989), or they could be associated with differences in the ways males and females are treated. Of course, it is possible that vulnerability to schizophrenia could be influenced by both biological sex and by the differential treatment of males and females, as well as by the interaction of these factors. Because there are cross-cultural variations in the socialization of males and females (e.g., Whiting & Edwards, 1988), cross-cultural research may help elucidate the precise reasons for the sex differences that have been observed in schizophrenia. For example, sex differences that are not found across cultures are likely to be at least partially influenced by sociocultural factors.

Culture

Although extensive cross-cultural research has been conducted in schizophrenia (e.g., Jablensky et al., 1992), there has been far less work examining potential cross-cultural differences in vulnerability to schizophrenia. Edell (Chapter 1) and Fernandes and Miller (Chapter 2) point out that research using the Wisconsin scales has been conducted almost exclusively with what Edell describes as "white, middle-class, academically advantaged, English-speaking college students." In contrast to the research using the Wisconsin scales, research examining EE has been conducted across multiple cultures. Hooley et al. (Chapter 3) point out that "the distribution of high- and low-EE relatives shows clear cultural variation." Hooley et al. also raise the possibility that the manner in which EE affects relapse may differ across cultures.

Specificity and Sensitivity

A reading of the schizophrenia literature suggests that what most schizophrenia researchers are searching for is a correlate that is specific to schizophrenia. Unfortunately, a correlate specific to schizophrenia has proved to be quite elusive. The issue of specificity to schizophrenia was discussed in all of these chapters. Edell (Chapter 1) points out that deviant scores on the Wisconsin scales can predict later psychosis but are not specific to schizophrenia (Chapman, Chapman, Kwapil, Eckblad, & Zinser, 1994). Fernandes and Miller (Chapter 2) state that "it is not clear whether anhedonics' or perceptual aberrators' psychophysiological abnormalities are associated with specific types of psychopathology or only with nonspecific risk for a range of disorders." Steinhauer and Friedman (Chapter 5) state that "the study of children of schizophrenic parents indicates that psychophysiological assessment may provide predictor variables for the development of global personality dysfunction." They express reservations, however, about the likelihood of the P300 being "a 'pure' vulnerability marker," in other words, "a marker for a specific psychopathological syndrome (e.g., schizophrenia)." Hooley et al. (Chapter 3) point out that high levels of EE can be found among the relatives of individuals with a wide range of psychiatric and medical disturbances. In fact, they refer to some unpublished work by Heckelman and Hooley in which high levels of EE were found in "normal mothers of healthy 3-year-old boys." Walker et al. (Chapter 4) come closest to describing a correlate that is specific to schizophrenia. They state that "there appears to be some diagnostic specificity of neuromotor abnormalities." However, it is not the case that signs of neuromotor abnormalities are unique to preschizophrenics or schizophrenics. Also, the degree to which neuromotor abnormalities appear to be specific to schizophrenia depends on the comparison group. Although preschizophrenic children may exhibit more neuromotor abnormalities than children who later develop mood disorders, neuromotor abnormalities would not appear to be specific to schizophrenia if the preschizophrenic children were compared to children with other neurodevelopmental disorders. Similarly, neuromotor abnormalities would not appear to be specific to schizophrenia if schizophrenic adults were compared with nonpsychotic adults with neurological disturbances such as Huntington's disease.

Although sensitivity is related to specificity, the former received less attention in these chapters than did the latter. Just as the phenomena described in these chapters are not unique to schizophrenia, neither are they present in all preschizophrenic or schizophrenic individuals. For example, some preschizophrenic and schizophrenic individuals do not exhibit neuromotor abnormalities. In fact, Chapman et al. (1994) found that not even all individuals who developed schizophrenia had deviant scores on the Wisconsin scales at the beginning of their study. Such results suggest that psychopathologists are unlikely to find a feature that is both highly sensitive and specific to schizophrenia, at least not using current operational definitions of the disorder.

Normal Functioning

Although the goal of these chapters is to discuss vulnerability to schizophrenia and its developmental course, the authors frequently refer to research on "normal" functioning and processes. In some cases, existing knowledge of normal functioning and processes is used to understand schizophrenia and vulnerability to the disorder. This is well illustrated in the chapter by Fernandes and Miller (Chapter 2), who use their understanding of different event-related brain potential (ERP) components gained from research with ordinary individuals to understand the results of research with hypothetically psychosis-prone individuals. Similarly, Walker et al. (Chapter 4) use their knowledge of the normal pattern of central nervous system maturation to develop hypotheses concerning the developmental course of schizophrenia. These examples suggest that our understanding of schizophrenia will ultimately depend on our understanding of normal processes and functioning. Indeed, progress in understanding schizophrenia and its symptoms can be impeded by our not understanding ordinary processes. Thus, Fernandes and Miller state that "obtaining psychophysiological evidence of anhedonia per se is somewhat constrained by the limited and controversial picture currently available of the normal psychophysiology of pleasure."

In all of the preceding examples, research on the ordinary was helpful in understanding the pathological. Conversely, in fields such as neuropsychology, there is a long history of research examining the pathological being used to improve understanding of normal functioning (Caramazza, 1992; but c.f. Kosslyn & Intriligator, 1992). The same could be true in the field of psychopathology. For example, the EE research that Hooley et al. (Chapter 3) describe has the potential to improve our understanding of interpersonal relationships and family functioning in *all* individuals and families, not just those with psychopathology. Similarly, Walker and coworkers' research examining the neuromotor and emotional disturbances of preschizophrenics may improve our understanding of the relationship between locomotor experience and emotional development (e.g., Campos, Bertenthal, & Kermoian, 1992). Just as theory and data should constantly influence one another, research examining the ordinary and research examining the psychopathological should continually complement one another. In fact, progress understanding both ordinary and psychopathological processes and outcomes might benefit from viewing the ordinary and the psychopathological as two sides of a single coin.

Theoretical Models of Schizophrenia

A theory of schizophrenia must explain how and why certain individuals develop the sort of characteristics most psychopathologists currently associate with the disorder, such as being out of touch with reality for more than a brief period of time. The evidence reviewed in these chapters indicates that schizophrenia should be viewed as a dynamic set of interacting phenomena that evolve over time, rather than as a single phenomenon that can be identified at a single point in time. A critical aspect of any theory of schizophrenia is the manner in which it accounts for heterogeneity in both cross-sectional symptom presentation (e.g., Andreasen & Olsen, 1982; Carpenter, Heinrichs, & Wagman, 1988) and longitudinal course (e.g., Bleuler, 1972/1978; Ciompi, 1980). A comprehensive theory of schizophrenia must be able to explain how and why the processes and outcomes described in these chapters vary between males and females and across cultures. Finally, a theory of schizophrenia must be able to explain how and why the disorder differs from other disturbances.

The fact that the features of schizophrenia evolve over time does not rule out the possibility that a single etiological factor may account for the developmental features of the disorder. For example, in addition to influencing which events occur, genes also influence when they occur. Thus, psychopathologists' desire to understand the etiology of schizophrenia, coupled with methodological advances in the field of molecular genetics, have led some investigators to search for a gene that causes schizophrenia. However, even if there were a single gene *necessary* for the development of schizophrenia, two findings from behavior genetic research demonstrate that no single gene can be *sufficient* to explain schizophrenia. The first critical finding is that concordance rates among monozygotic (MZ) twins are substantially below 100% regardless of how schizophrenia is operationalized (e.g., McGuffin, Farmer, Gottesman, Murray, & Reveley, 1984). This finding does not, by itself, rule out the possibility that genes are sufficient to explain schizophrenia. However, to maintain that genes are sufficient to explain schizophrenia in the face of the evidence demonstrating that MZ concordance rates are less than 100%, one would have to posit (a) that MZ twin pairs concordant for schizophrenia represent cases in which both twins possess the schizogene(s) and are therefore presumed to have "true" schizophrenia and (b) that MZ twin pairs discordant for schizophrenia represent cases in which the schizophrenic twin is a "phenocopy" and does not actually have a genetically defined form of schizophrenia. The second critical finding is that the rate of schizophrenia among the offspring of the normal MZ co-twins of schizophrenics is as high as the rate of schizophrenia among the offspring of the schizophrenic co-twins (Gottesman & Bertelsen, 1989). These results indicate that discordance among MZ twins cannot simply be attributed to the schizophrenic twin being a phenocopy but is more likely to be the result of unexpressed genetic vulnerability. Thus, the behavior genetic data demonstrate that genetic factors will never be sufficient to explain schizophrenia.

Although genes will never be sufficient to explain schizophrenia, it is conceivable that a gene or small set of genes will be found to be necessary for the development of schizophrenia. Even if this turns out to be the case, however, the identified gene or set of genes would not constitute an explanation of schizophrenia. The inability of a gene to explain schizophrenia is illustrated in the model of schizophrenia proposed by Meehl

(1990). Meehl posits that there is a single major gene that is necessary, but not sufficient, for the development of schizophrenia. Meehl, however, clearly distinguishes between (a) the schizogene; (b) schizotaxia, which he describes as the "genetically determined integrative defect, predisposing to schizophrenia" (Meehl, 1990, p. 35); (c) schizotypy, which he describes as a form of personality that develops in all or nearly all schizotaxic individuals; and (d) schizophrenia, which is the psychotic decompensation exhibited by a minority of schizotypes. In my view, the relationship between the hypothesized schizogene and schizophrenia is analogous, in at least some ways, to the relationship between the human immunodeficiency virus (HIV) and acquired immune deficiency syndrome (AIDS). Although HIV is necessary for the development of AIDS, it is not sufficient to explain all of the physiological facets of AIDS, let alone its sociopolitical aspects. Just as infection by the HIV virus is not sufficient to explain AIDS, no gene will ever be sufficient to explain schizophrenia.

Defining Schizophrenia

However schizophrenia will ultimately be defined, it is safe to say that a satisfactory definition will be embedded within an explicit theory of the disorder. Eventually, the definition of schizophrenia will be based on the manner in which psychopathologists explain a complicated set of developmental processes and outcomes. I predict that the definition of schizophrenia will turn out to be developmental in nature, which is consistent with the manner in which schizophrenia has been viewed ever since Kraepelin (1919/1971). Kraepelin did not define schizophrenia solely on the basis of cross-sectional information; rather, he emphasized its typical age of onset and progressive course. Issues concerning the nature of the onset and course of the disorder have been central to most definitions of schizophrenia since Kraepelin's time. For example, issues concerning onset and course are what psychopathologists have used to distinguish schizophrenia from other psychotic disorders such as "brief reactive psychosis" (e.g., Menuck, Legault, Schmidt, & Remington, 1989).

Some of the factors that contribute to schizophrenia may be categorical in nature, such as whether the individual possesses a particular gene or is exposed to a particular environmental insult. Other factors may be continuous in nature, such as the number of genes that contribute to a particular trait or the degree to which family members are emotionally supportive during difficult life transitions. Some continuous factors will be related to the development and course of schizophrenia in a linear fashion, whereas other continuous factors will be related to schizophrenia in a nonlinear manner. Some, but not all, factors will be independent of one another. Some factors will be correlated because they share common antecedents, whereas other factors will be correlated because one directly or indirectly influences the other. The different factors that influence the development and course of schizophrenia will differ in their specificity to schizophrenia and will also differ in terms of how common they are among schizophrenic and preschizophrenic individuals.

Ultimately, the manner in which schizophrenia will be defined will depend on (a) the factors that contribute to a developmental course characterized by nontransient psychotic symptoms accompanied by a deterioration in social and occupational functioning; (b) the relative contributions of the different factors; (c) the relative independence of the different factors; and (d) the specificity of the factors to a nontransient psychotic disturbance. Perhaps the simplest possible scenario would be the following. It is possible that some individuals currently considered to have schizophrenia share a single gene that contributes quite strongly to the development of the psychotic disorder but does not contribute to other types of disturbance, whereas other individuals currently considered to have schizophrenia do not possess the gene but rather develop the disorder as a result of different factors. If this were the case, what is currently considered to be a single disorder called schizophrenia would most likely be divided into two or more distinct disorders on the basis of their differing etiologies. At that point, the term *schizophrenia* would be used (a) to describe either the psychotic disturbance influenced by the single major gene, or the psychotic disturbance influenced by one of the other sets of factors; or (b) as a broad category encompassing two or more distinct disorders whose developmental outcomes happened to resemble each other. Meehl's (1990) model illustrates the first terminological revision. He proposed that only those individuals currently considered to have schizophrenia who possess the schizogene have "true" schizophrenia, whereas those individuals currently considered to have schizophrenia but who do not possess the schizogene have SHAITU syndrome.

The preceding scenario is a somewhat simple one because one etiological factor was specific to a single type of disturbance and also contributed quite strongly to its development. The picture will be more complicated if there are no factors specific to a single type of disturbance, or if those factors that are specific to a psychotic disturbance play only a relatively small role in its development and course. The following is an example of a complicated potential scenario. It is possible (a) that many factors contribute to the development and course of nontransient psychotic disturbances, (b) that none of the factors by themselves are specific to such disturbances, and (c) that the development and course of such disturbances result from a variety of *combinations* of factors. If this were the case, the definition of schizophrenia would depend largely on the relative independence of the different factors and the number of combinations of factors that contributed specifically to psychotic disturbances. The less independent the different factors or sets of factors, and the larger the number of combinations that contributed to psychotic disturbances, the broader the definition of schizophrenia would have to be. For example, if (a) most, if not all, of the factors were at least modestly correlated and (b) no single combination of factors could account for more than a small proportion of the individuals with psychotic disturbances, then schizophrenia would probably be defined as nontransient psychotic disturbances resulting from the interactions among a wide variety of correlated factors. If, on the other hand, (a) a single combination of factors or a small set of related combinations of factors were specifically associated with psychotic disturbances, (b) this combination or set of related combinations accounted for a modest proportion of the individuals with such problems, and (c) the factors that were constituents of these combinations were only weakly related to the remaining factors that contribute to psychotic disturbance, then what is currently considered to be a single disorder called *schizophrenia* would most likely be divided into two or more distinct disorders on the basis of the different combinations of factors that contribute to the development of psychotic disturbances. At that point, the term *schizophrenia* might be used to refer solely to individuals who develop psychotic disturbance as a result of a single or small set of related combinations of etiological factors. It is important to note that these examples represent only a small fraction of the possible scenarios.

Conclusions

Just as genes will never be sufficient to explain schizophrenia, I predict that no other single aspect of the disorder will ever be sufficient to explain it. For example, it will never be possible to explain schizophrenia solely in terms of neurotransmitters or interpersonal relationships. Rather, what will enable us to explain schizophrenia will be an understanding of how, when, and why the different features described in these chapters emerge, evolve, and are related to one another.

It remains to be seen exactly which factors contribute to the development and course of psychotic disturbances, how they are related to one another, and how specific they are to this particular form of psychopathology. However, the evidence reviewed in these chapters provides an excellent preview of the full picture that is likely to emerge. One can hope that the sort of behavioral high-risk research described in these chapters, coupled with psychopathology research that is not tied exclusively, if at all, to symptom-defined "syndromes" (e.g., Costello, 1992; Iacono & Clementz, 1993), along with research examining ordinary processes and outcomes, will lead to a satisfactory theory-based, comprehensive definition of schizophrenia.

Acknowledgments. I thank Megan McLaughlin and Greg Miller for their helpful comments on an earlier draft of this chapter.

References

Andreasen, N.C., & Olsen, S. (1982). Negative versus positive schizophrenia. *Archives of General Psychiatry, 39*, 789–794.

Barnard, P.J., & Teasdale, J.D. (1991). Interacting cognitive subsystems: A systemic approach to cognitive-affective interaction and change. *Cognition and Emotion, 5*, 1–39.

Berenbaum, H., & McGrew, J. (1993). Familial resemblance of schizotypic traits. *Psychological Medicine, 23*, 327–333.

Berenbaum, H., & Oltmanns, T.F. (1992). Emotional experience and expression in schizophrenia and depression. *Journal of Abnormal Psychology, 101*, 37–44.

Bleuler, M. (1978). *The schizophrenic disorders: Long-term patient and family studies* (S.M. Clemens, Trans.). New Haven: Yale University Press. (Original work published 1972.)

Caramazza, A. (1992). Is cognitive neuropsychology possible? *Journal of Cognitive Neuroscience, 4*, 80–95.

Carpenter, W.T., Heinrichs, D.W., & Wagman, A.M.I. (1988). Deficit and nondeficit forms of schizophrenia: The concept. *American Journal of Psychiatry, 145*, 578–583.

Campos, J.J., Bertenthal, B.I., & Kermoian, R. (1992). Early experience and emotional development: The emergence of wariness of heights. *Psychological Science, 3*, 61–64.

Chapman, L.J., Chapman, J.P., Kwapil, T.R., Eckblad, M., & Zinser, M.C. (1994). Putatively psychosis-prone subjects 10 years later. *Journal of Abnormal Psychology, 103*, 171–183.

Ciompi, L. (1980). Catamnestic long-term study on the course of life and aging of schizophrenics. *Schizophrenia Bulletin, 6*, 606–618.

Clarke, S., Assal, G., & de Tribolet, N. (1993). Left hemisphere strategies in visual recognition, topographical orientation and time planning. *Neuropsychologia, 31*, 99–113.

Costello, C.G. (1992). Research on symptoms versus research on syndromes: Arguments in favour of allocating more research time to the study of symptoms. *British Journal of Psychiatry, 160*, 304–308.

Cowen, E.L., & Work, W.C. (1988). Resilient children, psychological wellness, and primary prevention. *American Journal of Community Psychology, 16*, 591–607.

Crow, T.J. (1985). The two-syndrome concept: Origins and current status. *Schizophrenia Bulletin, 11*, 471–486.

DeLisi, L.E., & Crow, T.J. (1989). Evidence for a sex chromosome locus for schizophrenia. *Schizophrenia Bulletin, 15*, 431–440.

Erlenmeyer-Kimling, L., Cornblatt, B.A., Rock, D., Roberts, S., Bell, M., & West, A. (1993). The New York high-risk project: Anhedonia, attentional deviance, and psychopathology. *Schizophrenia Bulletin, 19*, 141–153.

Finn, S. (1982). Base rates, utilities, and DSM-III: Shortcomings of fixed-rule systems of psychodiagnosis. *Journal of Abnormal Psychology, 91*, 294–302.

Garmezy, N. (1987). Stress, competence, and development: Continuities in the study of schizophrenic adults, children vulnerable to psychopathology, and the search for stress-resistant children. *American Journal of Orthopsychiatry, 57,* 159–174.

Goldstein, J.M., & Tsuang, M.T. (1990). Gender and schizophrenia: An introduction and synthesis of findings. *Schizophrenia Bulletin, 16,* 179–183.

Gottesman, I.I., & Bertelsen, A. (1989). Confirming unexpressed genotypes for schizophrenia: Risks in the offspring of Fischer's Danish identical and fraternal discordant twins. *Archives of General Psychiatry, 46,* 867–872.

Iacono, W.G., & Beiser, M. (1992). Where are the women in first-episode studies of schizophrenia? *Schizophrenia Bulletin, 18,* 471–480.

Iacono, W.G., & Clementz, B.A. (1993). A strategy for elucidating genetic influences on complex psychopathological syndromes (with special reference to ocular motor functioning and schizophrenia). *Progress in Experimental Personality and Psychopathology Research, 16,* 11–65.

Jablensky, A., Sartorius, N., Ernberg, G., Anker, M., Korten, A., Cooper, J.E., Day, R., & Bertelsen, A. (1992). Schizophrenia: Manifestations, incidence and course in different cultures: A World Health Organization ten-country study. *Psychological Medicine,* Monograph Suppl. 20.

Johnson, M.K., & Multhaup, K.S. (1992). Emotion and MEM. In S.A. Christianson (Ed.), *The handbook of emotion and memory: Research and theory* (pp. 33–66). Hillsdale, NJ: Lawrence Erlbaum.

Kershner, J., & Micallef, J. (1992). Consonant-vowel lateralization in dyslexic children: Deficit or compensatory development? *Brain and Language, 43,* 66–82.

Kosslyn, S.M., & Intriligator, J.M. (1992). Is cognitive neuropsychology plausible? The perils of sitting on a one-legged stool. *Journal of Cognitive Neuroscience, 4,* 96–106.

Kraepelin, E. (1971). *Dementia praecox and paraphrenia* (R.M. Barclay, Trans.). New York: Krieger Huntington. (Original work published 1919.)

Leff, J., & Vaughn, C. (1985). *Expressed emotion in families.* New York: Guilford Press.

Lewine, R.R.J. (1981). Sex differences in schizophrenia: Timing or subtypes? *Psychological Bulletin, 90,* 432–444.

Mathews, A., & MacLeod, C. (1994). Cognitive approaches to emotion and emotional disorders. *Annual Review of Psychology, 45,* 25–50.

McGuffin, P., Farmer, A.E., Gottesman, I.I., Murray, R.M., & Reveley, A.M. (1984). Twin concordance for operationally defined schizophrenia: Confirmation of familiality and heritability. *Archives of General Psychiatry, 41,* 541–545.

Meehl, P.E. (1990). Toward an integrated theory of schizotaxia, schizotypy and schizophrenia. *Journal of Personality Disorders, 4,* 1–99.

Meehl, P.E. (1993). The origins of some of my conjectures concerning schizophrenia. *Progress in Experimental Personality and Psychopathology Research, 16,* 1–10.

Meehl, P., & Rosen, A. (1955). Antecedent probability and the efficiency of psychometric signs, patterns, or cutting scores. *Psychological Bulletin, 52,* 194–216.

Menuck, M., Legault, S., Schmidt, P., & Remington, G. (1989). The nosologic status of the remitting atypical psychoses. *Comprehensive Psychiatry, 30,* 53–73.

Rutter, M. (1985). Resilience in the face of adversity: Protective factors and resistance to psychiatric disorder. *British Journal of Psychiatry, 147*, 598–611.

Strauss, J.S., Carpenter, W.T., Jr., & Bartko, J.J. (1974). The diagnosis and understanding of schizophrenia: Part III. Speculations on the processes that underlie schizophrenic signs and symptoms. *Schizophrenia Bulletin, 1*(11), 61–69.

Whiting B.B., & Edwards, C.P. (1988). *Children of different worlds: The formation of social behavior.* Cambridge, MA: Harvard University Press.

Widiger, T.A., Hurt, S.W., Frances, A., Clarkin, J.F., & Gilmore, M. (1984). Diagnostic efficiency and DSM-III. *Archives of General Psychiatry, 41*, 1005–1012.

Part II
Mood Disorders

7
The Behavioral High-Risk Paradigm in the Mood Disorders

DANIEL N. KLEIN AND ROCHELLE L. ANDERSON

High-risk studies have represented a major approach to research in psychopathology since the paradigm was introduced by Pearson and Kley (1957) and Mednick and McNeil (1968). The rationale for the high-risk approach is based on both theoretical and methodological considerations. From a theoretical perspective, to develop comprehensive models of disorder, it is not only necessary to account for the processes underlying psychopathology but also to predict who is at risk for disorder, which factors predispose to illness, and under which circumstances the disorder will be manifested in vulnerable individuals (Depue, Monroe, & Shackman, 1979; Weiner, 1977). From a methodological standpoint, the study of individuals already exhibiting the disorder does not permit researchers to distinguish etiologically significant processes from the consequences of the disorder, which could include such factors as medication, institutionalization, and stigma (Garmezy & Streitman, 1974; Mednick & McNeill, 1968). In this chapter, we (a) outline the strengths and limitations of the three major approaches to identifying persons at risk for psychopathology; (b) discuss the available data on the validity of a continuum model of behavior disturbance as a basis for identifying persons at risk for the major mood disorders; and (c) review the literature on existing behavioral high-risk measures for affective disorders.

Major Approaches to Identifying Persons at Risk for Psychopathology

At present, there are three general approaches to the identification of persons at risk for psychopathology. These approaches use (a) genetic, (b) endophenotypic, and (c) exophenotypic indices of risk (Depue et al., 1981).

The Genetic Paradigm

The most widely used approach is the genetic, or familial, paradigm, which focuses on the first-degree relatives (generally offspring) of individuals with the disorder of interest. The rationale underlying this strategy is strictly empirical: Because most psychiatric disorders run in families, the children of an affected parent are at statistically higher risk for the disorder than their peers in the general population. In disorders with an established genetic component, the logic is even more compelling, because some proportion of the offspring will have inherited a diathesis for the disorder.

The major strength of the genetic paradigm lies in its established validity: Offspring of patients with most forms of psychopathology do, in fact, exhibit an increased risk for the disorder. In addition, samples selected on the basis of a family history of disorder may be more etiologically homogeneous than samples selected without regard to family history (Lewis, Reveley, Reveley, Chitkara, & Murray, 1987; Walker & Shaye, 1982), increasing the chances of detecting links between hypothesized risk factors and specific psychopathological outcomes.

Despite its considerable power, the genetic paradigm suffers from three major limitations. First, not all offspring will have inherited a diathesis, and, at present, it is not possible to distinguish between offspring who are and are not genetically predisposed to the disorder. As a result, a high proportion of false-positives will result in any offspring sample, substantially reducing the power to detect differences between high and low-risk samples (Hanson, Gottesman, & Meehl, 1977; Lewine, Watt, & Grubb, 1984). Second, many, if not most, individuals who develop major psychiatric disorders do not have a parent with the disorder. Hence, this strategy generates an unrepresentative sample of the risk population (Lewine et al., 1984). Third, the genetic high-risk approach is relatively uneconomical, because it requires expensive, time-consuming searches for patients with both the disorder of interest and appropriately aged offspring.

The Endophenotypic Paradigm

The second approach is the endophenotypic high-risk paradigm (Shields, Heston, & Gottesman, 1975), which uses a biological index to identify the risk group. Examples of biological variables used in this approach include monoamine oxidase activity (Buchsbaum, Coursey, & Murphy, 1976), electrodermal activity (Venables, Dalais, Mitchell, Mednick, & Schulsinger, 1983), and smooth-pursuit eye movements (Siever, Coursey, Alterman, Buchsbaum, & Murphy, 1984). The endophenotypic risk paradigm offers several potential advantages. First, because biological processes are more closely related to the action of genes than behavior is, endophenotypic indicators may provide more specific markers of the vulnerable genotype than behavior or symptomatology (Cloninger, Reich, & Yokoyama, 1983; Meehl, 1989; Shields et al., 1975). Second, to the extent that a disorder is biologically homogeneous, the endophenotypic strategy will yield a highly representative sample of the total risk pool. At the same time, to the degree that a disorder is etiologically heterogeneous, this approach has the potential to identify biologically homogeneous subtypes of the illness (Cromwell, 1984; Van Praag, 1990).

The endophenotypic high-risk paradigm has several limitations, however. First, few endophenotypic variables have been identified that satisfy the criteria for potential genetic markers (i.e., they are heritable, evident before the onset of illness, independent of clinical state, and associated with the disorder within families; Iacono, 1982; Reider & Gershon, 1978). Second, screening large groups with endophenotypic measures can be an expensive and cumbersome process. Finally, biological markers may interact with maturational processes in complex ways, obscuring their effects during particular developmental periods (Puig-Antich, 1986).

The Exophenotypic Paradigm

The final approach is the exophenotypic paradigm, which uses behavioral or personality trait indices to identify persons at risk for disorder. There are two general forms of this approach. The first employs relatively broad clusters of behavior, such as poor social adjustment (Cowen, Pederson, Babigian, Izzo, & Trost, 1973) or aggression and withdrawal (Schwartzman, Ledingham, & Serbin, 1985) as risk markers. Projects employing this approach have used teacher and peer ratings to assess risk in elementary schoolchildren (Schwartzman et al., 1985) and clinical assessments of patterns of behavior disturbance in clinically referred adolescents (M.J. Goldstein, Judd, Rodnick, Alkire, & Gould, 1968). In addition to the more general limitations of the exophenotypic approach discussed later, this particular form of the behavioral high-risk paradigm has two problems. First, the conceptual status of the risk indices is unclear. It is uncertain whether they should be viewed as causal factors or as early signs of disorder (Hanson, Gottesman, & Heston, 1990; Lewine et al., 1984; Richters & Weintraub, 1990). Second, because these risk indices are relatively nonspecific, it is not surprising that the high-risk subjects subsequently develop a broad range of psychopathological outcomes (Cowen et al., 1973; Jones, 1974).

The second form of the exophenotypic approach attempts to develop more specific indices of risk by focusing on traits and behaviors with a more specific relationship to the disorder of interest. These behaviors are assumed to be early signs of the disorder, rather than an independent set of predisposing factors. For example, Chapman and Chapman (1985) have developed self-report measures of a variety of schizotypal traits that are hypothesized to reflect a predisposition to psychotic disorders. In the mood disorders, Depue and his colleagues developed the General Behavior Inventory (GBI), which assesses chronic affective conditions throughout the full range of symptom severity (Depue & Klein, 1988; Depue et al., 1981; Depue, Krauss, Spoont, & Arbisi, 1989). Persons with milder cases of disorder identified by the GBI are presumed to be at risk for developing more severe forms of mood disorder.

The major assumption underlying recent exemplars of the behavioral high-risk paradigm is that full syndromal disorders can be conceptualized as the severe end of a continuum, at the other end of which lies a group of phenotypically similar but symptomatically milder conditions. These milder, "subsyndromal" conditions may or may not be continuous with normal personality variation.

In the area of schizophrenia, the continuum model derives from clinical observations that schizotypal features often precede the onset of psychosis (Bleuler, 1950; Kraepelin, 1911/1971; Parnas, Schulsinger, Schulsinger, Mcdnick, & Teasdale, 1982) and are more common in the relatives of schizophrenics than normals (Kendler, 1988). Similarly, in the mood disorders, the continuum model is based on Kraepelin's (1921) observation that premorbid manic (or hyperthymic), depressive (or dysthymic), cyclothymic, and irritable (or mixed state) temperaments were evident in at least 37% of manic–depressive patients and also appeared to be disproportionately common in their relatives.

The major advantage of the exophenotypic approach, particularly when it is based on self-report inventories, is its economy. Large samples from the general population can be screened at minimal cost, and subgroups of interest can be selected for more intensive study. In addition, if most individuals with the disorder exhibit a similar pattern of behavioral precursors, a second advantage would be its generalizability to the larger risk population. On the other hand, if the precursors of a disorder vary, the behavioral risk paradigm may be useful in identifying a subset of the risk population that is missed with the genetic and biological paradigms. Finally, the exophenotypic approach can be used in conjunction with other high-risk paradigms, such as the genetic, to identify subgroups of individuals that may be at particularly increased risk (e.g., cyclothmic offspring of patients with bipolar affective disorder; Klein, Depue, & Slater, 1986; Lewine et al., 1984).

The exophenotypic approach, however, also has several significant limitations. First, behavioral and personality processes operate at a considerable distance from the genetic level. As a result, the ability of exophenotypic indicators to separate risk and nonrisk populations in disorders in which there is a significant genetic component is likely to be modest (Depue et al., 1981; Meehl, 1973). The use of such fallible indicators is particularly likely to yield large numbers of false-positives, which is due to the low base rate of individuals at risk for severe psychopathology (Meehl, 1973).

Second, exophenotypic approaches based on continuum models of psychopathology are susceptible to circularity between the risk definition and the outcome disorder, particularly as the boundaries of the major psychiatric disorders are expanded to include milder forms. For example, before the classification of cyclothymia and dysthymia as forms of mood disorder, the meaning of the proposal that cyclothymics and dysthymics are "at risk" for mood disorders was relatively clear. However, now that cyclothymia and dysthymia are themselves regarded as forms of mood disorder, such a statement has become tautological. This problem can be mitigated to some degree by stipulating that cyclothymics and dysthymics are at risk for "more severe forms of" or "major" mood disorders. Even this claim remains problematic, however, because recent studies have suggested that cyclothymia and dysthymia can be relatively severe conditions in their own right (Klein, Riso, & Anderson, 1993; Klein, Taylor, Dickstein, & Harding, 1988), and the majority of cyclothymics and dysthymics have experienced superimposed major depressive episodes by young adulthood (Klein et al., 1993; Klein, Taylor, Harding, & Dickstein, 1990; Kocsis & Frances, 1987; Lewinsohn, Rohde, Seeley, & Hops, 1991). Hence, to avoid circularity, risk studies using exophenotypic indices need to ensure that the "risk" group has not already experienced the outcome disorder and to define the targeted outcome in such a manner that it represents a clinically significant increase in severity of psychopathology.

Third, even when phenomenological or etiological similarities between two conditions justify the use of a continuum model for nosological purposes, such a model is not necessarily appropriate from a longitudinal-risk perspective. Thus, although there may be evidence that condition X is a milder form of disorder Y from the standpoint of symptomatology and familial aggregation, it cannot be assumed that persons with condition X will necessarily develop disorder Y over time.

Fourth, the behavioral precursors of major psychiatric disorders may be relatively subtle and difficult to assess reliably. For example, the interrater reliability and test–retest stability of diagnoses of cyclothymia and bipolar 2 disorder are quite poor (Andreasen et al., 1981; Prusoff, Merikangas, & Weissman, 1988; Rice et al., 1986). Among the problems in making these diagnoses are raters' difficulty distinguishing hypomania from a normal "good" mood; subjects' problems recalling short-lived, mild periods of elated or depressed mood (Thomas & Diener, 1990); the effects of current mood state on recall (Aneshensel, Estrada, Hansell, & Clark, 1987); and individual, ethnic, and cultural differences in subjects' readiness to perceive or describe milder forms of mood disorder (Egeland, 1983; Gershon et al., 1982).

Finally, the behavioral high-risk approach, in and of itself, provides only limited information regarding the processes underlying risk. However, it may provide a method of selecting at-risk samples that can be studied more extensively in laboratory and naturalistic settings to elucidate these underlying processes (e.g., Depue, Kleiman, Davis, Hutchinson, & Krauss, 1985; Goplerud & Depue, 1985).

In the remainder of this chapter, we review the available data on the validity of the continuum model as a basis for identifying persons at risk for the major mood disorders and summarize the literature on existing behavioral high-risk measures for affective disorders.

The Continuum Model

Cyclothymia and Bipolar 2 Disorder

From the perspective of a continuum model, bipolar 1 disorder, which is characterized by full-blown manic and major depressive episodes, is viewed as the most severe end of a spectrum of bipolar conditions (Akiskal, 1983a; Depue & Monroe, 1978; Fieve & Duner, 1975; Klerman, 1981). Further down the continuum is bipolar 2 disorder, which is characterized by hypomanic and major depressive episodes. Finally, at the mildest end of the spectrum is cyclothymia, which is characterized by frequent hypomanic and depressive episodes that are phenomenologically similar to, but too brief in duration and associated with insufficient impairment to meet criteria for, mania or major depression. As would be expected by a continuum model, these conditions often shade into one another without any clear boundaries. Moreover, the definitions of these conditions vary significantly among diagnostic systems (Dunner, 1983). For example, most individuals meeting criteria for cyclothymia in the revised third edition of the *Diagnostic and Statistical Manual of Mental Disorders* (3rd ed., rev.) (DSM-III-R; American Psychiatric Association, 1987) would be diagnosed as bipolar 2 disorder, rather than cyclothymic personality, in the Research Diagnostic Criteria (RDC; Spitzer, Endicott, & Robins, 1978). Similarly, many patients with RDC bipolar 2 disorder would be diagnosed as cyclothymic in the DSM-III-R.

Current evidence links cyclothymia and bipolar 2 disorder to full-blown bipolar illness with regard to phenomenological, familial, and to a lesser degree, treatment response variables (Akiskal, Djenderedjian, Rosenthal, & Khani, 1977; Depue et al., 1981; Dunner, 1983; Endicott et al., 1985; Klein, Depue, & Slater, 1985; Weissman et al., 1984). Whereas these data are consistent with behavioral high-risk approaches based on a continuum model, the cornerstone of this paradigm rests on demonstrating that individuals with cyclothymia and bipolar 2 disorder are at increased risk for developing more severe forms of mood disorder over time. Unfortunately, follow-up studies of cyclothymia and bipolar 2 disorder are limited.

Two prospective follow-up studies examined bipolar 2 patients' risk of developing full-blown manic episodes. Dunner, Fleiss, and Fieve (1976) report that 3% of 71 bipolar 2 patients developed mania over the course of a 6-month to 5-year follow-up. Coryell et al. (1989) found that 11% of bipolar 2, compared to 60% of bipolar 1 and 4% of nonbipolar major depressive patients, developed manic episodes during a 5-year follow-up. During the follow-up period, however, 41% of the bipolar 2 patients experienced hypomanic episodes. Taken together, these studies suggest that bipolar 2 patients are at somewhat increased risk for developing manic episodes. However, the majority of bipolar 2 patients appear to have a relatively stable course that does not progress to mania.

Three follow-up studies of cyclothymics have been reported. Akiskal et al. (1977) followed 46 cyclothymic outpatients over a 2- to 3-year period. During this period, 7% developed mania and 15% developed hypomania. In addition, 13% of the remaining patients developed major depressive episodes.

Akiskal et al. (1985) identified 68 children and younger siblings of bipolar patients who had been referred for treatment for some form of behavioral disturbance in the previous year. Akiskal and his colleagues attempted to characterize the nature and type of onset of the disturbance and followed these subjects for a mean of 3 years. Of the 10 cyclothymics in the sample, 3 developed mania and 4 developed acute major depressive episodes during the follow-up period.

Finally, Klein and Depue (1984) conducted a 19-month follow-up with 45 cyclothymics and 64 controls identified from a college student population using the GBI. The follow-up evaluation was limited to a brief inventory assessing impairment and treatment. Paralleling the differences originally obtained in the index evaluation (Depue et al., 1981), the cyclothymics exhibited significantly greater impairment than the controls. In addition, two cyclothymics had been hospitalized after attempting suicide during the follow-up interval.

Overall, these data suggest that cyclothymics may be at increased risk for developing more severe forms of mood disorder. The extent of the risk, however, remains unclear. The samples in these studies have been modest in size, the follow-up periods have been short, and the number of cyclothymics who have developed full-blown manic episodes has been small. Although a number of subjects in the Akiskal et al. (1977, 1985) studies developed hypomanic and major depressive episodes, without more detailed characterizations of the samples, it is difficult to be certain that this represented significant increases in severity of psychopathology. Finally, while the Klein and Depue (1984) study indicated that the impairment associated with cyclothymia persists over the course of a short-term follow-up, it did not address whether this level of dysfunction increased or merely remained stable over time.

Dysthymia

Dysthymia, which is characterized by mild, chronic depressive symptomatology, can also be conceptualized within a continuum framework as an attenuated form of major depression. Although currently classified as a nonbipolar mood disorder in the DSM-III-R, some investigators believe that a subgroup of dysthymics may be related to bipolar disorder (Akiskal, 1983c).

Clinical, family, sleep neurophysiological, and pharmacological treatment studies have supported the link between dysthymia and the major mood disorders (Akiskal, 1983c; Klein et al., 1993; Kocsis & Frances, 1987; Kocsis et al., 1988). In addition, several follow-up studies demonstrated that dysthymics are at high risk for developing major depressive episodes (e.g., Gonzales, Lewinsohn, & Clarke, 1985; Keller, Lavori, Endicott, Coryell, & Klerman, 1983). Unfortunately, most of the dysthymics in these studies were in superimposed major depressive episodes at the time of the index assessment, and a high proportion of the others probably had past histories of major depression (Klein et al., 1993). Thus, these studies are not particularly informative with regard to risk for developing a first episode of major depression in dysthymics. Major depression is somewhat less frequent in younger and nonclinical samples of dysthymics (Lewinsohn et al., 1991; Weissman, Leaf, Bruce, & Florio, 1988). Hence, follow-up studies with these populations are more useful. Three studies have recently reported data that are relevant to this issue. In a study discussed previously, Akiskal et al. (1985) identified 12 dysthymics in a sample of referred offspring and younger siblings of bipolar patients. Two developed hypomania, and six developed acute major depressive episodes over the course of a 3-year follow-up.

Weissman and Johnson (1990) recently presented the preliminary results of a 1-year follow-up of dysthymics from the Epidemiological Catchment Area study. Limiting their analyses to dysthymics with no lifetime history of Axis-I disorders, they report that a significantly greater proportion of dysthymics than normal controls developed major depression within the 1-year follow-up interval.

Kovacs and her colleagues have been conducting a long-term follow-up study of 55 dysthymic children selected primarily from a psychiatric outpatient clinic (Kovacs et al., 1984; Kovacs & Gatsonis, 1989). Employing survival analysis and the Kaplan-Meier product-limit method, Kovacs (personal communication, May 1991) found that this group had an 80% cumulative probability of developing major depression within 168 months after the onset of dysthymia. The risk for developing major depression was significantly higher in the dysthymics than in children with adjustment disorders and nondepressed children with externalizing disorders. Interestingly, the dysthymic children were also at increased risk for developing bipolar disorders. Thus, 15% of the dysthymics, compared to 5% of the children with adjustment disorders and 2% of the nondepressed children with externalizing disorders, developed manic or hypomanic episodes during the follow-up period (Kovacs, 1991).

These data indicate that dysthymia is associated with a greatly increased risk for developing major depressive episodes. Importantly, the interval between the onset of dysthymia and the first major depressive episode is often short. In Kovacs et al.'s (1984) sample, for example, the median interval between the onsets of dysthymia and major depression was only 44 months. Hence, only a relatively narrow window may be available for behavioral risk identification before the onset of major depression.

Depressive Personality

The DSM-III-R defines cyclothymia and dysthymia in terms of chronic affective symptomatology (Akiskal, 1989, 1990; W.N. Goldstein & Anthony, 1988; Klein, 1990). This contrasts with the temperament and personality trait-based conceptualizations in the classical descriptive (Kraepelin, 1921; Kretschmer, 1936; Schneider, 1958) and psychoanalytic (Chodoff, 1972; Kernberg, 1988) literatures (see Phillips, Gunderson, Hirschfeld, & Smith, 1990, for a review). In a recent study, we found that although the DSM-III-R criteria for dysthymia and the criteria for depressive personality based on the work of Schneider (1958) and Akiskal (1983c) identified overlapping groups of patients, many dysthymics did not exhibit a depressive personality structure, any many individuals with depressive personality did not experience chronic depressive symptomatology (Klein, 1990). Hence, it appears that the constructs of dysthymia and depressive personality are not isomorphic. There was, however, a strong association between the depressive personality and a family history of bipolar disorder and hospitalization for affective disorder in first-degree relatives (Klein, 1990). This supports the view of depressive personality as part of the affective spectrum (Akiskal, 1989, 1990). Moreover, because it appears to be a milder, less symptomatic condition than dysthymia, it may present an opportunity for earlier risk identification, expanding the window between the appearance of subsyndromal manifestations and the onset of full syndromal disorder.

Hyperthymia

Hypomanic personality, or hyperthymia, is characterized by lifelong hypomanic symptoms, such as an optimistic, exuberant mood, high levels of energy, sociability, and confidence, sensation-seeking behavior, and a decreased need for sleep. Some investigators believe that hypomanic personality should also be viewed as part of the bipolar spectrum (Akhtar, 1988; Akiskal, 1989). This view, however, is controversial (Frances, Jacobsberg, Widiger, Manning, & Fyer, 1988; Morey & Ban, 1988), and hyperthymia may represent the upper end of a dimension of normal personality variation such as extraversion or positive affectivity.

Hyperthymic traits are generally adaptive in modern Western culture and may contribute to high levels of productivity and achievement (Akiskal, 1989). Thus, whereas hyperthymia is not uncommon in the general population (Depue et al., 1989), it is extremely rare in clinical settings (Akiskal et al., 1985; Klein et al., 1989).

Data addressing the relationship between hypomanic personality and the major mood disorders are scarce. Gershon et al. (1982) found similar rates of hypomanic personality in the relatives of bipolar 1, bipolar 2, and normal probands. In contrast, Akiskal (1984) reported that 12 hyperthymics seeking treatment for insomnia at a sleep disorders center tended to have family histories of mood disorder. In addition, they had significantly shorter rapid eye movement (REM) latencies than normal controls, a finding that is also characteristic of patients with major mood disorders. No data are available on hyperthymics' risk for developing manic and major depressive episodes over time.

Irritable Temperament

Kraepelin's (1921) fourth type of affective temperament, the irritable temperament, can be conceptualized as a subsyndromal mixed state (Akiskal, 1989). Many individuals with this temperamental style are probably classified as having cyclothymia, as several studies have described a subgroup of cyclothymics with predominantly irritable mood (Akiskal et al., 1977; Depue et al., 1981). Akiskal (1989) recently proposed criteria for the irritable temperament, but data regarding the validity of this subtype are not yet available.

Subthreshold Depressive Symptomatology

The continuum model consists of a series of affective syndromes ordered along a complex dimension based on a combination of severity and course. Alternatively, one could postulate a continuum model based simply on severity of cross-sectional symptomatology. In this approach, subthreshold levels of depressive symptoms, regardless of syndromal presentation, could be viewed as potential risk markers.

Several studies have provided evidence that persons with subthreshold levels of depressive symptoms, but with no lifetime history of mood disorder, are at increased risk for developing an affective condition. Weissman, Myers, Thompson, and Belanger (1986) administered a brief depression inventory to a large community sample in 1967 and 1969 and conducted structured diagnostic interviews with these subjects in 1975. Using lifetime diagnoses derived from the 1975 interview, they examined the relationship between inventory scores in 1967 and 1969 and the onset of a first lifetime episode of major depression between 1969 and 1975. Subjects with elevated symptom scores in both 1967 and 1969, or in only 1969, exhibited a three- to four-and-a-half-fold increase in the incidence of major depression compared to subjects with low depression scores at both assessments. Subjects with high depression scores in 1967, but not in 1969, did not exhibit an increased risk for major depression.

Brown, Bifulco, Harris, and Bridge (1986) conducted structured diagnostic interviews with a large community sample of women at two time points separated by a 1-year interval. They found that a significantly greater proportion of women with subclinical symptoms of depression and anxiety in the inital interview developed depressive episodes during the follow-up period than normal women. Interestingly, the increased risk for depressive episodes was limited to women whose subclinical symptoms had persisted for at least a year before the initial assessment. Brown et al. (1986) did not exclude subjects with a past history of depression from their analysis; hence, it is unclear how many of these cases were first episodes. They did report that previous psychiatric history was not associated with the onset of depression during follow-up. Their measure of previous history, however, consisted of prior psychiatric treatment. Hence, it is possible that some of Brown and co-workers' subclinical subjects may have been in partial remission from a previous untreated major depressive episode.

Finally, Murphy et al. (1989) examined the incidence of depression over a 16-year period in a community sample of individuals with no lifetime history of mood disorder at the time of the inital assessment. Although individuals with subthreshold depressive symptoms at baseline were at increased risk for developing a depressive disorder during the follow-up period, the results failed to reach statistical significance.

These studies are limited by the modest reliability of retrospective reports of depression in community samples (Bromet, Dunn, Connell, Dew & Schulberg, 1986). Moreover, misclassifications are most common in subjects closest to case thresholds (Robins, 1985). Hence, it is likely that some "subthreshold" subjects were misdiagnosed and had actually experienced full syndromal episodes before the initial assessment. Overall, however, these data suggest that individuals with subclinical levels of depressive symptoms are at increased risk for developing major affective episodes. The risk appears to be highest in those with chronic conditions (Brown et al., 1986; Weissman et al., 1986), many of whom are probably dysthymic. This suggests that rather than relying solely on cross-sectional symptomatology, one should consider the longitudinal course of mild mood disturbances in the identification of at-risk subjects.

Measures

As noted earlier, one of the major attractions of the behavioral high-risk paradigm is its efficiency and economy, particularly when the behaviors of interest can be assessed using self-report inventories. Two self-report inventories have been specifically developed within a behavioral high-risk framework for the major mood disorders: the GBI (Depue et al., 1981) and the Hypomanic Personality Scale (Eckblad & Chapman, 1986). In the final section of this chapter, we shall review the literature on the reliability and validity of these measures.

The General Behavior Inventory

The GBI was developed as a first-stage screening inventory for chronic and recurrent mood disorders (Depue et al., 1989; Depue & Klein, 1988). It was designed to identify the full range of affective spectrum cases. Hence, in nonclinical populations, the GBI may identify a group of "subsyndromal" cases who are at risk for developing more severe forms of mood disorder.

The GBI was originally designed to identify bipolar spectrum conditions, such as bipolar 2 disorder and cyclothymia (Depue et al., 1981). It was recently revised, however, to identify unipolar cases as well (Depue et al., 1989). The current version of the GBI consists of 73 items assessing the full range of hypomanic and depressive symptoms. The items are worded to reflect relatively low, but clinically significant, levels of intensity and duration. Items are answered on a 4-point frequency scale. Symptoms that are experienced "often" or "very often or almost constantly" each contribute 1 point to the total score.

The current version of the GBI has separate scales for depressive and hypomanic/biphasic (swings between hypomanic and depressive behaviors) items. Together, these scales delineate four affective patterns: bipolar (elevated GBI-H hypomania and GBI-D dysthymia scores); unipolar (elevated GBI-D only); hyperthymia (elevated GBI-H only); and noncase (both GBI-H and GBI-D scores in the noncase range).

Both the original and revised versions of the inventory exhibit high levels of internal consistency (Depue et al., 1981, 1989). In addition, both versions of the GBI have demonstrated good test–retest stability over 3- to 4-month intervals (Depue et al., 1981, 1989).

Like most self-report trait measures, the GBI appears to be somewhat influenced by subjects' clinical states (Hirschfeld et al., 1983). Dysthymics with superimposed major depressive episodes obtain higher scores on the GBI-D scale than dysthymics without major depression (Klein et al., 1989). Goodnick, Fieve, Peselow, Schlegel, and Filippi (1986) reported that patients in a major depressive episode exhibited significantly higher GBI scores than euthymic major depressives. Moreover, the GBI scores of patients with current major depression decreased significantly after treatment. Despite evidence of state effects, however, Goodnick et al. (1986) found that even when euthymic, patients with major affective disorders obtained significantly higher GBI scores than normal controls.

The GBI's case identification properties have been examined in six samples, including two large nonclinicial groups (Depue et al., 1981, 1989), three outpatient samples (Depue et al., 1981; Klein et al., 1988; Mallon, Klein, Bornstein, & Slater, 1986), and a group of offspring of patients with bipolar 1 disorder and nonaffective psychiatric conditions (Klein et al., 1986). In each of these studies, subjects completing the GBI were administered blind, structured diagnostic interviews, and diagnoses were derived blindly using specific diagnostic criteria. In all six studies, positive and negative predictive power and overall diagnostic efficiency for chronic and recurrent affective conditions were good to excellent. Moreover, the GBI distinguished between unipolar and bipolar affective cases with a high degree of accuracy. Not surprisingly, however, the GBI performed somewhat better in the nonclinical samples, where its major task was to discriminate affective cases from normals rather than from other psychiatric disorders.

The concurrent validity of the GBI has also been examined in several other ways. Blind diagnoses based on informants' reports were highly concordant with GBI classifications (Depue et al., 1981). In addition, GBI cases exhibited significantly greater between- and within-day variability in their moods and behavior than noncases on daily ratings over a 30-day period (Depue et al., 1981). Finally, GBI scores were significantly correlated with clinicians' ratings of subclinical affective symptoms in patients receiving lithium prophylaxis (Goodnick et al., 1986).

The discriminant validity of the GBI was addressed in the three outpatient studies cited previously (Depue et al., 1981; Klein et al., 1989; Mallon et al., 1986). Although the number and severity of nonaffective disorders in these studies were limited, the GBI consistently discriminated affective from nonaffective conditions with a high degree of accuracy. In addition, the GBI was reasonably effective in distinguishing chronic and recurrent mood disorders from nonrecurrent and transient affective conditions.

A number of findings have been reported that address the nomological validity of the GBI. The clinical characteristics of GBI-identified nonpatient cyclothymics were very similar to those reported for cyclothymic outpatients (Akiskal et al., 1977; Depue et al., 1981). GBI-identified cyclothymics reported significantly higher rates of mood disorders in their firs-degree relatives than noncases (Depue et al., 1981). In two studies, the offspring of bipolar patients obtained significantly higher GBI scores than the offspring of psychiatric and normal controls (Klein et al., 1986; Nurnberger et al., 1988). Compared to noncases, GBI-identified cyclothymics exhibited significant neuroendocrine dysregulation similar to that found in major depressives (Depue et al., 1985). GBI-identified cyclothymics and dysthymics took significantly longer than noncases to return to baseline mood levels following naturally occurring stressors (Goplerud & Depue, 1985). Finally, several studies have found that GBI cases and noncases differ significantly on a variety of psychophysiological tasks and measures (Lenhart, 1985; Lenhart & Katkin, 1986; Miller & Yee, 1994; Yee, Deldin, & Miller, 1992; Yee & Miller, 1988, 1994).

Only limited data on the predictive validity of the GBI are available. As discussed earlier, Klein and Depue (1984) found that GBI cases continued to exhibit significantly greater impairment than controls over the course of a 19-month follow-up study. In addition, Klein et al. (1989) reported that among outpatients with unipolar affective conditions, GBI-H scores were significantly elevated in those who subsequently developed RDC probable or definite hypomanic episodes over the course of a 6-month follow-up. Although it is possible that some of these patients were initially misdiagnosed, this is rendered less likely by the fact that only 15% of these subjects had scored in the bipolar range on the GBI at the index assessment, and, as a group, their mean GBI-H score was substantially below the cutoff for bipolar conditions. Unfortunately, long-term follow-up studies have not been conducted to address the critical question of whether GBI-identified cyclothymics and dysthymics are at increased risk for developing first episodes of mania and major depression.

Overall, the GBI appears to have accrued substantial concurrent, discriminant, and nomological validity. However, data on predictive validity are still limited. The measure appears to have a number of potential uses. First, it is an efficient means of screening nonclinical samples to select subjects with and without chronic and recurrent unipolar and bipolar affective conditions. Second, it may provide a means of identifying persons at risk for more severe forms of mood disorder. The joint use of the GBI and genetic high-risk paradigm to identify offspring of patients with major mood disorders who may be at particularly high risk is also a promising strategy that may reduce the high number of false-positives generated by both approaches (Klein et al., 1986; Lewine et al., 1984; Nurnberger et al., 1988). Third, given the poor reliability of affective spectrum diagnoses (Andreasen et al., 1981; Rice et al., 1986), and the likely variation across sites in the thresholds used to diagnose these conditions (White, Lewy, Sack, Blood & Wesche, 1990), the GBI could be used to provide convergent validation for clinical diagnoses and as a standard with which to compare samples across different studies and research centers. Finally, although further work is necessary, the GBI may be used as a continuous index of affective traits, rather than a means of identifying cases of discrete disorders. This has the advantage of offering investigators greater power and flexibility in data analyses.

The GBI also has several significant limitations. First, it is a first-stage screening inventory, rather than a diagnostic instrument. Hence, potential cases and noncases identified by the inventory should be confirmed with diagnostic interviews. Second, despite its selectivity for chronic and recurrent affective conditions, and its ability to distinguish between unipolar and bipolar disorders, the GBI case groups are still relatively heterogeneous, at least from a DSM-III-R standpoint. For example, unipolar cases include dysthymics, chronic and recurrent major depressives, double depressives, and probably some recurrent brief depressives, as described by Angst, Merikangas, Scheidegger, and Wicki (1990). Third, even in nonclinical samples, it cannot be assumed that most cases have not been treated or experienced major affective episodes. For example, Depue et al. (1989) reported that over 50% of their GBI cases, selected from a college student population, had a history of psychiatric treatment or counseling. Although this may not be a problem for investigators studying chronicity and relapse, it is problematic for high-risk studies that are interested in the onset of disorder in previously well individuals. Finally, because of the complex distinctions that it attempts to make, the GBI is longer and written in a more complex manner than most other depression inventories (Berndt, Schwartz, & Kaiser, 1983). As a result, it may be less useful in situations in which time is extremely limited, and in samples of less educated and lower socioeconomic status individuals.

The Hypomanic Personality Scale

Eckblad and Chapman's (1986) Hypomanic Personality Scale was also developed to identify persons at risk for bipolar disorder. It is a 48-item true–false questionnaire that includes a variety of hypomanic personality traits and symptoms. Unlike the GBI, it does not include depressive symptomatology. As a result, the scale does not attempt to distinguish between subjects with bipolar spectrum disorders (e.g., bipolar 2 disorder and cyclothymia) and hyperthymic personality. In fact, the majority of high-scoring subjects report having experienced significant depressive episodes (Eckblad & Chapman, 1986).

In a large sample of university students, the Hypomanic Personality Scale exhibited high internal consistency and test–retest stability over a 15-week interval (Eckblad & Chapman, 1986). In addition, the scale was uncorrelated with measures of social desirability and acquiesence. To assess concurrent validity, Eckblad and Chapman (1986) conducted blind structured interviews with groups of high- and low-scoring students. Compared to the low scorers, a significantly greater proportion of high scorers met RDC (Spitzer et al., 1978) for hypomanic episodes and affective personality disorders and reported week-long depressive episodes. In addition, the high scorers reported significantly higher levels of a number of hypomanic personality traits, substance abuse, and psychotic-like symptoms and exhibited a significantly lower level of social adjustment than the low-scoring group.

The Hypomanic Personality Scale was only moderately correlated with the original version of the GBI, and 25% of the high scorers scored above the case cutoff in Depue et al. (1981). This suggests that the two inventories may identify somewhat different, although overlapping, groups. It would be of interest to determine whether this also holds for the revised version of the GBI. If so, it would be important to directly compare the clinical characteristics of samples selected by each of these measures.

Apart from Eckblad and Chapman's (1986) initial validation study, further studies of the Hypomanic Personality Scale have not been published. Hence, data on the measure's discriminant and predictive validity are not available.

Conclusions

A decade after its initial formulation (Akiskal, 1983b; Depue et al., 1981), the behavioral high-risk approach to the major mood disorders remains a promising strategy for risk identification. To fulfill its initial promise, however, a number of critical tasks remain.

First, it is necessary to define the affective spectrum disorders more clearly and to delineate their boundaries more precisely. In particular, it is important to develop more reliable criteria and/or diagnostic interviews for the diagnosis of hypomania and to clarify the distinctions between cyclothymia and bipolar 2 disorder and between dysthymia and major depression. In addition, it is necessary to determine whether hypomanic and depressive personality are useful and valid constructs, and if so, whether they should be conceptualized as affective spectrum conditions.

Second, investigators must become more cognizant of the potential for circularity in the behavioral high-risk approach. In designing high-risk studies, target outcomes must be explicitly indicated, and samples of at-risk individuals must be carefully evaluated to screen out persons who have already experienced the outcome disorder.

Finally, and most importantly, prospective studies are necessary to test the core hypothesis that persons with affective spectrum conditions are indeed at risk for developing more severe forms of mood disorder and to evaluate the validity of measures purporting to identify high-risk groups.

References

Akhtar, S. (1988). Hypomanic personality disorder. *Integrative Psychiatry, 6,* 37–46.

Akiskal, H.S. (1983a). The bipolar spectrum: New concepts in classification and diagnosis. In L. Grinspoon (Ed.), *Psychiatry update: The American Psychiatric Association annual reivew* (Vol. 2, pp. 271–292). Washington, DC: American Psychiatric Press.

Akiskal, H.S. (1983b). Dysthymic and cyclothymic disorders: A paradigm for high-risk research in psychiatry. In J.M. Davis & J.W. Maas (Eds.), *The affective disorders* (pp. 211–232). Washington, DC: American Psychiatric Press.

Akiskal, H.S. (1983c). Dysthymic disorder: Psychopathology of proposed chronic depressive subtypes. *American Journal of Psychiatry, 140,* 11–20.

Akiskal, H.S. (1984). Characterologica manifestations of affective disorders: Toward a new conceptualization. *Integrative Psychiatry, 2,* 83–88.

Akiskal, H.S. (1989). Validating affective personality types. In L. Robins & J.E. Barrett (Eds.), *The validity of psychiatric diagnosis* (pp. 217–227). New York: Raven Press.

Akiskal, H.S. (1990). Towards a definition of dysthymia: Boundaries with personality and mood disorders. In S.W. Burton & H.S. Akiskal (Eds.), *Dysthymic disorder* (pp. 1–12). London: Gaskell.

Akiskal, H.S., Djenderedjian, A.H., Rosenthal, R.H., & Khani, M.K. (1977). Cyclothymic disorder: Validating criteria for inclusion in the bipolar affective group. *American Journal of Psychiatry, 134,* 1227–1233.

Akiskal, H.S., Downs, J., Jordan, P., Watson, S., Daugherty, D., & Pruitt, D.B. (1985). The prospective course of affective disturbances in the referred children and younger sibs of manic-depressives. *Archives of General Psychiatry, 42,* 996–1003.

American Psychiatric Association. (1987). *Diagnostic and statistical manual of mental disorders* (3rd ed., rev.). Washington, DC: Author.

Andreasen, N.C., Grove, W.M., Shapiro, R.W., Keller, M.B., Hirschfeld, R.M.A., & MacDonald-Scott, P. (1981). Reliability of lifetime diagnoses: A multicenter collaborative perspective. *Archives of General Psychiatry, 38,* 400–405.

Aneshensel, C.S., Estrada, A.L., Hansell, M.J., & Clark, V.A. (1987). Social psychological aspects of reporting behavior: Lifetime depressive episode reports. *Journal of Health and Social Behavior, 28,* 232–246.

Angst, J., Merikangas, K., Scheidegger, P., & Wicki, W. (1990). Recurrent brief depression: A new subtype of affective disorder. *Jounal of Affective Disorders, 19,* 87–98.

Berndt, D.J., Schwartz, S., & Kaiser, C.F. (1983). Readability of self-report depression inventories. *Journal of Consulting and Clinical Psychology, 51,* 627–628.

Bleuler, E. (1950). *Dementia praecox or the group of schizophrenias.* New York: International Universities Press. (Original work published, 1911)

Bromet, E.J., Dunn, L.O., Connell, M.M., Dew, M.A., & Schulberg, H.C. (1986). Long-term reliability of diagnosing lifetime major depression in a community sample. *Archives of General Psychiatry, 43,* 435–440.

Brown, G.W., Bifulco, A., Harris, T., & Bridge, L. (1986). Life stress, chronic subclinical symptoms and vulnerability to clinical depression. *Journal of Affective Disorders, 11,* 1–19.

Buchsbaum, M.S., Coursey, R.D., & Murphy, D.L. (1976). The biochemical high-risk paradigm: Behavioral and familial correlates of low platelet monoamine oxidase activity. *Science, 194,* 339–341.

Chapman, L.J., & Chapman, J.P. (1985). Psychosis proneness. In M. Alpert (Ed.), *Controversies in schizophrenia* (pp. 157–174). New York: Guilford Press.

Chodoff, P. (1972). The depressive personality: A critical review. *Archives of General Psychiatry, 27,* 666–667.

Cloninger, C.R., Reich, T., & Yokoyama, S. (1983). Genetic diversity, genome organization, and investigation of the etiology of psychiatric diseases. *Psychiatric Developments, 3,* 225–246.

Coryell, W., Keller, M., Endicott, J., Andreasen, N., Clayton, P., & Hirschfeld, R. (1989). Bipolar II illness: Course and outcome over a five-year period. *Psychological Medicine, 19,* 129–141.

Cowen, E.L., Pederson, A., Babigian, H., Izzo, L.D., & Trost, M.A. (1973). Long-term follow-up of early detected vulnerable children. *Journal of Consulting and Clinical Psychology, 41,* 438–446.

Cromwell, R.L. (1984). Preemptive thinking and schizophrenia research. In W.D. Spaulding & J.K. Cole (Eds.), *Theories of schizophrenia and psychosis* (pp. 1–46). Lincoln: University of Nebraska Press.

Depue, R.A., Kleiman, R.M., Davis, P., Hutchinson, M., & Krauss, S. (1985). The behavioral high risk paradigm and bipolar affective disorder: VIII. Serum free cortisol in nonpatient cyclothymic subjects selected by the General Behavior Inventory. *American Journal of Psychiatry, 142,* 175–181.

Depue, R.A., & Klein, D.N. (1988). Identification of unipolar and bipolar affective conditions in nonclinical and clinical populations by the General Behavior Inventory. In D.L. Dunner, E.S. Gershon, & J.E. Barrett (Eds.), *Relatives at risk for mental disorders* (pp. 179–202). New York: Raven Press.

Depue, R.A., Krauss, S., Spoont, M.R., & Arbisi, P. (1989). General Behavior Inventory identification of unipolar and bipolar affective conditions in a non-clinical university population. *Journal of Abnormal Psychology, 98,* 117–126.

Depue, R.A., & Monroe, S.M. (1978). The unipolar–bipolar distinction in the depressive disorders. *Psychological Bulletin, 85,* 1001–1030.

Depue, R.A., Monroe, S.M., & Shackman, S.L. (1979). The psychobiology of human disease: Implications for conceptualizing the depressive disorders. In R.A. Depue (Ed.), *The psychobiology of the depressive disorders: Implications for the effects of stress* (pp. 3–20). New York: Academic Press.

Depue, R.A., Slater, J.F., Wolfstetter-Kausch, H., Klein, D., Goplerud, E., & Farr, D. (1981). A behavioral paradigm for identifying persons at risk for bipolar depressive disorders: A conceptual framework and five validations studies (Monograph). *Journal of Abnormal Psychology, 90,* 381–437.

Dunner, D.L. (1983). Subtypes of bipolar affective disorder with particular regard to bipolar II. *Psychiatric Developments, 1,* 75–86.

Dunner, D.L., Fleiss, J.L., & Fieve, R.R. (1976). The course of development of mania in patients with recurrent depression. *American Journal of Psychiatry, 133,* 905–908.

Eckblad, M., & Chapman, L.J. (1986). Development and validation of a scale for hypomanic personality. *Journal of Abnormal Psychology, 95,* 214–222.

Egeland, J.A. (1983). Bipolarity: The iceberg of affective disorders? *Comprehensive Psychiatry, 24,* 337–344.

Endicott, J., Nee, J., Andreasen, N., Clayton, P., Keller, M., & Coryell, W. (1985). Bipolar II: Combine or keep separate? *Journal of Affective Disorders, 8,* 17–28.

Fieve, R.R., & Dunner, D.L. (1975). Unipolar and bipolar affective states. In F.F. Flach & S.C. Draghi (Eds.), *The nature and treatment of depression* (pp. 145–166). New York: John Wiley & Sons.

Frances, A., Jacobsberg, L., Widiger, T., Manning, D., & Fyer, M. (1988). Commentary. *Integrative Psychiatry, 6*, 47–48.

Garmezy, N., & Streitman, S. (1974). Children at risk: The search for the antecedents of schizophrenia: Part I. Conceptual models and research methods. *Schizophrenia Bulletin, 1*, 14–90.

Gershon, E.S., Hamovit, J., Guroff, J.J., Dibble, E., Leckman, J.F., Sceery, W., Targum, S.D., Nurnberger, J.I., Goldin, L.R., & Bunney, W.E. (1982). A family study of schizoaffective, bipolar I, bipolar II, unipolar, and normal control probands. *Archives of General Psychiatry, 39*, 1157–1167.

Goldstein, M.J., Judd, L.L., Rodnick, E.H., Alkire, A.A., & Gould, E. (1968). A method for the Study of social influence and coping patterns in the families of disturbed adolescents. *Journal of Nervous and Mental Disease, 147*, 233–251.

Goldstein, W.N., & Anthony, R.N. (1988). The diagnosis of depression and the DSMs. *American Journal of Psychotherapy, 42*, 180–196.

Gonzales, L.R., Lewinsohn, P.M., & Clarke, G.N. (1985). Longitudinal follow-up of unipolar depressives: An investigation of predictors of relapse. *Journal of Consulting and Clinical Psychology, 53*, 461–469.

Goodnick, P.J., Fieve, R.R., Peselow, E., Schlegel, A., & Filippi, A. (1986). General Behavior Inventory: Measurement of subclinical changes during depression and lithium prophylaxis. *Acta Psychiatrica Scandinavica, 73*, 529–532.

Goplerud, E., & Depue R.A. (1985). Behavioral response to naturally occurring stress in cyclothymia and dysthymia. *Journal of Abnormal Psychology, 94*, 128–139.

Hanson, D.R., Gottesman, I.I., & Meehl, P.E. (1977). Genetic theories and the validation of psychiatric diagnoses: Implications for the study of children of schizophrenics. *Journal of Abnormal Psychology, 86*, 575–588.

Hirschfeld, R.M.A., Klerman, G.L., Clayton, P.J., Keller, M.B., McDonald-Scott, P.L., & Larkin, B.H. (1983). Assessing personality: Effects of the depressed state on trait measurement. *American Journal of Psychiatry, 140*, 695–699.

Iacono, W. (1982). The genetics of psychopathology as a tool for understanding the brain: The search for a genetic marker of schizophrenia. In L. Leiblich (Ed.), *Genetics of the brain* (pp. 62–91). Amsterdam: Elsevier.

Jones, F.H. (1974). A four-year follow-up of vulnerable adolescents: The prediction of outcomes in early adulthood from measures of social competence, coping style, and overall level of psychopathology. *Journal of Nervous and Mental Disease, 159*, 20–39.

Keller, M.B., Lavori, P.W., Endicott, J., Coryell, W., & Klerman, G.L. (1983). "Double depression": Two-year follow-up. *American Journal of Psychiatry, 140*, 689–694.

Kendler, K.S. (1988). Familial aggregation of schizophrenia and schizophrenia spectrum disorders: Evaluation of conflicting results. *Archives of General Psychiatry, 45*, 377–383.

Kernberg, O.F. (1988). Clinical dimensions of masochism. *Journal of the American Psychoanalytic Association, 36*, 1005–1029.

Klein, D.N. (1990). Depressive personality: Reliability, validity, and relation to dysthymia. *Journal of Abnormal Psychology, 99*, 412–421.

Klein, D.N., & Depue, R.A. (1984). Continued impairment in persons at risk for bipolar affective disorder: Results of a 19-month follow-up study. *Journal of Abnormal Psychology, 93*, 345–347.

Klein, D.N., Depue, R.A., & Slater, J.F. (1985). Cyclothymia in the adolescent offspring of parents with bipolar affective disorder. *Journal of Abnormal Psychology, 94*, 115–127.

Klein, D.N., Depue, R.A., & Slater, J.F. (1986). Inventory identification of cyclothymia: IX. Validation in offspring of bipolar I patients. *Archives of General Psychiatry, 43*, 441–445.

Klein, D.N., Dickstein, S., Taylor, E.T., & Harding, K. (1989). Identifying chronic affective disorders in outpatients: Validation of the General Behavior Inventory. *Journal of Consulting and Clinical Psychology, 57*, 106–111.

Klein, D.N., Riso, L.P., & Anderson, R.L. (1993). DSM-III-R dysthymia: Antecedents and underlying assumptions. In L.J. Chapman, J.P. Chapman, & D.C. Fowles (Eds.), *Progress in experimental personality and psychopathology research* (Vol. 16). (pp. 222–253). New York: Springer-Verlag.

Klein, D.N., Taylor, E.T., Dickstein, S., & Harding, K. (1988). Primary early-onset dysthymia: Comparisons with primary nonbipolar nonchronic major depression on demographic, clinical, familial, personality, and socioenvironmental characteristics and short-term outcome. *Journal of Abnormal Psychology, 97*, 387–398.

Klein, D.N., Taylor, E.B., Harding, K., & Dickstein, S. (1990). The uniplar–bipolar distinction in the characterological mood disorders. *Journal of Nervous and Mental Disease, 178*, 318–323.

Klerman, G.L. (1981). The spectrum of mania. *Comprehensive Psychiatry, 22*, 11–20.

Kocsis, J.H., & Frances, A.J. (1987). A critical discussion of DSM-III dysthymic disorder. *American Journal of Psychiatry, 144*, 1534–1542.

Kocsis, J.H., Frances, A.J., Voss, C., Mann, J.J., Mason, B.J., & Sweeney, J. (1988). Imipramine treatment for chronic depression. *Archives of General Psychiatry, 45*, 253–257.

Kovacs, M. (1991). *The course of depressive disorders in childhood*. Grand Rounds, Department of Psychiatry, State University of New York at Stony Brook, Stony Brook, New York, March.

Kovacs, M., & Gatsonis, C. (1989). Stability and change in childhood-onset depressive disorders: Longitudinal course as a diagnostic validator. In L.N. Robins & J.E. Barrett (Eds.), *The validity of psychiatric diagnosis* (pp. 57–75). New York: Raven Press.

Kovacs, M., Feinberg, T.L., Crouse-Novak, M., Paulauskas, S.L., Pollock, M., & Finkelstein, R. (1984). Depressive disorders in childhood: II. A longitudinal study of the risk for a subsequent major depression. *Archives of General Psychiatry, 41*, 643–649.

Kraepelin, E. (1921). *Manic-depressive insanity and paranoia*. Edinburgh: E & S Livingstone.

Kraepelin, E. (1971). *Dementia praecox and paraphrenia*. Huntington, NY: Robert E. Krieger. (Original work published 1911)

Kretschmer, E. (1936). *Physique and character* (2nd ed.). London: Routledge.

Lenhart, R.E. (1985). Lowered skin conductance in a subsyndromal high-risk depressive sample: Amplitudes versus tonic levels. *Journal of Abnormal Psychology, 94*, 649–652.

Lenhart, R.E., & Katkin, E.S. (1986). Psychophysiological evidence for cerebral laterality effects in a high-risk sample of students with subsyndromal bipolar depressive disorder. *American Journal of Psychiatry, 143*, 602–607.

Lewine, R.R.J., Watt, N.F., & Grubb, T.W. (1984). High-risk-for-schizophrenia research: Sampling bias and its implications. In N.F. Watt, E.J. Anthony, L.C. Wynne, & J.E. Rolf (Eds.), *Children at risk for schizophrenia: A longitudinal perspective* (pp. 557–564). Cambridge: Cambidge University Press.

Lewinsohn, P.M., Rohde, P., Seeley, J.R., & Hops, H. (1991). Comorbidity of unipolar depression: I. Major depression with dysthymia. *Journal of Abnormal Psychology, 100*, 205–213.

Lewis, S.W., Reveley, A.M., Chitkara, B., & Murray, R.M. (1987). The familial/sporadic distinction as a strategy in schizophrenia research. *British Journal of Psychiatry, 151*, 306–313.

Mallon, J.C., Klein, D.N., Bornstein, R.F., & Slater, J.F. (1986). Discriminant validity of the General Behavior Inventory: An outpatient study. *Journal of Personality Assessment, 50*, 568–577.

Mednick, S.A., & McNeill, T. (1968). Current methodology in research on the etiology of schizophrenia. *Psychological Bulletin, 70*, 681–693.

Meehl, P.E. (1989). Schizotaxia revisited. *Archives of General Psychiatry, 46*, 935–944.

Miller, G.A., & Yee, C.M. (1994). Risk for severe Psychopathology: Psychometric screening and psychophysiological assessment. In J.R. Jennings, P.K. Ackles, & M.G.H. Coles (Eds.), *Advances in psychophysiology* (Vol. 5). (pp. 1–54). London: Jessica Kingsley.

Morey, L.C., & Ban, T.A. (1988). Commentary. *Integrative Psychiatry, 6*, 50–52.

Murphy, J.M. Sobol, A.M., Olivier, D.C., Monson, R.R., Leighton, A.H., & Pratt, L.A. (1989). Prodromes of depression and anxiety: The Stirling County study. *British Journal of Psychiatry, 155*, 490–495.

Nurnberger, J.I., Hamovit, J., Hibbs, E.D., Pellegrini, D., Guroff, J.J., Maxwell, M.E., Smith, A., & Gershon, E.S. (1988). A high-risk study of primary affective disorder: Selection of subjects, initial assessment, and 1- to 2-year follow-up. In D.L. Dunner, E.S.. Gershon, & J.E. Barrett (Eds.), *Relatives at risk for mental disorder* (pp. 161–177). New York: Raven Press.

Parnas, J., Schulsinger, F., Schulsinger, H., Mednick, S.A., & Teasdale, T.W. (1982). Behavioral precursors of the schizophrenia spectrum: A prospective study. *Archives of General Psychiatry, 39*, 658–664.

Pearson, J.S., & Kley, I.B. (1957). On the application of genetic expectancies as age-specific base rates in the study of human behavior disorders. *Psychological Bulletin, 54*, 406–420.

Phillips, K.A., Gunderson, J.G., Hirschfeld, R.M.A., & Smith, L.E. (1990). A review of the depressive personality. *American Journal of Psychiatry, 147*, 830–837.

Prusoff, B.A., Merikangas, K.R. & Weissman, M.M. (1988). Lifetime prevalence and age of onset of psychiatric disorders: Recall 4 years later. *Journal of Psychiatric Research, 22*, 107–117.

Puig-Antich, J. (1986). Psychobiological markers: Effects of age and puberty. In M. Rutter, C.E. Izard, & P.B. Read (Eds.), *Depression in young people: Clinical and developmental perspectives* (pp. 341-381). New York: Guilford Press.

Reider, R.O., & Gershon, E.S. (1978). Genetic strategies in biological psychiatry. *Archives of General Psychiatry, 35*, 866-873.

Rice, J.P., McDonald-Scott, P., Endicott, J., Coryell, W., W.M., Keller, M.B., & Altis, D. (1986). The stability of diagnosis with an application to bipolar II disorder. *Psychiatry Research, 19*, 286-296.

Richters, J., & Weintraub, S. (1990). Beyond diathesis: Toward an understanding of high-risk environments. In J. Rolf, A.S. Masten, D. Cicchetti, K.H. Neuchterlein, & S. Weintraub (Eds.), *Risk and protective factors in the development of psychopathology* (pp. 67-96). Cambridge, England: Cambridge University Press.

Robins, L.N. (1985). Epidemiology: Reflections on testing the validity of psychiatric interviews. *Archives of General Psychilatry, 42*, 918-924.

Schneider, K. (1958). *Psychopathic personalities.* London: Cassell.

Schwartzman, A.E., Ledingham, J.E., & Serbin, L.A. (1985). Identification of children at risk for adult schizophrenia: A longitudinal study. *International Review of Applied Psychology, 34*, 363-380.

Shields, J., Heston, L.L., & Gottesman, I.I. (1975). Schizophrenia and the schizoid: The problem for genetic analysis. In R.R. Fieve, D. Rosenthal, & H. Brill (Eds.), *Genetic research in psychiatry* (pp. 167-197). Baltimore: Johns Hopkins University Press.

Siever, L.J., Coursey, R.D., Alterman, I.S., Buchsbaum, M.S., & Murphy, D.L. (1984). Smooth pursuit eye movement impairment: A vulnerability marker for shizotypal personality disorder in a volunteer population. *American Journal of Psychiatry, 141*, 1560-1566.

Spitzer, R.L., Endicott, J., & Robins, E. (1978). *Research Diagnostic Criteria (RDC) for a selected group of functional disorders* (3rd ed.). New York: New York State Psychiatric Institute, Biometrics Research.

Thomas, D.L., & Diener, E. (1990). Memory accuracy in the recall of emotions. *Journal of Personality and Social Psychology, 59*, 291-297.

Van Praag, H.M. (1990). Two-tier diagnosing in psychiatry. *Psychiatry Research, 34*, 1-11.

Venables, P.H., Dalais, J.C., Mitchell, D.A., Mednick, S.A., & Schulsinger, F. (1983). Outcome at age nine of psychophysiological selection at age three for risk of schizophrenia: A Mauritian study. *British Journal of Developmental Psychology, 1*, 21-30.

Walker, E., & Shaye, J. (1982). Familial schizophrenia: A predictor of neuromotor and attentional abnormalities in schizophrenia. *Archives of General Psychiatry, 39*, 1153-1156.

Weiner, H. (1977). *Psychobiology and human disease.* New York: Elsevier.

Weissman, M.M., Gershon, E.S., Kidd K.K., Prusoff, B.A., Leckman, J.F., Dibble, E., Hamovit, J., Thompson, W.D., Pauls, D.L., & Guroff, J.J. (1984). Psychiatric disorders in the relatives of probands with affective disorders: The Yale University-National Institute of Mental Health collaborative study. *Archives of General Psychiatry, 41*, 13-21.

Weissman, M.M., & Johnson, J. (1990). *Epidemiology of dysthymia.* Paper presented at the 29th annual meeting of the American College of Neuropsychopharmacology, San Juan, Puerto Rico, December.

Weissman, M.M., Leaf, P.J., Bruce, M.L., & Florio, L. (1988). Epidemiology of dysthymia in five communities: Rates, risks, comorbidity, and treatment. *American Journal of Psychiatry, 145,* 815–819.

Weissman, M.M., Myers, J.K., Thompson, W.D., & Belanger, A. (1986). Depressive symptoms as a risk factor for mortality and for major depression. In L. Erlenmeyer-Kimling & N.E. Miller (Eds.), *Life-span research on the prediction of psychopathology* (pp. 251–260). Hillsdale, NJ: Lawrence Erlbaum.

White, D.M., Lewy, AJ., Sack, R.L., Blood, M.L., & Wesche, D.L. (1990). Is winter depression a bipolar disorder? *Comprehensive Psychiatry, 31,* 196–204.

Yee, C.M., Deldin, P.J., & Miller, G.A. (1992). Early stimulus processing in dysthymia and anhedonia. *Journal of Abnormal Psychology.*

Yee, C.M., & Miller, G.A. (1988). Emotional information processing: Modulation of fear in normal and dysthymic subjects. *Journal of Abnormal Psychology, 97,* 54–63.

Yee, C.M., & Miller, G.A. (1994). A dual-task analysis of resource allocation in dysthymia and anhedonia. *Journal of Abnormal Psychology, 103,* 625–636.

8
Selected Psychophysiological Measures in Depression: The Significance of Electrodermal Activity, Electroencephalographic Asymmetries, and Contingent Negative Variation to Behavioral and Neurobiological Aspects of Depression

Scott R. Sponheim, John J. Allen, and William G. Iacono

Although the phenomenon of depression has been described in one form or another since the time of Hippocrates, major advances in understanding the etiology, course, and treatment of depression have come only recently. These advances include an increased understanding of the genetics and epidemiology of depression, as well as the development of pharmacological and cognitive interventions for the disorder. Psychophysiologically-based studies hold potential for further elucidating the nature and biological mechanisms of depression. Well-chosen psychophysiological paradigms, which measure the physiological manifestations of psychological processes and conditions, can link the underlying pathophysiology of depression to observable facets of the disorder. Psychophysiological studies can also provide a means for identifying unseen characteristics that may signal a liability for the development of depression in individuals (Iacono, 1991; Iacono & Ficken, 1989).

The traditional approach of identifying psychophysiological indices that simply differentiate depressed from nondepressed persons, although useful for cataloging features of the depressed state, does not provide information that will necessarily lead to an understanding of the underlying risk for the development of depression or a clarification of the neurobiological underpinnings of depressive symptomatology. A stronger finding than a simple differentiation of groups is that a psychophysiological index characterizes depressed persons not only when depressed but also when euthymic (either before or after episodes). Such an index taps a feature of the disorder that is not merely a characteristic of a depressive state but is an enduring trait of an individual that is potentially related to a risk of developing depression. Another finding stronger than simple differentiation of depressed from nondepressed individuals is that a psychophysiological index demonstrates a relationship with a particular set of depressive symptoms. Such a finding may point to a biologically distinct subtype of depression and could be linked to underlying disturbances involved in the genesis or maintenance of depression. Last, psychophysiological measures associated with another index of anomalous biological functioning (e.g., a blunted response to a neurotransmitter agonist) may suggest specific neurobiological disturbances underlying clinical depression.

In this chapter, we highlight several lines of psychophysiological investigation that appear relatively promising for identifying neurobiological involvement in depression. The lines of research to be reviewed have either identified trait characteristics of depressed individuals' psychophysiology or have shown direct associations between a psychophysiological index and aspects of depressive symptomatology or measures of biological functioning.

Electrodermal Activity in Depression

Perhaps the best replicated psychophysiological correlates of depression come from studies of autonomic variables, especially those examining electrodermal activity. Typically, these investigations have employed simple information-processing paradigms focused on the orienting response and its habituation. In this section, we review these studies, paying special attention to those reports using electrodermal measures. Our review shows that individuals with mood disorders are electrodermally hypoactive. This reduced activation appears to be a stable trait. It is present during symptom remission and in persons with subclinical manifestations of disorder. Why mood-disorder patients show depressed electrodermal activity is uncertain. Though hardly conclusive, the evidence to date suggests that the electrodermal system itself is dysfunctional, perhaps as a result of a defect in peripheral cholinergic or central noradrenergic mediation.

Electrodermal Studies of Depression

A substantial number of studies show that depressed patients have low levels of electrodermal activity, including reduced tonic levels, high rates of failure to respond to stimuli, attenuated response amplitudes, and few spontaneous responses. For instance, in investigations of orienting response habituation, 68% of mood-disorder patients, on average, have been shown not to respond at all to innocuous auditory stimuli (see Iacono, Ficken, & Beiser, 1993, for a review). This rate of electrodermal nonresponding is more than twice that seen in the normal comparison subjects assessed in these same studies. Electrodermal hypoactivation has been reported in studies using other paradigms as well, including those involving quiet relaxation (Carney, Hong, Kulkarni, & Kapila, 1981; Lapierre & Butter, 1980; Storrie, Doerr, & Johnson, 1981; Ward, Doerr, & Storrie, 1983), word association and classical conditioning (Dawson, Schell, & Catania, 1977), reaction time (Giedke, Bolz, & Heimann, 1980), exposure to emotional and cognitive stressors (Donat & McCullough, 1983), and respiratory maneuvers (Iacono et al., 1983; Storrie et al., 1981). With one exception (Levinson, 1991), investigations of the effects of stimulus intensity and meaningfulness have shown that depressed electrodermal activation is present across stimulus conditions (Bernstein et al., 1988; Iacono et al., 1984).

Although many of the subjects in these studies were receiving anticholinergic medications (antidepressants), it is unlikely that the observed electrodermal hypoactivity was simply a drug effect. Contrasts of the electrodermal functioning of patients on and off these medications have failed to find differences (e.g., Bernstein et al., 1988; Dawson et al., 1977; Iacono et al., 1983, 1984; Ward et al., 1983), and several studies have found reduced electrodermal activity in drug-free depressed patients (Bernstein et al., 1988; Donat & McCullough, 1983; Lapierre & Butter, 1980; Levinson, 1991; Storrie et al., 1981).

Electrodermal hypoactivity appears to be a trait, perhaps one with potential to identify subclinical manifestations of disorder. Iacono et al. (1984) showed that various electrodermal measures yielded relatively stable values over time, with mood-disorder patients generating 1-year retest stability correlations ranging from .45 to .65. Several investigations have indicated that electrodermal measures remain unchanged in depressed patients despite clinical improvement following drug therapy (Storrie et al., 1981) and electroconvulsive therapy (Dawson et al., 1977; Noble & Lader, 1971). In addition, low levels of electrodermal activity have been observed in depressed and manic patients during symptom remission (Iacono et al., 1983, 1984; Janes & Stock, 1982). Lenhart (1985) found a nonclinical sample of college students identified as cyclothymic to be electrodermally hyporesponsive. Similar results have been reported by McCarron (1973) in nonhospitalized, unmedicated college students with high scores on the Minnesota Multiphasic Personality Inventory depression scale.

Although an association between electrodermal hypofunctioning and depression has been reported in over a dozen investigations, the relationship between electrodermal functioning and bipolar disorder or mania has received far less attention. Williams, Iacono, and Remick (1985) reported that acutely depressed bipolar and unipolar patients did not differ on various tonic and phasic skin conductance measures. Iacono et al. (1983, 1984) showed that during remission, bipolar patients with many prior episodes had reduced levels of electrodermal activity compared to normal subjects. Iacono et al. (1993), in a study of bipolar manics experiencing their first episode of disorder, demonstrated that these patients were electrodermally equivalent to patients with major depression and less responsive than normal subjects. For many of the electrodermal variables in these studies by Iacono and associates, compared to depressives, bipolar participants tended to be somewhat less hypoactive. More investigations of bipolar disorder are needed to determine the significance of this apparent difference between depressed and manic patients in the degree of electrodermal hypoactivity. It should also be noted that diminished electrodermal arousal is not specific to mood disorders. It is well established that schizophrenic patients show a similar dysfunction (see Iacono, Ficken, & Beiser, 1993, for a review).

Various types of evidence suggest that, among depressives, those with the lowest electrodermal levels are those with the most severe affliction. Compared to other types of depressed patients, diminished electrodermal activation has been observed in endogenous (Mirken & Coppen, 1980; Noble & Lader, 1971), retarded (Bernstein et al., 1988; Lader & Wing, 1969; Williams et al., 1985) and psychotic (Byrne, 1975) depression. Low levels of electrodermal responsivity have also been associated with violent suicidal behavior in depressed patients (Edman, Asberg, Levander, & Schelling, 1986; Keller, Wolfersdorf, Straub, & Hole, 1991). Finally, Dawson et al. (1977) showed that the number of clinical symptoms was inversely related to electrodermal activity in depressed patients.

The Significance of Electrodermal Hypoactivity in Depression

The literature reviewed previously indicates that depressed patients, whether acutely ill or in remission, show electrodermal hypoactivation. This psychophysiological anomaly is present in both unipolar and bipolar disorder, is independent of changes in clinical state, is stable over time, and does not appear to depend on experimental conditions or stimulus characteristics. It is seen in subsyndromal individuals in nonclinical populations. Among depressed patients, low levels of electrodermal activity are evident in those with more severe disorders and with the type of phenomenology often associated with depressions believed to have a strong biological component. It would be premature to suggest that low electrodermal activity identifies risk for depression because no data indicate that it predicts the development of disorder in asymptomatic individuals, it has not been observed in the biological relatives of depressed people, nor is it specific to depression. Taken together, the evidence points to the conclusion that for at least a subgroup of depressed patients, electrodermal hypoactivation is an index of a trait reflecting a psychobiological dysfunction.

The nature of the dysfunction is difficult to pinpoint, but it may reflect a defect in the peripheral cholinergic mediation of the electrodermal system. The strongest support for this argument derives from the work of Bernstein et al. (1988). These investigators showed that although depressed patients fail to respond electrodermally to orienting stimuli, they do generate normal finger-pulse amplitude responses. The response dissociation seen for these two variables indicates that the central processing system of depressed individuals is intact because they can emit normal orienting responses when these responses are indexed through an adrenergically mediated cardiovascular variable. Their failure to emit an electrodermal orienting response thus is unlikely to reflect an orienting deficit. Rather, Bernstein et al. (1988) assert that it represents a cholinergic deficiency. This interpretation is supported by studies reviewed by Bernstein et al. (1988), which indicate that other cholinergically mediated autonomic measures appear to be underactive in depression (e.g., salivation, forearm blood flow), while adrenergic responses (e.g., pupillary dilation) seem to be unimpaired.

It is also possible, however, that diminished electrodermal activation reflects the involvement of central noradrenergic control over electrodermal activity. Animal studies conducted by Yamamoto and colleagues (1990, 1991) have demonstrated that destruction of noradrenergic fibers (by injecting a catecholamine neurotoxin into the ascending noradrenergic bundle of the pons) eliminated skin conductance responses and reduced spontaneous electrodermal fluctuations and tonic skin conductance level. Destruction of dopaminergic fibers in the ventral tegmentum area resulted in no electrodermal changes. Additionally, Yamamoto et al. (1991) showed that injection of a different neurotoxin, selective for noradrenergic nerve terminals, into the fourth ventricle reduced spontaneous electrodermal fluctuations and skin conductance responses in rats.

Electroencephalographic Asymmetries in Depression

A substantial body of research supports the premise that the two cerebral hemispheres subserve very different emotional functions and may be involved differentially in clinical mood disorders. The early observations of Goldstein (1939) of brain-damaged patients indicated that left-hemisphere damage produces a "catastrophic reaction," whereas comparable damage to the right hemisphere would leave affected individuals with an indifference to their impairments or inappropriate jocularity (Gainotti, 1972; Robinson, 1983). Although clear differences exist between the two hemispheres with respect to emotion, the simple right–left dichotomization provides an insufficient account. Compelling evidence from Robinson and colleagues (Robinson, 1983; Robinson & Benson, 1981; Robinson, Kubos, Starr, Reo, & Price, 1984; Robinson & Price,

1982; Robinson & Szetele, 1981) indicates that the severity of depressive symptomatology is strongly and positively correlated with the proximity of a left-hemisphere lesion to the frontal pole, whereas anterior right-hemisphere lesions more often produce undue cheerfulness or an emotional indifference. These findings are further corroborated by studies of the pharmacological inactivation of one or the other hemisphere. Amytal injections to the left side typically produce the catastrophic reaction whereas right-sided injections produce euphoric behavior (Alema, Rosadini, & Rossi, 1961; Perria, Rosadini, & Rossi, 1961; Terzian, 1964).

However, the finding that damage to or inactivation of a brain locus causes an emotional disturbance does not necessarily imply that the affected area subserves the emotion in intact brains or that it is necessarily involved in functional psychopathology. Naturally occurring brain lesions seldom respect anatomical boundaries, affecting not only local structures but interconnecting pathways as well. Moreover, although damage to the frontal lobes certainly can produce emotional sequelae, it does not invariably do so (e.g., Gainotti, 1972). Methods that identify relationships between mood, behavior, and brain function are required to specify what neurobehavioral systems may be disrupted by lesions or pharmacological inactivation. Electrophysiological studies are important in that they provide an opportunity to examine the functioning of brain regions in intact humans.

Electroencephalographic Lateralization Studies of Depression and Depressed Mood

Electroencephalographic (EEG) studies of the lateralization of emotion typically rely on collecting a sample of EEG activity from several symmetrically placed scalp electrodes (with a common reference, usually the vertex) during an eyes-closed or sometimes eyes-open resting condition. The EEG data are then examined for specific frequencies of interest— almost exclusively alpha band (8 to 12.5 Hz) activity—at the different loci. A robust and replicable finding from these investigations is *relative* excess of alpha-band activity in the left frontal region when compared to the homologous right frontal region in persons experiencing clinical depression (Allen, Iacono, Depue, & Arbisi, 1993; Henriques & Davidson, 1991), in psychometrically defined depressed college students (Schaffer, Davidson, & Saron, 1983), and in nondepressed subjects temporarily experiencing sad mood (Bennett, Davidson, & Saron, 1981; Davidson, Schwartz, Saron, Bennett, & Goleman, 1979). This pattern of asymmetry also holds prognostic value in that it predicts infant crying to maternal separation (using a broader spectrum <12 Hz, Davidson & Fox, 1988, 1989). Additionally, it may also serve as a vulnerability indicator, in that formerly depressed persons show this relative excess of left frontal alpha power even when euthymic (Allen et al., 1993; Henriques & Davidson, 1990). The only study that fails to confirm the finding (Tucker & Dawson, 1984) used method actors without controlling the strategies used to evoke emotions and without a neutral baseline against which to compare each subject's "depressive" EEG recordings.

Finally, two studies suggest that the pattern of frontal asymmetry may serve as a state-independent marker of depression in both unipolar (Henriques & Davidson, 1990) and bipolar depressions (Allen et al., 1993). Whether or not this asymmetry indexes a vulnerability factor remains to be determined by prospective studies of high-risk populations. Family studies are also important in this respect, because such data could assess whether the asymmetries may index a genetically transmitted liability for depression.

The Significance of Electroencephalographic Asymmetries in Depression and Depressed Mood

Although near-consensus exists with respect to the finding of greater relative left frontal alpha activity during depression, there exists less agreement about the meaning of these data. Increases in alpha activity have been interpreted as reflecting a decrease in activation (e.g., Shagass, 1972), and the greater relative left frontal alpha activity has been interpreted (see Davidson & Fox, 1988, for a review) as indicating that the left frontal region is less activated than the homologous right region during depression. This interpretation closely parallels the data from the studies of lesions and pharmacological inactivations and suggests that the left frontal region is involved in the experience of positive mood and that inactivation of, or damage to, this region results in depressed mood. This interpretation is similar to others' conceptualization of depression as reflecting an absence of positive emotion (Clark & Watson, 1991; Tellegen, 1985).

More specifically, Davidson and Tomarken (1989) suggest that the fundamental left–right frontal hemispheric difference may involve the dimension of approach and avoidance. From this perspective, left frontal systems subserve the functions of behavioral approach and engagement whereas right frontal systems mediate withdrawal and avoidance; this interpretation is consistent with the neurobiological conceptualization of depression and seasonal affective disorder of Depue and his colleagues (Depue, Arbisi, Spoont, Leon, & Ainsworth, 1989; Depue Krauss, & Spoont, 1987; Depue & Iacono, 1990). Depue et al. view depressions as reflecting a deficit in systems of reward incentive and behavioral facilitation. The EEG asymmetries, interpreted as indexing left frontal hypoactivity, are also consistent with neurochemical evidence suggesting that dopamine, an excitatory neurotransmitter hypothesized to mediate the system of behavioral facilitation (Depue et al., 1987, 1989; Depue & Iacono, 1989), exhibits denser concentration in the left limbic structures in the rat (Denenberg, 1981). Additionally, a converging line of evidence from pharmacological, imaging, electrophysiological, and postmortem studies suggests preferential dopaminergic mediation of left-hemisphere function in humans (for a review see Flor-Henry, 1986; Tucker & Williamson, 1984). In sum, left frontal underactivation may reflect a dopaminergically mediated deficiency in the behavioral facilitation system in depression.

A competing interpretation of these EEG findings (Tucker, 1981) holds that right-hemisphere subcortical structures generate positive emotions when released from right frontal cortical inhibitory functions. Thus, the excess right-hemisphere activity (relatively less alpha) putatively reflects cortical suppression of positive emotion centers, thus producing depression. Although this conceptualization captures the cortical–subcortical dimension of complexity, it cannot reconcile that subcortical lesions on the right side produce the same effects as cortical lesions (Cummings & Mendez, 1984; Whitlock, 1982).

A major limitation in assessing the meaning of the frontal EEG asymmetries is that in most cases only data from the alpha band are reported and available. Before the advent of the personal computer, the computational complexities of decomposing EEG data by Fourier transform into all of its constituent frequencies made simple alpha extraction a logical choice. Yet as Ahren and Schwartz point out, "Alpha blocking is not a unitary phenomenon," and "the pre-eminence of alpha as the sole indicator of cerebral activation needs to be re-evaluated" (1985, p. 753). One study that specifically examined beta activity found increased right-frontal beta activity correlated with depressed mood, but increased left frontal beta activity correlated with depressive ideation (Perris & Monakhov, 1979).

Moreover, whereas Henriques and Davidson (1990) found frontal *alpha* asymmetries, no consistent effects were found in the other bands; the claim that increased alpha reflects a hypoactivity would be bolstered by a concomitant finding of decreased beta activity. Other studies have also observed alpha increases without concomitant decreases in other bandwidths, most notably beta (Ahren & Schwartz, 1985; Pockberger, Petsche, Rappelsberger, Zidek, & Zapotoczky, 1985). Thus, it cannot be ubiquitously accepted that a relative excess of alpha activity at one site reflects lesser activation at that site (Ray, 1990). More importantly, however, scalp potentials are spatiotemporally low-pass filtered signals that can originate from diverse locations, both cortical and subcortical. The electroencephalographic data bearing on the brain systems underlying the mood disorders have thus far yielded a two-dimensional picture (left–right, anterior–posterior).

Contingent Negative Variation in Depression

Since the studies of Walter and colleagues (Walter, 1964; Walter et al., 1964), researchers have examined the brain's electrical responses to foster an understanding of biological and behavioral aspects of depression. Because they are records of the ephemeral processes of the brain, scalp recordings of stimulus-related brain potentials (termed *event–related potentials*) have been thought to hold particular promise for linking biological and behavioral phenomena.

The contingent negative variation (CNV) is an event-related potential identified as relevant to depression and is associated with the learned contingency between a pair of stimuli (Walter et al., 1964)—typically, a warning stimulus followed by an unpleasant imperative stimulus. The CNV has been shown to be diminished in depressed patients (e.g., Small & Small, 1971). Attempts to clarify the functional significance of CNV abnormalities have provided evidence suggesting that the potential is an electrocortical manifestation of a process involving orientation to the warning stimulus (Loveless & Sanford, 1974) and preparation for the imperative stimulus (McCarthy & Donchin, 1978; Rockstroh, Elbert, Canavan, Lutzenberger, & Birbaumer, 1989; Rohrbaugh & Gaillard, 1983). Investigations have also shown that the process of orientation and preparation captured in the CNV is influenced by the emotional and motivational content of the stimuli (Howard, Fenton, & Fenwick, 1982; Klorman & Ryan, 1980; Simons, Ohman, & Lang, 1979). Taken together, these studies raise the possibility that CNV abnormalities found in depression reflect aberrations in the orientation and preparation process in which depressed individuals engage when confronted with stimuli. In this section, we describe a typical CNV paradigm, review CNV studies of depressed individuals, and summarize studies examining the neurobiological basis of the CNV.

The Contingent Negative Variation Paradigm

The CNV is elicited in a classic conditioning paradigm. The subject is first presented with a conditional stimulus (called S1, or the warning stimulus) which is typically neutral in character (i.e., possessing no inherent associations; most frequently a tone, click, or flash of light). After a fixed delay (usually several seconds but ranging from 0.5 to 15 s across studies; for a review see Rockstroh et al., 1989), the subject is presented with the imperative stimulus (called S2, or the unconditional or indicative stimulus), which often has an aversive quality (e.g., a bright flashing light or a tone that has been paired with a shock). S2 stimuli of a neutral or positive nature have also been used by researchers. After S2's onset, the subject terminates the stimulus as quickly as possible with a behavioral response, usually a button press. The negative shift in brain electrical activity that occurs between S1 and S2 composes the CNV. S1 elicits a series of relatively short-lived positive and negative scalp potentials ending between 0.2 and 0.6 s after S1 onset (depending on S1 characteristics). After these brief potentials, the CNV develops as a slow negative potential shift of approximately 15 microvolts on average. The onset of S2 terminates the CNV with a large and immediate positive potential change back to pre-S1 voltage levels. A number of S1–S2 trials (usually from 10 to 50) are presented and then averaged together to cancel bioelectrical signals not time-locked to the stimuli. The result is a discernable image of the CNV. The CNV is typically recorded over frontal and central aspects of the brain.

Contingent Negative Variation Studies of Depression

A considerable number of investigations have shown that depressed individuals exhibit small CNVs in comparison to nonpsychiatric control subjects (Claverie, Brun, Nizard, Brenot, & Paty, 1984; Giedke et al., 1980; Giedke & Heimann, 1987; Nakamura, Iida, Fukui, & Takahashi, 1982; Small & Small, 1971; Small, Small, & Perez, 1971; Timsit-Berthier et al., 1984). Overall, reductions in the CNV during depression appear to be relatively independent of the physical characteristics of S1 and S2. Despite using clicks, tones, or light flashes of varying intensities and duration for S1 and S2, researchers continue to find the effect. It appears, however, that the subject's prepared behavioral response to S2 is important in uncovering CNV abnormalities because all the studies that found CNV reductions in depression required the subject to perform motor movements. Decremented CNVs in depressed individuals are also exclusive to those investigations that implement relatively brief intervals between S1 and S2 (from 0.5 to 2 s).

In her 1986 review of the CNV in psychiatry, Timsit-Berthier addressed state–trait characteristics associated with CNV abnormalities. She concluded that diminished CNV amplitudes in affective disorders depend on clinical state and that normal amplitudes return when patients become euthymic. Two studies published since 1986 further confirm Timsit-Berthier's conclusion (Ashton et al., 1988; Giedke & Heimann, 1987). Ashton et al. (1988) found that severely depressed patients had smaller CNV amplitudes than patients with milder depressions. Also, Giedke and Heimann (1987) measured a significant association between increased CNV size and clinical improvement under the administration of amitriptyline and oxaprotiline. They attributed the augmented CNV amplitudes, however, to a drug effect rather than clinical improvement. The identification of reduced CNV amplitudes in drug-free patients (Small & Small, 1971; Small et al., 1971) weighs against Giedke and Heimann's interpretation and indicates that the phenomenon is unlikely to be solely the consequence of medications.

A number of studies suggest that diminished CNVs are more common in individuals suffering from severe forms of depression. Decremented CNVs are more prominent in patients with retarded and endogenous depressions than in individuals with milder depression (Elton, 1984; Verhey, Lomers, & Edmonds, 1984). Timsit-Berthier et al. (1987) found that 12 of 24 (50%) retarded and psychotic depressives had small CNV amplitudes (<12 microvolts), while 8 of 30 (27%) neurotic depressives (as defined by the Present State Exam CATEGO classes, Wing, Cooper, & Sartorius, 1974) fell into the small CNV category. Specifically, this study demonstrated that depressed patients with delusions, hallucinations, psychomotor retardation, or agitation had small CNV amplitudes on average. In contrast, patients free of these more severe symptoms, and who experienced prominent anxiety, in addition to their depression, tended toward normal and large CNV amplitudes. The large CNV amplitude group also differed from the small and normal CNV groups in other event-related potentials preceding and following the CNV (i.e., smaller P300s, larger N100s, and larger postimperative-negative variation). These findings, in conjunction with an earlier study showing similar results (Timsit-Berthier & Timsit 1981), suggest that CNV abnormalities may, in addition to being associated with subtypes of severe depression, occur together with other electrocortical abnormalities. Additionally, studies have often shown that samples of depressed patients have more variation in their CNV amplitudes than either normal comparison or other psychiatric samples (Bolz & Giedke, 1981; Giedke, Bolz, & Heimann, 1980; Knott & Lapierre, 1987; Nakamura et al., 1982; Timsit-Berthier et al., 1984). This increased variability of CNV amplitudes in depressed samples may be due in part to the relationship between the severity of depressive subtype and CNV amplitude. Claverie et al. (1984) is the only study that has failed to find an association between a subtype of severe depression and greater reductions in CNV amplitude. This study, however, had limited power to detect group differences (only eight endogenous and eight neurotic depressed patients were examined) and implemented a ratio measure of the CNV instead of amplitude and area measures used in most other investigations.

Studies that have failed to identify diminished CNVs in depressed patients when compared to nonpsychiatric subjects (Elton, 1984; Knott & Lapierre, 1987, Knott et al., 1991; Thier, Axmann, & Giedke, 1986) have altered the CNV paradigm. Specifically, these investigations have reduced the role of S1 as a warning for a required behavioral response to S2. Elton (1984) preceded S1 with a green or red feedback light, which may have functioned as a warning stimulus for the subject. In this case, S1 would carry little value as a warning stimulus, conceivably limiting the development of a CNV between S1 and S2. In a study by Thier et al. (1986), S2 (a tone) required the subject to decide whether or not to respond behaviorally based on S2's pitch. Thus, S1 did not function as a warning of an upcoming motor movement but instead warned of a stimulus that would determine if a motor movement was required. Thier et al. (1986) point out that the use of a choice reaction-time task instead of a conventional simple reaction-time task could have affected their results. Knott and Lapierre (1987) and Knott et al. (1991) also used choice reaction-time tasks, similarly reducing the role of S1 as purely a warning stimulus for a required motor movement. Knott et al. (1991) found no significant differences between depressed patients and nonpsychiatric subjects in a simple reaction-time condition. This study, however, failed to record the CNV from a midline frontal site such as Fz, a site commonly used in CNV studies. Bolz and Giedke (1981), the only other published study that has failed to show significant depressed group-control group differences, found that in a simple reaction-time condition, a depressed patient group exhibited a nonsignificant trend toward smaller CNV amplitudes.

The notion that failures to find CNV abnormalities in depression occur only in studies that alter the experimental paradigm is consistent with a report from an international pilot study of CNV in mental disorders involving laboratories in Belgium, England, and the Netherlands (Abraham et al., 1980). This report stressed the importance of standardizing experimental procedures for the collection of CNV data, contending that even slight changes would likely result in discrepant findings among researchers.

Although researchers have shown depressed patients to have small CNV amplitudes, this finding is not specific to depression. Individuals with schizophrenia, dementia, head trauma, and psychopathy have all been reported to exhibit diminished CNV amplitudes (Timsit-Berthier, 1986).

The Significance of Diminished Contingent Negative Variation in Depression

Contingent Negative Variation and Psychomotor Retardation in Depression

Depressed individuals commonly experience psychomotor retardation as part of their depressive symptomatology. Within the CNV paradigm, a depressed subject's reaction time to S2 provides a measure of psychomotor retardation (Byrne, 1976; Hoffman, Gonz, & Mendlewicz, 1985). Contingent negative variation studies have demonstrated that reaction times to S2 are longer for depressed individuals than nondepressed comparison subjects (Claverie et al., 1984; Giedke et al., 1980; Giedke & Heimann, 1987; Knott & Lapierre, 1987; Thier et al., 1986) and have been directly associated with small CNV amplitudes in depressed patients (Ashton et al., 1988; Timsit-Berthier et al., 1987). Data from nonpsychiatric samples, however, indicate that the concurrence between longer reaction times and decremented CNVs is not exclusive to depression but may be a general electrophysiological–behavioral relationship (Arito & Oguri, 1990; Mackay & Bonnet, 1990). Rockstroh, Elbert, Lutzenberger, and Birbaumer (1982) presented compelling evidence of a CNV–reaction time association by demonstrating that subjects who learn to increase CNV negativity through feedback training (displaying to the subject ongoing frontal voltage levels) shorten their reaction times to S2. Taken together, these studies suggest that CNV abnormalities during depression may reflect in part an aberration in the cortical processes activated when an individual anticipates a required motor movement. In depression, these anomalous cortical processes may be manifested in changes of psychomotor activity.

Contingent Negative Variation and Dopaminergic Functioning

Timsit-Berthier, Mantanus, Ansseau, Doumont, and Legros (1983), relying in part on Marczynski's conceptualization of CNV genesis (Marczynski, 1978, 1986), proposed a complex model suggesting that catecholaminergic activity is critically involved in the development of the CNV. Of special interest here is their hypothesis that CNV amplitude is directly related to dopaminergic and noradrenergic (i.e., norepinephrine) activity in the brain and that decreased activity in these two neurotransmitter systems is associated with decreased CNV amplitudes in psychiatric patients.

In a series of studies, Timsit-Berthier tested catecholaminergic influences on CNV amplitude in psychiatric and normal samples (Timsit-Berthier et al., 1987; Timsit-Berthier et al., 1983; Timsit-Berthier, Mantanus, Marissiaux, et al., 1986; Timsit-Berthier, Mantanus, Poncelet, et al., 1986). In these studies, apomorphine and clonidine were used to measure the sensitivity of dopamine postsynaptic and norepinephrine receptors, respectively, as indicated by the release of growth hormone into blood plasma after the administration of either drug. (Apomorphine is a dopaminergic agonist that acts on, among other sites, dopaminergic receptors in the hypothalamus resulting in the dispersal of growth hormone-releasing factor. Clonidine is similar to norepinephrine in that they are both alpha-2 adrenoceptor agonists, which, like apomorphine, act on receptors in the hypothalamus, leading to the release of growth hormone-releasing factor.) If plasma growth hormone increases markedly after apomorphine or clonidine administration, then receptor sensitivity is high for the targeted neurotransmitter.

In her investigations, Timsit-Berthier found support for dopaminergic involvement and more tenuous support for the influence of norepinephrine in CNV determination. In a sample of 24 psychiatric patients diagnosed with affective disorder, neurosis, or schizophrenia, growth hormone response after apomorphine administration correlated .83 with

CNV amplitude, indicating that increased sensitivity of dopaminergic postsynaptic receptors was associated with larger CNV amplitudes (Timsit-Berthier et al., 1983). Growth hormone response after clonidine administration correlated .48 with CNV amplitude. In a second study, Timsit-Berthier and co-workers examined 28 psychiatric inpatients (17 depressed, 2 neurotic, 7 schizophrenic, and 2 manic) and found CNV amplitude correlated .56 with growth hormone response to apomorphine (Timsit-Berthier, Mantanus, Marissiaux, et al., 1986). For the control group of 20 normal males, however, growth hormone response to apomorphine was not significantly related to CNV amplitude. Timsit-Berthier et al. suggested that the simplicity of the task reduced the alertness of normal subjects, thereby removing the association between dopamine receptor sensitivity and CNV amplitude. In another study, Timsit-Berthier, Mantanus, Poncelet, et al. (1986) administered clonidine to 19 normal males and apomorphine to 17 other normal males. Growth hormone response after apomorphine administration correlated .60 with CNV amplitude. The correlation between CNV amplitude and growth hormone response after clonidine administration was not significant.

Adding to the studies of Timsit-Berthier, Tecce (1991) reviewed evidence supporting the role of dopamine in CNV genesis. Tecce found that shortly after the administration of dopamine agonists, CNV amplitude increased (Kopell, Wittner, Lunde, Wolcott, & Tinklenberg, 1974; Tecce, Cattanach, Hill, & Cole, 1987; Tecce & Cole, 1974) and that, after dopamine antagonist administration, the potential decreased (Tecce, Cole, & Savignano-Bowman, 1975; Thompson et al., 1978), implicating dopamine as playing a modulatory role in CNV determination. To account for diminished CNV amplitudes found in schizophrenia, a disorder posited as associated with excess dopamine, Tecce proposed that dopamine and CNV amplitude are related by an inverted U function. Collectively, Timsit-Berthier and Tecce's work suggests that the role of dopamine in CNV determination is not unique to depression but cuts across diagnostic categories. This may, in part, account for reduced CNV amplitudes noted in other forms of psychopathology (Timsit-Berthier, 1986). The role of norepinephrine in CNV genesis is not clear, given several failures to identify a relationship between growth hormone response after clonidine administration and CNV amplitude.

In an investigation targeting depression, Timsit-Berthier et al. (1987) found support for dopaminergic and noradrenergic dysfunction in the disorder. They showed depressed patients to have low levels of growth hormone response to apomorphine and clonidine as well as marked variability in their CNV amplitudes. Depressed individuals with *large* CNV amplitudes, however, had the most significantly blunted growth hormone responses to clonidine, indicating, contrary to prediction, that more dramatic decreases in noradrenergic activation were associated with larger CNV amplitudes. Because the large CNV group's mean reaction time was significantly shorter, one might expect these individuals to generate larger CNV amplitudes and possibly to exhibit less psychomotor retardation in their depressions. Nevertheless, this finding calls into question any simple relationship between noradrenergic activity and CNV amplitude in depression. In conjunction with evidence cited previously, this study suggests that the primary catecholaminergic dysfunction resulting in decremented CNV amplitudes during depression more likely involves dopaminergic functioning than noradrenergic systems.

Electrodermal, Electroencephalographic, and Contingent Negative Variation Abnormalities in Relation to Behavioral and Neurobiological Aspects of Depression

In this chapter we highlighted electrodermal hypoactivity, increased left frontal EEG alpha activity, and decremented CNVs as significant findings in the psychophysiological study of depression. Our review suggests that, in addition to being anomalous in depressed individuals, electrodermal activity, frontal EEG asymmetries, and CNV may provide insights into the neurobiological bases of depression through identification of a trait defect in biological functioning and from an association with specific aspects of depressive symptomatology or an index of neurobiological activity (e.g., growth hormone response to apomorphine). Taken together, investigations implementing these psychophysiological measures provide a complex picture of depression. Nevertheless, consistencies can be identified in this literature that may guide an understanding of selected behavioral and neurobiological aspects of depression and, consequently, suggest several testable hypotheses. First, electrodermal hypoactivation, increased left-frontal EEG alpha activity, and diminished CNVs may occur most prominently and conjunctively in particularly severe forms of depression (e.g., melancholic, retarded, psychotic, and endogenous depressions). Second, electrodermal abnormalities may in part measure trait disturbances in central noradrenergic processes that predispose an individual to experience depressions characterized by either high or low levels of arousal. Consistent with a review by Zahn (1986), individuals with electrodermal hypoactivity (suggesting noradrenergic underactivation) may tend to experience depressions characterized by underarousal (e.g., retarded depression), whereas people who are electrodermally hyperactive (suggesting noradrenergic overactivation) may be predisposed to have depressions with high levels of arousal (e.g., agitated or anxious depressions). Finally, whereas increased left-frontal EEG alpha activity may be interpreted to signify a tonic disruption of dopaminergic functioning (because the asymmetry does not change with clinical remission), diminished CNV amplitudes appear to be state dependent, suggesting that phasic changes in dopaminergic functioning may underlie the expression of aspects of depressive symptomatology. Phasic dysregulation of dopaminergic systems (possibly indexed by the CNV and expressed as clinical depression) may be superimposed on a tonic vulnerability in the dopaminergic functioning (possibly indexed by EEG asymmetries).

Investigations examining the associations between depressive symptomatology and electrodermal activity, frontal EEG asymmetries, or CNV, generally support the hypothesis that psychophysiological abnormalities occur most prominently, and possibly together, in severe forms of depression. As noted, electrodermal hypoactivity (Lader & Wing, 1969; Mirken & Coppen, 1980) and diminished CNV amplitudes occur more often in severe depressions (Elton, 1984; Verhey et al., 1984). Specifically, psychomotor retardation, a symptom often associated with severe depression, appears to be related to electrodermal hypoactivity (Williams et al., 1985) and to reductions in CNV amplitude (Ashton et al., 1988; Timsit-Berthier et al., 1987). Although not yet directly tested, frontal EEG asymmetries may also be more prominent in severe forms of depression. Consistent with the hypothesis that severe depression is psychophysiologically distinct from mild depression are studies showing that patients suffering from severe depression exhibit disturbances in biological functioning while those suffering mild depressions do not (e.g., Maes, Maes et al., 1990; Zimmerman, Stangl, & Coryell, 1985).

Studies using quantitative approaches to analyze the nature of depressive symptoms in patient populations also lend credence to the idea that severe depression is a distinct subtype. Cluster analyses based on depressive symptoms have recovered groups of individuals experiencing prominent characteristics of severe depression (e.g., Andreasen & Grove, 1982: pervasiveness of loss of interest, lack of reactivity of mood, loss of appetite, and weight loss; Maes, Cosyns et al., 1990: loss of interest, nonreactivity of mood, distinct quality of mood, early morning awakening, psychomotor disturbances, and anorexia and weight loss; Paykel, 1971). Blashfield and Morey (1979) reviewed 11 cluster-analytic studies of depressed patients' symptomatology and found all 11 studies extracted a group of individuals with a severe form of depression (usually labeled retarded, endogenous, or psychotic). Also, factor-analytic studies of depressed inpatient symptomatology have found a prominent retarded-psychotic depression factor with high loadings of symptoms associated with severe depression (see Gersh & Fowles, 1979).

Although investigations of electrodermal activity, frontal EEG asymmetries, and CNVs, as well as quantitative studies of depressive phenomenology, can be seen as promising, these studies carry only limited value for exploring the concurrence of psychophysiological anomalies in severe depression. Multivariate psychophysiological examinations of depressed patients are clearly necessary to directly study whether or not the psychophysiological abnormalities of interest occur together in patients afflicted with a particularly severe form of depression. To date, no studies have examined electrodermal activity, frontal EEG asymmetries, and CNV in the same depressed subjects. However, three studies by Giedke and colleagues identified both CNV and electrodermal anomalies in patients with primary depression (Bolz & Giedke, 1981; Giedke et al., 1980; Giedke & Heimann, 1987). Across the three studies, which together included 95 patients and 75 control subjects, electrodermal activity was found to be significantly hypoactive in patients when measured by mean level of activity or by frequency and amplitude of electrodermal responses. Patients also exhibited significantly lower mean CNV area than control subjects in two of the studies (Giedke et al., 1980; Giedke & Heimann, 1987). The high Hamilton Depression Scale (Hamilton, 1960) scores of subjects in these three studies (means = 25.5, 24.6, and 23.6) indicate that the patients were severely depressed. Geidke et al. (1980), the only investigators to report correlations between psychophysiological indices in depressed patients, found CNV area (recorded midline over the frontal lobes) to correlate .52 with the amplitude of skin resistance-orienting response. This finding suggests a coupling between diminished CNV amplitudes and small electrodermal responses.

Several lines of research relate to the hypothesis that electrodermal activity may index a predisposition for the level of arousal in depression and, in part, be a manifestation of central noradrenergic functioning. First, depressed patients with anxiety or agitation have been shown to exhibit higher rates of spontaneous electrodermal fluctuations and slower rates of habituation than nonpsychiatric control subjects and individuals with retarded depressions (see Boucsein, 1992; Zahn, 1986 for reviews). These findings, in conjunction with the evidence that electrodermal hypoactivity is a trait characteristic of depressed individuals, are consistent with the possibility that electrodermal activity indexes a trait predisposition to experience depressions with high or low levels of arousal. At a neurobiological level, evidence suggests that norepinephrine mediates electrodermal activity in the reticular formation (Boucsein, 1992). Animal studies indicate that electrodermal activity is influenced by norepinephrine at the level of the ascending noradrenergic bundle, which passes through the reticular formation of the pons (Bernthal & Koss, 1984; Yamamoto et al., 1990, 1991). Additionally, the absence of lateralized electrodermal abnormalities in depression may suggest that defects in the cerebral controls of electrodermal activity do not occur above the level of the reticular formation. In a review of animal and human studies, Boucsein (1992) identified ipsilateral hypothalamic effects on sweat secretion as well as contralateral influences by the premotor cortex and basal ganglia. However, reticular formation stimulation was followed by bilateral responses, possibly indicating that bilateral electrodermal anomalies originate at, or below, the level of the reticular formation.

A body of evidence bears on the hypothesis that individuals experiencing some forms of depression undergo a phasic dysregulation of dopaminergic functioning (as indexed by the CNV) that may be superimposed on a tonic vulnerability in dopamine systems (as indexed by EEG asymmetries). Our review suggests that decremented CNVs are state phenomena occurring mostly during active depressions and are associated with postsynaptic dopamine receptor subsensitivity as measured by growth hormone response to pharmacological challenges (Timsit-Berthier & Timsit, 1981; Timsit-Berthier, Mantanus, Marissiaux et al., 1986; Timsit-Berthier, Mantanus, Poncelet et al., 1986). More specifically, Timsit-Berthier (1986) and Marczynski (1978, 1986), reviewing animal studies, found evidence that dopamine played a modulatory role in cortical processes leading to the initiation of motor movements and the development of the CNV. Additionally, our review shows that frontal EEG asymmetries are likely to be a trait phenomenon, because people who are at risk for depression (i.e., previously depressed individuals during euthymic periods and people with subthreshold levels of depression) exhibit increased left frontal EEG alpha activity. In turn, this EEG abnormality has been associated with low levels of positive affect, a construct thought in part to characterize depression (Clark & Watson, 1991; Tellegen, 1985). Specifically, low levels of positive affect have been related to depressive symptoms (Watson, Clark, & Carey, 1989) and can be seen as encompassing two criteria of depression: (a) loss of pleasure in all, or almost all, activities, and (b) lack of reactivity of mood to usually pleasurable stimuli (Watson & Kendall, 1989). Studies by Wheeler et al. (1993) and Tomarken, Davidson, Wheeler, and Doss (1992) have demonstrated in nonpsychiatric populations an association between low levels of positive affect (i.e., low levels of approach) and increased left frontal EEG alpha activity. Trait levels of positive affect have also been shown to correlate .75 to .89 with measures of dopaminergic functioning in a nonpsychiatric sample (Depue et al., 1992). Thus, evidence is provided for dopaminergic involvement underlying CNV and EEG abnormalities in depressions characterized by low levels of positive affect. However, it remains to be shown directly that this is a neurochemical factor linking increased left frontal EEG alpha activity and reductions in CNV. A direct way to study state–trait associations of EEG and CNV in relation to dopaminergic functioning would be to longitudinally examine all three variables in a sample of depressed individuals. With such a study, it would be fruitful to examine CNV laterality effects to determine whether reductions in CNV amplitude are lateralized to the left during depression, thereby providing more specific anatomical agreement with increased levels of left frontal EEG alpha activity.

The empirical evidence that CNV and EEG abnormalities are dopaminergically based is consistent with proposals calling attention to disturbances in dopaminergic functioning as contributing to depressive symptomatology (Depue & Iacono, 1989; Silverstone, 1985; Swerdlow & Koob, 1987). Depue and Iacono (1989) posited that hypoactivity of the dopamine system is manifested in a depression characterized by low positive affect and expressed as (a) low incentive-reward motivation to participate in social, vocational, recreational, and sexual stimuli, and (b) delayed and slowed motor responses. As evidence, Depue and Iacono cite the antidepressant effects of dopamine agonists, the propensity for tricyclic antidepressants to trigger hypomanic and manic disturbances, and the enhancement of dopamine-mediated behaviors in rats as a result of chronic tricyclic antidepressant administration.

Conclusion

In this chapter we focused on electrodermal hypoactivity, increased left frontal EEG alpha activity, and decremented CNVs as significant psychophysiological findings in depression. Our review shows electrodermal hypoactivition is likely a trait characteristic of depressed patients that is present during remission and in individuals with subclinical manifestations of depression. Electrodermal activity appears to be lowest in individuals with severe forms of depression, and evidence suggests that the anomaly may be due to a defect in peripheral cholinergic functioning. Studies have also shown electrodermal hypoactivity to reflect an underactivation of central noradrenergic systems, particularly in the ascending noradrenergic bundle of the pons. The extant literature supports reductions in left frontal EEG alpha to be a trait characteristic of depressed individuals that is also found in people who are not depressed but are experiencing a sad mood. Alpha asymmetries may reflect a dopaminergically mediated deficiency in the left frontal brain system subserving the functions of behavioral approach and positive affect. Evidence suggests that diminished CNVs in depressed patients are a state phenomenon and that CNV amplitudes return to normal levels when patients are euthymic. Investigations have also shown CNV amplitudes to be more diminished in severe depressions and to be associated with psychomotor retardation. Researchers using pharmacological challenges have demonstrated that small CNV amplitudes in depression are related to the desensitization of postsynaptic dopamine receptors. Consequently, decremented CNV amplitudes may be the manifestation of a defect in dopaminergically mediated brain processes leading to psychomotor activation. Last, electrodermal hypoactivity, increased left frontal EEG alpha activity, and reductions in CNV amplitudes have each been identified in drug-naive or drug-free depressed subjects, indicating that these psychophysiological abnormalities are not purely the result of medications.

From consistencies in the literature, several testable hypotheses can be offered regarding psychophysiological abnormalities in relation to behavioral and neurobiological aspects of depression. First, electrodermal hypoactivation, increased left frontal EEG alpha activity, and diminished CNVs may occur most prominently and together in particularly severe forms of depression (e.g., melancholic, retarded, psychotic, and endogenous depressions). Second, electrodermal abnormalities may in part measure trait disturbances in central noradrenergic processes that predispose an individual to experience depression characterized by either high or low levels of arousal. Individuals with electrodermal hypoactivity (suggesting noradrenergic underactivation) may tend to experience depressions characterized by underarousal (e.g., retarded depression), whereas people who are electrodermally hyperactive (suggesting noradrenergic overactivation) may be predisposed to have depressions with high levels of arousal (e.g., agitated or anxious depressions). Finally, whereas increased left frontal EEG alpha activity may be interpreted to signify a tonic disruption of dopaminergic functioning (because the asymmetry does not change with clinical remission), diminished CNV amplitudes appear to be state dependent, suggesting that phasic changes in dopaminergic functioning may underlie the expression of aspects of depressive symptomatology. Phasic dysregulation of dopaminergic systems (possibly indexed by the CNV and expressed as clinical depression) may be superimposed on a tonic vulnerability in the dopaminergic functioning (possibly indexed by EEG asymmetries). Although these hypotheses, based predominantly on psychophysiological research, offer a framework for linking behavioral and neurobiological aspects of depression, they remain to be tested directly.

References

Abraham, P., Docherty, T.B., Spencer, S.C., Verhey, F.H., Lamers, T.B., Emonds, P.M., Timsit-Berthier, M., Gerono, A., & Rousseau, J.C. (1980). An international pilot study of CNV in mental illness. In H.H. Kornhuber & L. Deecke (Eds.), *Motivation, motor, and sensory processes of the brain* (pp. 535–542). Amsterdam/New York: Elsevier.

Ahern, G.L., & Schwartz, G.E. (1985). Differential lateralization for positive and negative emotion in the human brain: EEG spectral analysis. *Neuropsychologia, 23*, 745–455.

Alema, G.C., Rosadini, G., & Rossi, G.F. (1961). Psychic reactions associated with intracarotid Amytal injection and relation to brain damage. *Excerpta Medica, 37*, 154–155.

Allen, J.J., Iacono, W.G., Depue, R.A., & Arbisi, P. (1993). Regional electroencephalographic asymmetries in bipolar seasonal affective disorder before and after exposure to bright light. *Biological Psychiatry, 33*, 642–646.

Andreasen, N.C., & Grove, W.M. (1982). The classification of depression: Traditional versus mathematical approaches. *American Journal of Psychiatry, 139*, 45-52.

Arito, H., & Oguri, M. (1990). Contingent negative variation and reaction time of physically-trained subjects in simple and discriminative tasks. *Industrial Health, 28*, 97-106.

Ashton, H., Golding, J.F., Marsh, V.R., Thompson, J.W., Hassanyeh, F., & Tyrer, S.P. (1988). Cortical evoked potentials and clinical rating scales as measures of depressive illness. *Psychological Medicine, 18*, 305-317.

Bennett, J., Davidson, R.J., & Saron, C. (1981). Patterns of self-rating in response to verbally elicited affective imagery: Relation to frontal vs. parietal EEG asymmetry. *Psychophysiology, 18*, 158.

Bernstein, A.S., Riedel, J.A., Graae, F., Siedman, D., Steele, H., Connolly, J., & Lubowsky, J. (1988). Schizophrenia is associated with altered orienting activity; depression with electrodermal (cholinergic?) deficit and normal orienting response. *Journal of Abnormal Psychology, 97*, 3-12.

Bernthal, P.J., & Koss, M.C. (1984). Evidence for two distinct sympathoinhibitory bulbo-spinal systems. *Neuropharmacology, 23*, 31-36.

Blashfield, R.J., & Morey, L.C. (1979). The classification of depression through cluster analysis. *Comprehensive Psychiatry, 20*, 516-527

Bolz, J., & Geidke, H. (1981). Controllability of an aversive stimulus in depressed patients and healthy controls: A study using slow brain potentials. *Biological Psychiatry, 16*, 441-452.

Boucsein, W. (1992). *Electrodermal activity*. New York: Plenum Press.

Byrne, D.G. (1975). A psychophysiological distinction between types of depressive states. *Australian and New Zealand Journal of Psychiatry, 9*, 181-185.

Byrne, D.G. (1976). Choice reaction time in depressive states. *British Journal of Social and Clinical Psychology, 15*, 149-156.

Carney, R.M., Hong, B.A., Kulkarni, S., & Kapila, A. (1981). A comparison of EMG and SCL in normal and depressed subjects. *Pavlovian Journal of Biological Science, 16*, 212-216.

Clark, L.A., & Watson, D. (1991). Tripartite model of anxiety and depression: Psychometric evidence and taxonomic implications. *Journal of Abnormal Psychology, 100*, 316-336.

Claverie, B., Brun, A., Nizard, A., Brenot, P., & Paty, J. (1984). Multiparametric outlines with CNV: Application to depressive syndromes. In R. Karrer, J. Cohen, & P. Teuting (Eds.), Brain and information. *Annals of the New York Academy of Sciences, 425*, 556-563.

Cummings, J.L., & Mendez, M.P. (1984). Secondary mania with focal cerebrovascular lesions. *American Journal of Psychiatry, 141*, 1084-1087.

Davidson, R.J., & Fox, N.A. (1988). Cerebral asymmetry and emotion: Developmental and individual differences. In D.L. Molfese & S.J. Segalowitz (Eds.), *Brain lateralization in children* (pp. 191-206). New York: Guilford Press.

Davidson, R.J., & Fox, N.A. (1989). Frontal brain asymmetry predicts infants' response to maternal separation. *Journal of Abnormal Psychology, 98*, 127-131.

Davidson, R.J., & Tomarken, A.J. (1989). Laterality and emotion: An electrophysiological approach. In F. Boller & J. Grafman (Eds.), *Handbook of*

Neuropsychology (Vol. 3, pp. 419-441). Amsterdam, The Netherlands: Elsevier.

Dawson, M., Schell, A., & Catania, J. (1977). Autonomic correlates of depression and clinical improvement following electroconvulsive shock therapy. *Psychophysiology, 14*, 569-578.

Denenberg, V.H. (1981). Hemispheric laterality in animals and the effects of early experience. *Behavioral and Brain Sciences, 4*, 1-49.

DePue, R.A., Arbisi, P., Spoont, M.R., Leon, A., & Ainsworth, B. (1989). Dopamine functioning in the behavioral facilitation system and seasonal variation in behavior: Normal population and clinical studies. In N.E. Rosenthal & M.C. Blehar (Eds.), *Seasonal affective disorders and phototherapy* (pp. 230-259). Guilford Press: New York.

Depue, R.A., & Iacono, W.G. (1989). Neurobehavioral aspects of affective disorders. *Annual Review of Psychology, 40*, 457-492.

Depue, R.A., Krauss, S., & Spoont, M.R. (1987). A two-dimensional threshold model of seasonal bipolar affective disorder. In D. Magnuson & A. Ohman (Eds.), *Psychopathology: An interactionist perspective* (pp. 95-123). New York: Academic Press.

Depue, R.A., Luciana, M., Arbisi, P., & Collins, P. (1994). Relation of agonist-induced dopamine activity to personality. *Journal of Personality and Social Psychology, 67*, 485-498.

Donat, D.C., & McCullough, J.P. (1983). Psychophysiological discriminants of depression at rest and in response to stress. *Journal of Clinical Psychology, 39*, 315-320.

Edman, G., Asberg, M., Levander, S., & Schelling, D. (1986). Skin conductance habituation and cerebrospinal fluid 5-hydroxyindoleacetic acid in suicidal patients. *Archives of General Psychiatry, 43*, 586-592.

Elton, M. (1984). A longitudinal investigation of event-related potentials in depression. *Biological Psychiatry, 19*, 1635-1649.

Flor-Henry, P. (1986). Observations, reflections, and speculations on the cerebral determinants of mood and on the bilaterally asymmetrical distributions of the major neurotransmitter systems. *Acta Neurologica Scandinavica, 74* (Suppl. *109*), 75-89.

Gainotti, G. (1972). Emotional behavior and hemispheric side of lesion. *Cortex, 8*, 41-55.

Gersh, F., & Fowles, D.C. (1979). Neurotic depression: The concept of anxious depression. In R.A. Depue (Ed.), *The psychobiology of the depressive disorders: Implications for the effect of stress* (pp. 55-80). New York: Academic Press.

Giedke, H., Bolz, J., & Heimann, H. (1980). Evoked potentials, expectancy wave, and skin resistance in depressed patients and healthy controls. *Pharmakopsychiatrie Neuro-Psychopharmakologie, 13*, 91-101.

Giedke, H., & Heimann, H. (1987). Psychophysiological aspects of depressive syndromes. *Pharmakopsychiatrie Neuro-Psychopharmakologie, 20*, 177-180.

Goldstein, K. (1939). *The organism: An holistic approach to biology, derived from pathological data in man.* New York: American Book.

Hamilton, M. (1960). A rating scale for depression. *Journal of Neurology, Neurosurgery, and Psychiatry, 23*, 56-62.

Henriques, J.B., & Davidson, R.J. (1990). Regional brain electrical asymmetries discriminate between previously depressed and healthy control subjects. *Journal of Abnormal Psychology, 99*, 22–31.

Henriques, J.B., & Davidson, R.J. (1991). Left frontal hypoactivation in depression. *Journal of Abnormal Psychology, 100*, 535–545.

Hoffman, G.M.A., Gonze, J.C., & Mendlewicz, J. (1985). Speech pause rate as a method for evaluation of psychomotor retardation in depressive illness. *British Journal of Psychiatry, 146*, 535–538.

Howard, R.C., Fenton, G.W., & Fenwick, P.B.C. (1982). *Event-related brain potentials in personality and psychopathology: A pavlovian approach*. Chichester, England: Research Studies Press.

Iacono, W.G. (1991). Psychophysiological assessment of psychopathology. *Psychological Assessment: A Journal of Consulting and Clinical Psychology, 3*, 309–320.

Iacono, W.G., & Ficken, J. (1989). Psychophysiological research strategies. In G. Tarpin (Ed.), *Handbook of clinical psychophysiology* (pp. 45–70). London: John Wiley.

Iacono, W.G., Ficken, J.W., & Beiser, M. (1993). Electrodermal nonresponding in first-episode psychosis as a function of stimulus significance. In J.C. Roy (Ed.), *Progress in electrodermal research* (pp. 239–255). New York: Plenum.

Iacono, W., Lykken, D., Haroian, K., Peloquin, L., Valentine, R., & Tuason, V. (1984). Electrodermal activity in euthymic patients with affective disorders: One-year retest stability and the effects of stimulus intensity and significance. *Journal of Abnormal Psychology, 93*, 304–311.

Iacono, W., Lykken, D., Peloquin, L., Lumry, A., Valentine, R., & Tuason, V. (1983). Electrodermal activity in euthymic unipolar and bipolar affective disorders. *Archives of General Psychiatry, 40*, 557–568.

Janes, C., & Strock, B. (1982). Skin conductance responding following major depressive episode remission. *Psychophysiology, 19*, 566. (Abstract)

Keller, F., Wolfersdorf, M., Straub, R., & Hole, G. (1991). Suicidal behaviour and electrodermal activity in depressive inpatients. *Acta Psychiatrica Scandinavica, 83*, 324–328.

Klorman, R., & Ryan, R.M. (1980). Heart rate, contingent negative variation, and evoked potentials during the anticipation of affective stimulation. *Psychophysiology, 17*, 513–523.

Knott, V.J., & Lapierre, Y.D. (1987). Electrophysiological and behavioral correlates of psychomotor responsivity in depression. *Biological Psychiatry, 22*, 313–324.

Knott, V.J., Lapierre, Y.D., Lugt, D.L., Griffiths, L., Bakish, D., Browne, M., & Horn, E. (1991). Preparatory brain potentials in major depressive disorder. *Progress in Neuro-Psychopharmacology and Biological Psychiatry, 15*, 257–262.

Kopell, B.S., Wittner, W.K., Lunde, D.T., Wolcott, L.J., & Tinklenberg, J.R. (1974). The effects of methamphetamine and secobarbital on the contingent negative variation amplitude. *Psychopharmacologia, 34*, 55–62.

Lader, M.H., & Wing, L. (1969). Physiological measures in agitated and retarded depressed patients. *Journal of Psychiatric Research, 7*, 89–100.

Lapierre, Y.D., & Butter, H.G. (1980). Agitated and retarded depression: A clinical psychophysiological investigation. *Neuropsychobiology, 6*, 217–223.

Lenhart, R.E. (1985). Lowered skin conductance in a subsyndromal high-risk depressive sample: Response amplitudes versus tonic levels. *Journal of Abnormal Psychology, 94*, 648–652.

Levinson, D.F. (1991). Skin conductance orienting response in unmedicated RDC schizophrenic, schizoaffective, depressed, and control subjects. *Biological Psychiatry, 30*, 663–683.

Loveless, N.E., & Sanford, A.J. (1974). Slow potential correlates of preparatory set. *Biological Psychology, 1*, 303–314.

MacKay, W.A., & Bonnet, M. (1990). CNV, stretch reflex and reaction time correlates of preparation for movement direction and force. *Electroencephalography and clinical Neurophysiology, 76*, 47–62.

Maes, M., Cosyns, P., Maes, L., D'Hondt, P., & Schotte, C. (1990). Clinical subtypes of unipolar depression: Part I. A validation of the vital and nonvital clusters. *Psychiatry Research, 34*, 29–41.

Maes, M., Maes, L., Schotte, C., Vandewoude, M., Martin, M., D'Hondt, P., Blockx, P., Scharpe', S., & Cosyns, P. (1990). Clinical subtypes of unipolar depression: Part III. Quantitative differences in various biological markers between the cluster-analytically generated nonvital and vital depression classes. *Psychiatry Research, 34*, 59–75.

Marczynski, T.J. (1978). Neurochemical mechanisms in the genesis of slow potentials: A review and some clinical implications. In D.A. Otto (Ed.), *Multidisciplinary perspectives in event-related brain potential research* (pp. 25–35). Washington, DC: U.S. Government Printing Office.

Marczynski, T.J. (1986). A model of the brain. *Electroencephalography and Clinical Neurophysiology* (Suppl. 38), 351–367.

McCarron, L.T. (1973). Psychophysiological discriminants of reactive depression. *Psychophysiology, 10*, 223–230.

McCarthy, G., & Donchin, E. (1978). Brain potentials associated with structural and functional visual matching. *Neuropsychologia, 16*, 571–585.

Mirkin, A.M., & Coppen, A. (1980). Electrodermal activity in depression: Clinical and biochemical correlates. *British Journal of Psychiatry, 137*, 93–97.

Nakamura, M., Iida, H., Fukui, Y., & Takahashi S. (1982). Melancholia and excessive CNV recovery after nonresponse condition. *Folia Psychiatrica et Neurologica Japonica, 36*, 81–88.

Noble, P., & Lader, M. (1971). The symptomatic correlates of the skin conductance changes in depression. *Journal of Psychiatric Research, 9*, 61–69.

Paykel, E.S. (1971). Classification of depressed patients: A cluster analysis derived grouping. *British Journal of Psychiatry, 118*, 275–288.

Perria, P., Rosadini, G., & Rossi, G.F. (1961). Determination of side of cerebral dominance with Amobarbital. *Archives of Neurology, 4*, 175–181.

Perris, C., & Monakhov, K. (1979). Depressive symptomatology and systemic structural analysis of the EEG. In J. Gruzelier & P. Flor-Henry (Eds.), *Hemisphere asymmetries of function in psychopathology*. Amsterdam/New York/Oxford: Elsevier North-Holland.

Pockberger, H., Petsche, H., Rappelsberger, P., Zidek, B., & Zapotoczky, H.G. (1985). On-going EEG in depression: A topographic spectral analytical pilot study. *Electroencephalography and Clinical Neurophysiology, 61*, 349–358.

Ray, W.J. (1990). The electrocortical system. In J.T. Cacioppo & L.G. Tassinary (Eds.), *Principals of psychophysiology: Physical, social, and inferential elements* (pp. 385–412). New York: Cambridge University Press.

Robinson, R.G. (1983). Poststroke affective disorders. In M. Reivich (Ed.), *Cerebrovascular disease, 13th Princeton Conference* (pp. 137–152). New York: Raven Press.

Robinson, R.G., & Benson, D.F. (1981). Depression in aphasic patients: Frequency, severity, and clinical-pathological correlations. *Brain and Language, 14*, 282–291.

Robinson, R.G., Kubos, K.L., Starr, L.B., Reo, K., & Price, T.R. (1984). Mood disorders in stroke patients: Importance of location of lesion. *Brain, 107*, 81–93.

Robinson, R.G., & Price, T.R. (1982). Post-stroke depressive disorders: A follow-up study of 103 patients. *Stroke, 13*, 635–641.

Robinson, R.G., & Szetele, D. (1981). Mood changes following left-hemisphere brain injury. *Annals of Neurology, 9*, 447–453.

Rockstroh, B., Elbert, T., Canavan, A., Lutzenberger, W., & Birbaumer, N. (1989). *Slow cortical potentials and behaviour*. Munich: Urban & Schwarzenberg.

Rockstroh, B., Elbert, T., Lutzenberger, W., & Birbaumer, N. (1982). The effects of slow cortical potentials on response speed. *Psychophysiology, 19*, 211–217.

Rohrbaugh, J.W., & Gaillard, A.W.K. (1983). Sensory and motor aspects of the contingent negative variation. In A.W.K. Gaillard & W. Ritter (Eds.), *Tutorials in ERP research: Endogenous components* (pp. 269–310). Amsterdam, The Netherlands: North-Holland.

Schaffer, C.E., Davidson, R.J., & Saron, C. (1983). Frontal and parietal electroencephalogram asymmetry in depressed and nondepressed subjects. *Biological Psychiatry, 18*, 753–762.

Shagass, C. (1972). Electrical activity of the brain. In N.S. Greenfield & R.H. Sternbach (Eds.), *Handbook of psychophysiology* (pp. 263–328). New York: Holt, Reinhart & Winston.

Silverstone, T. (1985). Dopamine in manic depressive illness. *Journal of Affective Disorders, 8*, 225–231.

Simons, R.F., Ohman, A., & Lang, P.J. (1979). Anticipation and response set: Cortical, cardiac, and electrodermal correlates. *Psychophysiology, 16*, 222–233.

Small, J.G., & Small, I.F. (1971). Contingent negative variation (CNV) correlations with psychiatric diagnosis. *Archives of General Psychiatry, 25*, 550–554.

Small, J.G., Small, I.F., Perez, H.C. (1971). EEG, Evoked potential, and contingent negative variations with lithium in manic depressive disease. *Biological Psychiatry, 3*, 47–58.

Storrie, M., Doerr, H., & Johnson, M. (1981). Skin conductance characteristics of depressed subjects before and after therapeutic intervention. *Journal of Nervous and Mental Disease, 169*, 176–179.

Swerdlow, N.R., & Koob, G.F. (1987). Dopamine, schizophrenia, and depression: Toward a unified hypothesis of cortico-striato-pallido-thalamic function. *Behavior and Brain Sciences, 10*, 197–208.

Tecce, J.J. (1991). Dopamine and CNV: Studies of drugs, disease and nutrition. *Electroencephalography and Clinical Neurophysiology* (Suppl. 42), 153–164.

Tecce, J.J., Cattanach, L., Hill, C.D., & Cole, J.O. (1987). Dextroamphetamine effects on CNV magnitude in type A and B individuals. *Electroencephalography and Clinical Neurophysiology* (Suppl. 40), 549–555.

Tecce, J.J., & Cole, J.O. (1974). Amphetamine effects in man: Paradoxical drowsiness and lowered electrical brain activity (CNV), *Science, 185*, 451–453.

Tecce, J.J., Cole, J.O., & Savignano-Bowman, J. (1975). Chlorpromazine effects on brain activity (contingent negative variation) and reaction time in normal women. *Psychopharmacologia, 43*, 293–295.

Tellegen, A. (1985). Structures of mood and personality and their relevance to assessing anxiety, with an emphasis on self-report. In A.H. Tuma & J.D. Maser (Eds.), *Anxiety and the anxiety disorders* (pp. 681–706). Hillsdale, NJ: Lawrence Erlbaum.

Terzian, H. (1964). Behavioral and EEG effects of intracarotid sodium amytal injections. *Acta Neurochirurgica, 12*, 230–240.

Thier, P., Axmann, D., & Giedke, H. (1986). Slow potentials and psychomotor retardation in depression. *Electroencephalography and Clinical Neurophysiology, 63*, 570–581.

Thompson, J.W., Newton, P., Pocock, P.V., Cooper, R., Crow, H., McCallum, W.C., & Papakostopoulos, D. (1978). Preliminary study of pharmacology of contingent negative variation in man. In D.A. Otto (Ed.), *Multidisciplinary perspectives in event-related potentials research* (pp. 51–55). Washington, DC: U.S. Government Printing Office.

Timsit-Berthier, M. (1986). Contingent negative variation (CNV) in psychiatry. *Electroencephalography and Clinical Neurophysiology* (Suppl. 38), 429–434.

Timsit-Berthier, M., Gerono, A., Rousseau, J.C., Mantanus, H., Abraham, P., Verhey, F.H.M., Lamers, T., & Emonds, P. (1984). An international pilot study of CNV in mental illness. In R. Karrer, J. Cohen, & P. Teuting (Eds.), Brain and information. *Annals of the New York Academy of Sciences, 425*, 629–637.

Timsit-Berthier, M., Mantanus, H., Ansseau, M., Devoitille, J., Dal Mas, A., & Legros, J.J. (1987). Contingent negative variation in major depressive patients. *Electroencephalography and Clinical Neurophysiology* (Suppl. 40), 762–771.

Timsit-Berthier, M., Mantanus, H., Ansseau, M., Doumont, A., & Legros, J.J. (1983). Methodological problems raised by contingent negative variations interpretation in psychopathological conditions. In C. Perris, D. Kemali, & M. Koukkou-Lehmann (Eds.), *Advances in biological psychiatry: Vol. 13. Neurophysiological correlates of normal cognition and psychopathology* (pp. 80–92). Basel: Karger.

Timsit-Berthier M., Mantanus, P., Marissiaux, P., Ansseau, M., Doumont, A., Geenen, V., & Legros, J.J. (1986). CNV and dopamine receptor reactivity: Correlations with the apomorphine test: *Electroencephalography and Clinical Neurophysiology* (Suppl. 40), 403–405.

Timsit-Berthier, M., Mantanus, H., Poncelet, M., Marissiaux, P., & Legros, J.J. (1986). Contingent negative variation as a new method to assess the catecholaminergic systems. In V. Gallai (Ed.), *Maturation of the CNS and evoked potentials* (pp. 260–268). New York: Elsevier Science.

Timsit-Berthier, M., & Timsit, M. (1981). Toward a neurochemical interpretation of CNV in psychiatry. In C. Perris, D. Kemali, & L. Vacca (Eds.), *Advances in* biological psychiatry: Vol. 6. Electroneurophysiology and psychopathology (pp. 165–172). Basel, Switzerland: Karger.

Tomarken, A.J., Davidson, R.J., Wheeler, R.E., & Doss, R.C. (1992). Individual differences in anterior brain asymmetry and fundamental dimensions of emotion. *Journal of Personality and Social Psychology, 62*, 676–687.

Tucker, D.M. (1981). Lateral brain function, emotion, and conceptualization. *Psychological Bulletin, 89*, 19–46.

Tucker, D.M., & Dawson, S.L. (1984). Asymmetric EEG power and coherence as method actors generated emotions. *Biological Psychology, 19*, 63–75.

Tucker, D.M., & Williamson, P.A. (1984). Asymmetric neural control systems in human self-regulation. *Psychological Review, 91*, 185–215.

Verhey, F., Lamers, T., & Edmonds, P. (1984). A second baseline in relating ERP and measured psychopathology. In R. Karrer, J. Cohen, & P. Teuting (Eds.) Brain and information. *Annals of the New York Academy of Sciences, 425*, 638–644.

Walter, W.G. (1964). The contingent negative variation: An electrical sign of the significance of association in the human brain. *Science, 146*, 434.

Walter, W.G., Cooper, R., Aldridge, V.J., McCallum, W.C., & Winter, A.L. (1964). Contingent negative variation: An electric sign of sensorimotor association and expectancy in the human brain. *Nature, 203*, 380–384.

Ward, N.G., Doerr, H.O., & Storrie, M.C. (1983). Skin conductance: A potentially sensitive test for depression. *Psychiatry Research, 10*, 295–302.

Watson, D., Clark, L.A., & Carey, G. (1988). Positive and negative affect and their relation to anxiety and depressive disorders. *Journal of Abnormal Psychology, 97*, 346–353.

Watson, D., & Kendall, P.C. (1989). Understanding anxiety and depression: their relation to negative and positive affective states. In P.C. Kendall & D. Watson (Eds.), *Anxiety and depression: Distinctive and overlapping features* (pp. 3–25). New York: Academic Press.

Watson, D., Clark, L.A., & Carey, G. (1988). Positive and negative affect and their relation to anxiety and depressive disorders. *Journal of Abnormal Psychology, 97*, 346–353.

Wheeler, R.E., Davidson, R.J., & Tomarken, A.J. (1993). Frontal brain asymmetry and emotional reactivity: A biological substrate of affective style. *Psychophysiology, 30*, 82–89.

Whitlock, F.A. (1982). *Symptomatic affective disorders*. New York: Academic Press.

Williams, K., Iacono, W., & Remick, R. (1985). Electrodermal activity among subtypes of depression. *Biological Psychiatry, 20*, 158–162.

Wing, J.K., Cooper, J.E., & Sartorius, N. (1974). *Measurement and classification of psychiatric symptoms: An instruction manual for the PSE and CATEGO program*. Cambridge, England: Cambridge University Press.

Yamamoto, K., Arai, H., & Nakayama, S. (1990). Skin conductance response after 6-hydroxydopamine lesion of central noradrenergic system in cats. *Biological Psychiatry, 28*, 151–160.

Yamamoto, K., Hoshino, T., Takahashi, Y., Kaneko, H., & Ozawa, N. (1991). Skin conductance activity after intraventricular administration of 6-hydroxydopa in rats. *Biological Psychiatry, 29*, 365–375.

Zahn, T.P. (1986). Psychophysiological approaches to psychopathology. In M.G.H. Coles, E. Donchin, S.W. Porges (Eds.), *Psychophysiology: Systems, processes, and applications* (pp. 508–610). New York: Guilford Press.

Zimmerman, M., Stangl, D., & Coryell, W. (1985). The research criteria for endogenous depression and the dexamethasone suppression test: A discriminant function analysis. *Psychiatry Research, 14*, 197–208.

9
The Cognitive Diathesis for Depression

Michael W. O'Hara

Cognitive models dominate theorizing and research on the psychological etiology of depression (Alloy, 1988). These models have two major common characteristics. First, each of these models postulates the existence of a psychological characteristic that puts an individual at risk for depression. These characteristics are often called *diatheses* and are usually presumed to be rather stable and to have developed as early as childhood. Examples of diatheses include depressive self-schema (Beck, Rush, Shaw, & Emery, 1979), dysfunctional attributional style (Abramson, Seligman, & Teasdale, 1978), and maladaptive self-control (Rehm, 1977). Second, cognitive theories usually imply that some sort of stress (often conceptualized as a stressful life event) is necessary to activate the diathesis and bring on an episode of depression. The theories vary in the extent to which the nature of the stressful event should match in some way the depression diathesis (e.g., a loss schema matching with an "exit" event).

One of the important advances in testing the validity of cognitive models of depression over the past 10 years has been the use of the longitudinal design. This design represents an advance over earlier (and still commonly used) cross-sectional designs. Particularly in the area of depression research, where some of the presumed etiological factors are similar to the symptoms of the disorder, it is important to separate in time the measurement of the cause and the outcome. In contrast to longitudinal studies of risk factors in schizophrenia, populations that are followed in depression studies often are not considered to be at high risk for depression. As will become clear later in this chapter, only a few investigators (most notably, Constance Hammen) have followed, for example, offspring of parents with major depression to test cognitive theories of depression. The only other high-risk context in which cognitive theories have been tested is childbearing (O'Hara & Zekoski, 1988). Recent studies, however, have suggested that childbearing women may not be at increased risk for nonpsychotic depression (O'Hara, Zekoski, Philipps, & Wright, 1990; Troutman & Cutrona, 1990).

The two cognitive models that have had the greatest influence in research on depression are the helplessness model of depression (Abramson et al., 1978) and Beck's cognitive model of depression (Beck et al., 1979). Prospective research that bears on the validity of these two models will be the focus of this chapter. In the first section, the helplessness model and Beck's cognitive model of depression are briefly described. The role of stressful life events (the activating component in these theories) in the etiology of depression are addressed next. Third, the various designs that have been used in prospective research are discussed. Fourth, the prospective research bearing on the helplessness model and Beck's cognitive model are reviewed. Finally, the significance of this work in illuminating the etiology of depression is discussed.

Cognitive Models

Learned Helplessness

The reformulated learned helplessness model of depression (Abramson et al., 1978)[1] posits that internal, stable, and global attributions for past or present noncontingency (often negative events such as a failure or loss) should lead to the expectation of future noncontingency and consequently to symptoms of helplessness/depression. Abramson et al. allowed that many factors influence the attributions that individuals might make for instances of past or present noncontingency. One psychological factor was labeled attributional style or the predisposition to make certain types of attributions for events. A dysfunctional attributional style (e.g., internal, stable, and global attributions for negative outcomes) was regarded as a potential diathesis for depression. That is, individuals who habitually

---

[1] The most recent version of the helplessness model of depression (Abramson, Metalsky, & Alloy, 1989; Abramson, Seligman, & Teasdale, 1978) posits that hopelessness is a proximal sufficient cause of depression. Hopelessness in this context is defined as "the expectation that highly desired outcomes are unlikely to occur or that highly aversive outcomes are likely to occur and that no response in one's repertoire will change the likelihood of the occurrence of these outcomes" (Abramson, et al., 1988, p. 7). The path to this state of hopelessness is often through stable and global attributions for important negative life events (e.g., loss of a relationship or job, poor performance in some valued activity). These attributions are described as proximal contributory causes; that is, in and of themselves, they will not lead directly to a depressive episode. Abramson and her colleagues argue that the attributions made by an individual for a particular negative life event are influenced by factors such as a depressogenic attributional style and situational cues that are likely to affect specific attributions (Abramson et al., 1989). These factors influencing specific attributions, and the negative events themselves are considered distal contributory causes of depression (Abramson et al., 1989).

make internal, stable, and global attributions for negative events are at increased risk for depression. Research addressing the validity of the reformulated learned helplessness model has largely focused on the association between attributional style and concurrent or later depression. Relatively less research has examined attributions for specific events and their association with later depression (Cutrona, 1983).

Beck's Cognitive Model

A central element of the cognitive model is the cognitive triad, which includes beliefs reflecting a negative view of self, world, and future (Beck, Rush, Shaw, & Emery, 1979). The cognitive model posits that these beliefs lead directly to the symptoms of depression. For example, a negative view of self would lead directly to low self-esteem. A negative view of the world might result in disturbances in interpersonal relationships that could have a direct impact on social adjustment. A negative view of the future may lead to pessimism or in the extreme, suicidal ideation.

Beck and his colleagues (Beck et al., 1979) posit the existence of idiosyncratic schemas, often formed at an early age, which organize current life experiences in a manner that makes an individual vulnerable to depression. For example, an individual who experienced a painful loss as a child might have a schema regarding loss that is very broad, such that any sort of a loss, no matter how trivial, is encoded as a significant event. The schema might also encode losses as having catastrophic consequences so that the individual on experiencing even a small loss (e.g., being passed over for a promotion) is subject to cognitions and feelings associated with a major loss (e.g., death of a loved one). The model further states that activation of depressive schemas is also associated with an increased likelihood of faulty information processing or cognitive distortions. These systematic errors in thinking such as arbitrary inference and selective abstraction are thought to be directly responsible for the cognitive triad.

Similar to the helplessness model, Beck's cognitive model posits that some sort of negative life event usually starts this chain. The major difference in the models with respect to the role of life events is that the cognitive model seems to allow more clearly for an individual to misinterpret the significance or importance of a life event.

Stressful Life Events

Adverse life circumstances have long been recognized as a potential cause of depression (Paykel, 1979). The concept of reactive depression as a diagnosis, in contrast to endogenous depression, recognizes the importance of stressful life events. Despite a number of methodological difficulties with life events research (Monroe & Peterman, 1988), studies have consistently found a link between stressful life events and depression defined in terms of both clinical diagnosis and high symptom levels (Monroe & Peterman, 1988; Paykel, 1982). However, the proportion of variance in depression symptom level or diagnosis accounted for by negative life events often is relatively small. The most potent life events with respect to causing depression appear to be those that involve severe long-term threat or loss (G.W. Brown & Harris, 1989; Paykel, 1982).

The role of stressful life events in cognitive models is to activate a "pathological" psychological process that results in symptoms of depression. In the view of the learned helplessness model, this process involves the expectation of future noncontingency, and in the cognitive model, this process involves the negative cognitive triad. Within the framework of the learned helplessness model, any major negative event or series of negative events of lesser magnitude could be perceived as significant and as having causes that are internal, stable, and global, thus leading to a helplessness depression. Within the framework of the cognitive model, the same sort of analysis would apply. The cognitive model, however, does imply that there should be some sort of matching between the nature of an individual's depressogenic schema and the type of negative life event. That is, a negative life event involving loss should be more likely to provoke a depression in an individual who has a loss-related negative self-schema than an individual who has an achievement-related negative self-schema. Although many of the studies of cognitive models of depression have included a consideration of negative life events and their interaction with cognitive diatheses (at least in a statistical sense), in only a few recent studies has the specific nature of the negative events received much attention (e.g., Hammen, Ellicott, & Gitlin, 1989).[2]

Prospective Designs

Prospective studies of the cognitive etiology of depression have taken several different approaches. These approaches include (a) measuring a cognitive construct and depression at two points in time, (b) measuring a cognitive construct and depression and following subjects through a predictable stressful life event, and (c) measuring a cognitive construct, depression, and wide range of stressful life events (which may or may not have specific etiological relevance in the depression model) and following subjects for a predetermined amount of time. The logic of most of this

---

[2] An excellent discussion of conceptual issues related to the role of life stressors in diathesis–stress theories of depression is contained in Monroe and Simons (1991).

work is that once the relation between depression at time 1 and time 2 has been accounted for, such as in a hierarchical multiple regression, a significant association still should be found between the cognitive construct at time 1 and depression at time 2. In some studies, it is also predicted that a concurrent measure of negative life events and the interaction of the level of stressful life events and the cognitive construct should also be associated with depression at time 2.

Review of the Literature

Attributional Reformulation of the Learned Helplessness Model of Depression

The majority of work on the cognitive high-risk paradigm has tested predictions from the helplessness model of depression (Abramson et al., 1978). Almost all studies reviewed in this section used self-report measures of depressive symptomatology such as the Beck Depression Inventory (BDI) (Beck, Ward, Mendelson, Mock, & Erbaugh, 1961) as the outcome measure.

Time 1 to Time 2

In an early study, Golin, Sweeney, and Shaeffer (1981) assessed level of depressive symptomatology and attributional style at two points in time, 1 month apart, in a sample of 180 college students. The investigators performed a cross-lagged panel correlational analysis and found that, much as the helplessness model would predict, more stable and global attributions (and the composite measure of attributional style) for bad outcomes and less stable attributions for good outcomes at time 1 were significantly related to higher levels of depressive symptomatology at time 2. These correlations were significantly greater than the associations between depression at time 1 and the attributional measures at time 2. Although the requirements for a cross-lagged panel analysis are rather stringent (Kenny & Harackiewicz, 1979), the data in this study met most of the statistical assumptions of this procedure.

Lewinsohn and his colleagues (Lewinsoh, Hoberman, & Rosenbaum, 1988; Lewinsohn, Steinmetz, Larson, & Franklin, 1981) followed a community sample ($N = 998$) over about an 8-month period. They obtained self-report (Center for Epidemiological Studies Depression Scale [CES-D], Radloff, 1977) and interview-based assessments of depression (Research Diagnostic Criteria [RDC]-defined depression). With respect to attributions, only the locus of control dimension for success and failure was assessed. Only external attributions for negative events predicted time 2 CES-D scores after controlling for time 1 CES-D scores. The attributional measures were not associated with time 2 diagnostic status.

Seligman et al. (1984) assessed depression and attributional style in 96 third through sixth graders at two points in time, 6 months apart, using the Children's Depression Inventory (CDI; Kovacs & Beck, 1977) and the Children's Attributional Style Questionnaire (CASQ). In a hierarchical multiple regression, they found that the CASQ composite (of the locus of control, stability, and globality dimensions) for negative events accounted for significant variance in time 2 depression over and above the variance accounted for by time 1 depression and other demographic variables. Depression at time 1 was not associated with attributional style at time 2.

Each of the studies described in this section followed subjects prospectively; however, subjects were not followed through any sort of high-risk period, and attributions for concurrent negative life events were not assessed. The two studies using the adult or child version of the Attributional Style Questionnaire (ASQ) obtained the predicted results. The findings from the Lewinsohn et al. (1988) study, in which only the locus of control dimension was assessed, were somewhat inconsistent with the predictions of the reformulated helplessness model (Abramson et al., 1978). This inconsistency, however, does not cause much difficulty for the revised version of the helplessness model, the hopelessness model (Abramson, Metalsky, & Alloy, 1989). The revision asserts that the locus of control dimension is not related to risk for depression; rather, it is related only to the likelihood that self-esteem deficits will be part of the picture of depression if it develops.

CHILDBEARING

Childbearing has been an important context within which attributional predictions have been tested. Two assumptions have underlain much of this research: women are at increased risk for depression after childbirth, and the social and physical adjustments (many of which could be considered negative events, e.g., colicky baby) related to childbirth are stressful. The basic prediction of most of these studies has been that attributional style characterized by internal, stable, and global attributions for negative events (and the opposite for positive events) would put women at increased risk for depression after childbirth.

In an early study, Manly, McMahon, Bradley, and Davidson (1982) followed 50 women from the third trimester of pregnancy until 3 days postpartum. Women were given the ASQ and the BDI, the Depression Adjective Check List, and the McLean-Hakstian depression scale during pregnancy and after delivery. In the context of a hierarchical multiple regression, neither attributions for negative or positive events were associated with level of depressive symptomatology after controlling for initial level of depressive symptomatology for each of the depression measures. In a very similar study, however, O'Hara, Rehm, and Campbell (1982) followed 170 women from the second trimester of pregnancy until, on the average, 12 weeks postpartum. Attributional style (assessed by the ASQ) and depression level (the BDI) were assessed during pregnancy; only depression level was assessed postpartum. In the context of a hierarchical multiple regression, the set of cognitive-behavioral variables that included a composite index of attributional style accounted for significant variance in postpartum depression level. The measure of attributional style was the only significant variable in that set.

Cutrona (1983) followed 85 women from the third trimester of pregnancy until 2 and 8 weeks postpartum. Subjects completed the ASQ (only items for negative events) and the BDI during pregnancy. After delivery at 2 and 8 weeks postpartum, subjects completed a BDI and a measure of subjects' actual attributions for stressful childcare-related events. Limiting the analysis to those subjects who were not depressed during pregnancy (BDI < 9), it was found that attributional style during pregnancy was significantly associated with level of depressive symptomatology at 2 and 8 weeks postpartum (after controlling for initial depression level). Moreover, attributional style was also predictive of depression level at 8 weeks postpartum among women who were depressed at 2 weeks postpartum after controlling for both level of depression during pregnancy and at 2 weeks postpartum. Finally, Cutrona found that attributions for actual stressful events did not mediate the significant relation between attributional style and level of postpartum depressive symptomatology.

Later prospective studies have not yielded much support for attributional style or attributions for actual events as etiologically related to postpartum depression. O'Hara, Neunaber, and Zekoski (1984) followed 99 women from the second trimester of pregnancy until 9 weeks postpartum. Attributional style measured during pregnancy was not associated with level of postpartum depressive symptomatology or postpartum depression diagnosis. Similarly, Whiffen (1988) found no association between either pre- or postpartum attributions for actual negative life events and level of postpartum symptomatology or depression diagnosis.

EXAMINATION PERFORMANCE

Metalsky and colleagues (Metalsky, Abramson, Seligman, Semmel, & Peterson, 1982; Metalsky, Halberstadt, & Abramson, 1987) have reported on two prospective studies of college students. Students completed measures of attributional style and mood before a midterm examination and indicated the grades on the examination with which they would be happy and unhappy. In the first study (Metalsky et al., 1982), the attributional dimensions of locus of control and globality were predictive of post-test feedback mood scores (after controlling for mood scores at the time attributional style was assessed) for students who performed poorly (by their own standards). Attributional style was not predictive of mood among students who did not do poorly on the examination. The authors argued that the study constituted strong support for the helplessness model because attributional style in the context of a specific negative event (but not in the absence of a specific negative event) was predictive of later mood. This position was later challenged by Williams (1985), who showed that the correlations between measures of attributional style and later mood in the subjects who had done poorly and those who had not done poorly on the examination did not differ significantly.

Metalsky et al. (1987) conducted a second study to replicate and extend the results of their earlier work. In this study, generality of attributions (stability and globality combined, consistent with the hopelessness reformulation, Abramson et al., 1989) for achievement-related events and the interaction of examination outcome and generality attributions did not predict depressive mood at the time the students received the results of their examination. However, the examination outcome itself was significantly related to depressive mood. When mood was reexamined 2 days later, the interaction of examination outcome and generality attributions significantly predicted depressive mood. The authors also found that generality attributions for the students' performance on the examination were significantly correlated with generality attributions for hypothetical achievement-related events. Further, the specific attributions that students made for their performance on the examination appeared to mediate the relation between attributional style and mood 2 days after students received their examination grade. Finally, attributional style assessed for negative interpersonal outcomes was not related to mood at any point.

Hunsley (1989) also examined the relation between attributional style for negative events and mood following return of examination grades. Interestingly, performance satisfaction and the interaction of performance satisfaction and attributional style for negative events were significant predictors of mood immediately after the examination. However, only performance satisfaction was a significant predictor of mood immediately after grades were returned. This latter finding was consistent with that of Metalsky et al. (1987). Unlike Metalsky et al. (1987), Hunsley did not assess mood a second time after grades were returned.

Follette and Jacobson (1987) used methodology similar to Metalsky et al. (1982) and Hunsley (1989). They found that attributional style for negative events was not predictive of depressive mood either alone or in the context of an attributional style X satisfaction with performance interaction following the receipt of an examination grade. Time 1 Multiple Affect Adjective Check List (MAACL) depression scores and satisfaction with performance were significant predictors of MAACL depression scores following receipt of examination grades. Attributions made for examination performance at the time the grades were received were also significantly correlated with MAACL depression scores measured at the same time.

Life Events

Adults

Hammen and her colleagues have conducted a number of studies evaluating elements of the helplessness model of depression. In an early study (Hammen, Krantz, & Cochran, 1981), 315 students were followed over a 7- to 8-week period. They completed the BDI and a measure of attributions (including the control, locus of control, stability, and globality dimensions) for five of the most stressful life events occurring in the past 3 months. Depression and the other measures were obtained again at the second assessment. Low controllability and globality were associated with later depression; however, initial level of depression appeared not to have been entered in the hierarchical multiple regression.

Cochran and Hammen (1985) followed 409 college students over a 2-month period. Depression was represented at both points in time by a latent variable composed of the BDI, Profile of Mood States, and the Costello and Comrey Depression Scale. At both points in time, subjects made attributions for the causes of up to five personally upsetting life events and then rated the causes of these events on the dimensions of degree of upset, control, globality, stability, locus of control, and intentionality. In the context of testing a causal model linking depression and attributions for upsetting events at time 1 with depression at time 2, only depression at time 1 showed a significant relation with depression at time 2.

Johnson and Miller (1990) assessed attributional style, undesirable life events, and depression in a sample of 80 undergraduate students on two occasions separated by a 1-month interval. After accounting for time 1 depression, the effects for attributional style, undesirable life events, their interaction were not significant in accounting for time 2 depression.

Metalsky and Joiner (1992) followed 152 college students over a 5-week period. Their interest was in testing the diathesis stress and the causal mediation components of the hopelessness model (see footnote 1). These cognitive diatheses included (a) attributional style (generality attributions); (b) a generalized tendency to make negative inferences about the self, given that a negative life event has occurred; and (c) a generalized tendency to infer that negative life events are going to lead to dire consequences. They found that the interactions of attributional style and stress, negative inferences about self and stress, and maladaptive inferences about negative life events and stress were all significant predictors of time 2 depression, after accounting for time 1 depression and (in separate regressions) the individual components of the interaction terms (e.g., attributional style). They also found evidence that hopelessness was a partial mediator of the association between the depression and the interactions of stress and generality attributions and stress and negative inferences about the self. Finally, no evidence indicated that these effects were observed in predicting anxiety, suggesting a specific relation between several components of the hopelessness model and depression.

Pagel, Becker, and Coppel (1985) followed 38 caretakers of spouses with Alzheimer's disease patients over a 10-month period. Their principal concern was with the extent to which the caretakers' judgments regarding the amount of control they had over several important current stressful situations (i.e., their spouses' unpredictable, upsetting behavior, the life changes they experienced as a result of their spouses' illness, and their personal reactions to their spouses' unpredictable behavior) was predictive of later depression. The authors also assessed caretakers' locus of control regarding the patients' initial symptoms before the Alzheimer's diagnosis. Finally, they assessed dispositional optimism. The measure of depression was a composite of the BDI, the depression subscale of the Symptom Check List (SCL-90-R), and the Hamilton Depression Rating Scale. They found that initial depression, perceived loss of control over the patients' symptoms, the interaction of internal attributions for patients' symptoms and perceived loss of control over patients' symptoms, and perceived loss of control of personal reactions to patients' symptoms were significantly related to follow-up depression level in the context of a hierarchical multiple regression.

Children

Hammen, Adrian, and Hiroto (1988) followed 79 children (average age = 12.5 years) of unipolar and bipolar depressive mothers over a 6-month period. The initial assessment included current and lifetime history of psychiatric disorder, the CDI, and attributional style using the CASQ. At the 6-month follow-up, diagnostic status was assessed again, as was the degree of threat of recent stressful life events. Depression at time 1 and degree of threat were the significant predictors of depression diagnosis at time 2. Interestingly, the attribution X degree of threat interaction was a significant predictor of nonaffective diagnoses at time 2.

Nolen-Hoeksema, Girgus, and Seligman (1986) followed 168 children in grades 3 through 5 for 1 year. Five times during the year, the children completed the CDI, CASQ, and a life events questionnaire. These workers found that attributional style significantly predicted later depression for each adjacent occasion after controlling for concurrent depression level. In a similar manner, however, depression level predicted later attributional style for each adjacent occasion. Also, for two of the four occasions, the interaction of the negative life events measure and attributional style was predictive of later depression (time 2 to time 3 and time 4 to time 5).

J.D. Brown and Siegel (1988) followed 176 females in grades 7 through 11 over a period of 8 months. At the initial assessment, depression was measured using the CES-D. Subjects also completed a life stress inventory and were asked to select the most upsetting event. Subjects made ratings about how upsetting the event was and how much control they had over the event and ratings of the locus of control, stability, and globality of the cause of the event. The authors predicted that the interaction of attribution style and controllability judgments would predict future depression. This prediction was confirmed. After controlling for time 1, depression, internal, stable, and global attributions for *uncontrollable* events were related to depression 8 months later. For outcomes ascribed to controllable causes, internal and global attributions were associated with lower levels of depressed mood. Interestingly, the stressful event for which students made the attributions occurred at least 8 months before the follow-up depression assessment.

Beck's Cognitive Model

The studies to be reviewed in this section are similar to those reviewed in the learned helplessness section with respect to design and subjects, with one important exception. Several recent tests of the cognitive model have included depressed patients in remission and have evaluated elements of the cognitive model in predicting depression relapse and remission.

Time 1 to Time 2

The Lewinsohn et al. (1988) study described earlier (see section entitled, "Attributional Reformulation of the Learned Helplessness Model of Depression") also included measures of cognitive constructs related to the cognitive model. Several of these constructs measured at time 1 were predictive of level of depression (CES-D scores) 8 months later (time 2) after controlling for time 1 depression scores. These constructs included expectancies of positive and negative outcomes and irrational beliefs. None of the cognitive variables were associated with diagnostic status at the time 2 assessment.

Childbearing

Two studies have evaluated the cognitive model in the context of a specific life event (Gotlib, Whiffen, Wallace, & Mount, 1991; O'Hara, Rehm, & Campbell, 1982). The O'Hara et al. prospective study of postpartum depression was described earlier (see section entitled,

"Childbearing"). The Dysfunctional Attitude Scale (DAS; Weissman & Beck, 1978) was the measure of the cognitive diathesis, and it was administered during pregnancy. Although the DAS was significantly correlated with postpartum depression level, in the hierarchical multiple regression, it did not show a significant association with postpartum depression level after accounting for level of depression during pregnancy and the other cognitive-behavioral variables (e.g., attributional style and self-control). Similar findings were obtained by Gotlib et al. (1991), who found that the DAS obtained during pregnancy was not associated with postpartum depression diagnostic status in a sample of 435 women.

LIFE EVENTS

Hammen, Marks, deMayo, and Mayol (1985) followed 83 college students over 4 months who were classified into one of four groups: depressed with a depressive self-schema, depressed without a depressive self-schema, nondepressed with a depressive self-schema, and nondepressed without a depressive self-schema. No association was found between schema status and depression (based on self-report or clinical interview) over a 4-month period. Also, the interaction of schema status and negative life events was not significantly related to later depression. Initial depression status and negative life events, however, were related to later depressive symptomatology. The subjects in this study also completed an additional assessment of memories of times within the past month when they felt good and bad about themselves on two occasions, 2 months apart. The ratio of bad/good memories (measure of schema relevant recall) was the dependent variable. In summary, the authors found that the bad/good memory ratio was related to current depression level and not to schema classification.

Barnett and Gotlib (1988) followed 57 female undergraduates over a 3-month period. At time 1, subjects completed the BDI, the DAS, a life events measure, and a measure of social support. In the context of a hierarchical multiple regression, Barnett and Gotlib found that depression measured at time 1 and life events and social support measured at time 2 and the interaction of dysfunctional attitudes measured at time 1 and social support measured at time 2 were significant predictors of depression at time 2. Dysfunctional attitudes alone and their interaction with life events were not significant in the regression.

Barnett and Gotlib (1990) replicated their earlier study with a larger sample of female undergraduates ($N = 177$), and they included 63 male undergraduates as well. Once again, they found that depression measured at time 1 and life events and social support measured at time 2 and the interaction of dysfunctional attitudes measured at time 1 and social support measured at time 2 were significant predictors of depression at time 2 for female subjects. The same main effects were significant for the male subjects as well as the interaction of social support and life events, both measured at time 2. Two subscales of the DAS based on an earlier factor analysis (Cane, Olinger, Gotlib, & Kuiper, 1986), performance evaluation (15 items), and approval by others (10 items), were substituted for the total DAS in the regression analyses. Only the interaction of the performance evaluation subscale and social support at time 2 was significant in predicting time 2 depression and only for the female subjects.

Zuroff, Igreja, and Mongrain (1990) followed 46 college women over a 12-month period. At the time of the initial assessment, subjects were given the DAS, the Depressive Experiences Questionnaire (to assess dependency and self-criticism), and the BDI. Approximately 1 year later, subjects completed the measures again, focusing on the period when their mood was the worst in the preceding year. Subjects also completed an adjective-rating form that assessed anaclitic and introjective feelings and cognitions during the worst period. Controlling for time 1 depression, the DAS was significantly associated with level of depression during the worst period; however, the DAS was not associated with the anaclitic or introjective scales. The measures of dependency and self-criticism were not associated with later depression; however, the dependency measure was significantly associated with anaclitic depression, and the self-criticism measure was associated with introjective depression.

LIFE EVENTS AND SCHEMA MATCHING

Hammen and her colleagues (Hammen, Marks, Mayol, & deMayo, 1985) have conducted several studies assessing the role of self-schemas, stressful life events, and their interaction as etiological factors in depression. College students were classified as having dependent or self-critical schemas (or neither) based on an assessment of the types of responses they gave to a questionnaire assessing their recall of positive and negative interpersonal and achievement related events over the previous month. They had three depression outcome measures: mean BDI across the four follow-ups, number of times (0 to 4) the BDI score was >14 across the follow-ups, and the number of times (0 to 4) that subjects met criteria for major or minor depression across the follow-ups. The authors predicted that, for the subjects with dependent schemas, significant correlations should be noted between number of interpersonal-related life events and depression and that, for the subjects with self-critical schemas, significant correlations should be noted between number of achievement-related life events and depression. The results of the study partially confirmed the hypotheses. For example, the correlation between interpersonal events and the number of times the BDI score was >14 was significantly greater for the dependent schema subjects ($r = .41$) than for the self-critical schema subjects ($r = -.08$). Conversely, the correlation between number of achievement events and the number of times the BDI score was >14 was significantly greater for the self-critical schema subjects ($r = .51$) than for the dependent schema subjects ($r = -.10$). Although these results were exactly in line with predictions, other results involving other measures of depression and other measures of life events were not as consistently in line with predictions.

Several studies have evaluated the role of life events and schema matching in the context of relapse/remission or exacerbation of symptoms in an already depressed sample. For example, Hammen, Ellicott, and Gitlin (1989) followed 15 unipolar depressives who could be classified as either sociotropic or autonomous (Beck, Epstein, Harrison, & Emery, 1983) from a period of remission through a clear and significant onset of depressive symptoms. The prediction was that the life events that occurred in the 3 months before onset of symptoms would tend to match the cognitive subtype; that is, a preponderance of interpersonal events would be noted for patients classified as a sociotropic subtype and a preponderance of achievement/autonomy events would be noted for patients classified as an autonomous subtype. Overall, more matching events than nonmatching events occurred before relapse in the subjects. In a series of regression analyses, only the factors of achievement events and the achievement events X autonomy score were significantly associated with severity of symptoms of depression at the time of relapse.

In a similar study, Hammen, Ellicott, Gitlin, and Jamison (1989) followed 22 unipolar depressives who were in remission or who were experiencing chronic symptoms but not in an acute episode, for at least 6 months. Subjects were classified as sociotropic or autonomous subtypes. Among the six unipolar patients who experienced an exacerbation or onset of depression during the follow-up period, five had a preponderance of matching events (one had no events), a significant association. With respect to level of symptomatology over the follow-up period, unipolar patients who had more congruent objective stress had higher levels of depressive symptomatology compared to subjects with equal levels of noncongruent stress.

Hammen, Ellicott, and Gitlin, (1992) followed a sample of 49 bipolar depressives over a period of 6 months to 2 years. Similar to the study by Hammen, Ellicott, Gitlin, and Jamison (1989), in this study, patients were characterized as sociotropic or autonomous subtypes (or neither). Life events were also classified as interpersonal or achievement related (or other). The authors found no evidence of an association between relapse and matching of cognitive subtypes and life event types. A hierarchical multiple regression did reveal a significant association between severity of symptoms in bipolar patients and the measures of sociotropy, interpersonal events, and the interaction of these two measures. However, the authors failed to control for initial level of symptom severity in their regression, leaving open the possibility that the significant effects of their measures of sociotropy, interpersonal events, and their interaction were due to their association with initial levels of symptom severity.

The DAS (Weissman & Beck, 1978) was used to classify remitted depressives as either self-critical ($N = 16$) or as dependent ($N = 10$) in a 6-month longitudinal study (Segal, Shaw, & Vella, 1989). Life events were also assessed, and a subset were classified as either representing self-critical concerns or dependent concerns. The mean BDI across the follow-ups and the presence or absence of relapse were the dependent variables. The results suggested that matching of schema and life event type was important for subjects characterized as having dependent schemas. The authors found significant associations between levels of interpersonal events and follow-up BDI scores. In the context of a multivariate analysis of variance (MANOVA), subjects characterized as dependent relapsed more often after the experience of interpersonal life events than after achievement life events.

In a study similar to Segal et al. (1989), Segal, Shaw, Vella, and Katz (1992) followed 59 remitted depressed subjects over a 1-year period, conducting assessments every 2 months. In contrast to their earlier findings, the DAS self-criticism by number of achievement events interaction was significant in predicting relapse. Only when events occuring in the 2 months before relapse and only when interpersonal stress ratings were used rather than number of interpersonal events was the interaction between DAS dependency and interpersonal stress (events) a significant predictor of relapse.

In only one study has the cognitive model been tested in a prospective study of children. Hammen and Goodman-Brown (1990) followed up a sample of children (ages 8 to 16) of women who had participated in the two studies described earlier (Hammen, Ellicott, & Gitlin, 1989; Hammen, Ellicott, Gitlin, & Jamison, 1989). Schemas and life events were assessed in a way similar to Hammen et al. (1989). The Kiddie-SADS (Schedule for Affective Disorders and Schizophrenia) was used to determine if and when each child experienced the onset or exacerbation of a depressive episode. The authors found a significant association between onset or exacerbation of depression and whether or not the children had experienced a preponderance of schema congruent life events. Most of the congruency (in 9/10 cases where a child became depressed during the follow-up period) was accounted for by children who had dependent self-schema and experienced interpersonally related life events.

Significance of Prospective Research

It is probably a mistake to view much of the work reviewed in this chapter as "high-risk" in any sense. Most of the studies followed college students or in some cases, schoolchildren, who were not at any particular risk for depression, over a period of time. There were two major exceptions to this rule. A number of investigators followed women from pregnancy through the postpartum period. Women were thought to be at increased risk for depression during the postpartum period because of the major physical, psychological, and social adjustments required after delivery. The advantage of studying childbearing women is that all subjects are undergoing a predictable and stressful event that can be dated rather precisely. It was expected that women who possessed a cognitive diathesis (e.g., depressive self-schema or dysfunctional attributional style) would be especially prone to experience a depressive episode during this period. Some support for this hypothesis was forthcoming (Cutrona, 1983; O'Hara et al., 1982); however, other similar studies provided little support (O'Hara et al., 1984). The second high-risk context involved the study of depressed patients in remission and children of depressed patients (Hammen, Ellicott, & Gitlin, 1989; Hammen & Goodman-Brown, 1990). These studies were most commonly done to test the cognitive model and showed evidence that schema-congruent life events may be particularly important in precipitating depression onset, relapse, or exacerbation.

A second characteristic of this work is that changes in the level of depressive symptomatology were studied in the vast majority of cases. In those studies that used depression diagnosis as an outcome, little support was found for the attributional predictions (Lewinsohn et al., 1988; O'Hara et al., 1984); however, as described in the previous paragraph, Hammen and her colleagues found evidence that schema-congruent life events were predictive in adults and children of depression onset and relapse. Despite these encouraging results, these studies had relatively small samples ($N = 12$ to 64) and need replication.

The advantage of studying changes in levels of depressive symptomatology over time rather than changes in diagnostic status (i.e., nondepressed to depressed) is that severity measures are much more sensitive to change in the affective state of subjects than are binary diagnostic indices, which are rather crude indicators of change. That is to say, a small change in a subject's psychological state probably would be picked up by a measure such as the BDI and probably would not be detected in the context of a binary diagnostic judgment. This characteristic increases the likelihood that statistically significant associations between cognitive predictor variables and severity measures would be observed, compared to the case of cognitive predictor variables and diagnostic measures. Despite the advantages of studying easily accessible populations with rather sensitive measures of depressive symptomatology, clinically significant depression should increasingly become the focus of prospective studies of depression. Clinically significant depression is the phenomenon that the helplessness and cognitive models of depression were developed to explain. Moreover, in terms of public health, it is important to examine the validity of cognitive models with respect to major depression.

Lewinsohn et al. (1988) have argued, based on the results of their prospective study of a very large community sample, that cognitive diatheses increase risk for negative affect (e.g., measured by instruments such as the BDI; see Watson & Tellegen, 1985) but that cognitive diatheses do not increase risk for clinically significant depression (e.g., diagnoses based on the RDC and the *Diagnostic and Statistical Manual of Mental Disorders*, 3rd ed., rev., American Psychiatric Association, 1987). Negative affect is viewed as a common core of an elevated BDI score and a diagnosable episode of depression, but each has its unique characteristics. Lewinsohn et al. argue that negative affect in combination with other specific risk factors (e.g., being female, younger age, and previous history of depression) are responsible for clinically significant depression. With respect to the unique predictors of negative affect, they argue that cognitive factors such as a dysfunctional attributional style and irrational beliefs are most important. This view suggests that cognitive diatheses may play only an indirect role in causing clinically significant depression by operating through negative affect. Although this perspective is in accord with many of the findings summarized above, the more recent work by Hammen and her colleagues belies this view. Once again, studies with large samples of subjects at risk for clinically significant depression are needed to test the validity of the cognitive models of depression.

A final perspective on prospective studies of cognitive models of depression comes from Abramson, Alloy, and Metalsky (1988), who have argued that the various research strategies used do not provide adequate tests of the basic postulates of the theories. For example, both the cognitive and helplessness theories emphasize the importance of a stressful event(s) interacting with the cognitive diathesis to produce depression or the conditions necessary for depression from the perspective of the specific theory (i.e., expectation of future noncontingency or negative cognitive triad). Few studies evaluating either model have focused on theoretically relevant stressful events (Hammen's work being the outstanding exception) or theoretically relevant intervening variables (e.g., expectation of future noncontingency).

Despite the inadequacies of current research strategies (Abramson et al., 1988) and the possibility that cognitive models may have more relevance for understanding in development of negative affect than of clinical depression (Lewinsohn et al., 1988), the research reviewed in this chapter does provide limited support for both the helplessness and cognitive models of depression. Future studies will have the best chance of illuminating the role of cognitive diatheses in clinical depression if they (a) use a large sample of subjects at high risk for clinical depression, (b) incorporate multiple measures of cognitive diatheses, (c) consider the types of relevant life stressors that are likely to occur in subjects' lives, and (d) follow subjects over a long period with frequent assessments of depression status. Undoubtedly, these studies will be expensive and require multiple centers to be carried out successfully.

References

Abramson, L.Y., Alloy, L.B., & Metalsky, G.I. (1988). The cognitive diathesis-stress theories of depression: Toward an adequate evaluation of the theories' validities. In L.B. Alloy (Ed.), *Cognitive processes in depression* (pp. 3–30). New York: Guilford Press.

Abramson, L.Y., Metalsky, G.I., & Alloy, L.B. (1989). Hopelessness depression: A theory-based subtype of depression. *Psychological Review, 96*, 358–372.

Abramson, L.Y., Seligman, M.E.P., & Teasdale, J.D. (1978). Learned helplessness in humans: Critique and reformulation. *Journal of Abnormal Psychology, 87*, 49–74.

Alloy, L.B. (Ed.). (1988). *Cognitive processes in depression*. New York: Guilford Press.

American Psychiatric Association. (1987). *Diagnostic and statistical manual of mental disorders* (3rd ed., rev.). Washington, DC: Author.

Barnett, P.A., & Gotlib, I.H. (1988). Dysfunctional attitudes and psychosocial stress: The differential prediction of future psychological symptomatology. *Motivation and Emotion, 12*, 251–270.

Barnett, P.A., & Gotlib, I.H. (1990). Cognitive vulnerability to depressive symptoms among men and women. *Cognitive Therapy and Research, 14*, 47–61.

Beck, A.T., Epstein, N., Harrison, R., & Emery, G. (1983). *Development of the sociotropy-autonomy scale: A measure of personality factors in psychopathology.* Unpublished manuscript, University of Pennsylvania.

Beck, A.T., Ward, C.H., Mendelson, M., Mock, J., & Erbaugh, J. (1961). An inventory for measuring depression. *Archives of General Psychiatry, 4*, 561–569.

Beck, A.T., Rush, A.J., Shaw, B.F., & Emery, G. (1979). *Cognitive therapy of depression*. New York: Guilford Press.

Brown, G.W., & Harris, T.O. (1989). Depression. In G.W. Brown & T.O. Harris (Eds.), *Life events and illness*. New York: Guilford Press.

Brown, J.D., & Siegel, J.M. (1988). Attributions for negative life events and depression: The role of perceived control. *Journal of Personality and Social Psychology, 54*, 316–322.

Cane, D.B., Olinger, L.J., Gotlib, I.H., & Kuiper, N.A. (1986). Factor structure of the Dysfunctional Attitudes Scale in a student population. *Journal of Clinical Psychology, 42*, 307–309.

Cochran, S.D., & Hammen, C.L. (1985). Perceptions of stressful life events and depression: A test of attributional models. *Journal of Personality and Social Psychology, 48*, 1562–1571.

Cutrona, C.E. (1983). Causal attributions and perinatal depression. *Journal of Abnormal Psychology, 92*, 161–172.

Follette, V.M., & Jacobson, N.S. (1987). Importance of attributions as a predictor of how people cope with failure. *Journal of Personality and Social Psychology, 52*, 1205–1211.

Golin, S., Sweeney, P.D., & Shaeffer, D.E. (1981). The causality of causal attributions in depression: A cross-lagged panel correlational analysis. *Journal of Abnormal Psychology, 90*, 14–22.

Gotlib, I.H., Whiffen, V.E., Wallace, P.M., & Mount, J.H. (1991). Prospective investigation of postpartum depression: Factors involved in onset and recovery. *Journal of Abnormal Psychology, 100*, 122–132.

Hammen, C., Adrian, C., & Hiroto, D. (1988). A longitudinal test of the attributional vulnerability model in children at risk for depression. *British Journal of Clinical Psychology, 27*, 37–46.

Hammen, C., Ellicott, A., & Gitlin, M. (1989). Vulnerability to specific life events and prediction of course of disorder in unipolar depressed patients. *Canadian Journal of Behavioural Science, 21*, 377–388.

Hammen, C., Ellicott, A., & Gitlin, M. (1992). Stressors and sociotropy/autonomy: A longitudinal study of their relationship to the course of bipolar disorder. *Cognitive Therapy and Research, 16*, 409–418.

Hammen, C., Ellicott, A., Gitlin, M., & Jamison, K.R. (1989). Sociotropy/autonomy and vulnerability to specific life events in patients with unipolar depression and bipolar depression. *Journal of Abnormal Psychology, 98*, 154–160.

Hammen, C., & Goodman-Brown, T. (1990). Self-schemas and vulnerability to specifc life stress in children at risk for depression. *Cognitive Thereapy and Research, 14*, 215–227.

Hammen, C., Krantz, S.E., & Cochran, S.D. (1981). Relationships between depression and causal attributions about stressful life events. *Cognitive Therapy and Research, 5*, 351–358.

Hammen, C., Marks, T., deMayo, R., & Mayol, A. (1985). Self-schemas and risk for depression: A prospective study. *Journal of Personality and Social Psychology, 49*, 1147–1159.

Hammen, C., Marks, T., Mayol, A., & deMayo, R. (1985). Depressive self-schemas, life stress, and vulnerability to depression. *Journal of Abnormal Psychology, 94*, 308–319.

Hunsley, J. (1989). Vulnerability to depressive mood: An examination of the temporal consistency of the reformulated learned helplessness model. *Cognitive Therapy and Research, 13*, 599–608.

Johnson, J.G., & Miller, S.M. (1990). Attributional, life-event, and affective predictors of onset of depression, anxiety, and negative attributional style. *Cognitive Therapy and Research, 14*, 417–430.

Kenny, D.A., & Harackiewicz, J.M. (1979). Cross-lagged panel correlation: Practice and promise. *Journal of Applied Psychology, 64*, 372–379.

Kovacs, M., & Beck, A.T. (1977). An empirical-clinical approach toward a definition of childhood depression. In J.G. Schulterbrandt & A. Raskin (Eds.), *Depression in Childhood: Diagnosis, treatment and conceptual models* (pp. 1–25). New York: Raven Press.

Lewinsohn, P.M., Hoberman, H.M., & Rosenbaum, M. (1988). A prospective study of risk factors for unipolar depression. *Journal of Abnormal Psychology, 97*, 251–264.

Lewinsohn, P.M., Steinmetz, J.L., Larson, D.W., & Franklin, J. (1981). Depression-related cognitions: Antecedent or consequence? *Journal of Abnormal Psychology, 90*, 213–219.

Manly, P.C., McMahon, R.J., Bradley, C.F., & Davidson, P.O. (1982). Depressive attributional style and depression following childbirth. *Journal of Abnormal Psychology, 91*, 245–254.

Metalsky, G.I., Abramson, L.Y., Seligman, M.E.P., Semmel, A., & Peterson, C. (1982). Attributional styles and life events in the classroom: Vulnerability and invulnerability to depressive mood reactions. *Journal of Personality and Social Psychology, 43*, 612–617.

Metalsky, G.I., Halberstadt, L.J., & Abramson, L.Y. (1987). Vulnerability to depressive mood reactions: Toward a more powerful test of the diathesis-stress and causal mediation components of the reformulated theory of depression. *Journal of Personality and Social Psychology, 52*, 386–393.

Metalsky, G.I. & Joiner, T.E. (1992). Vulnerability to depressive symptomatology: A prospective test of the diathesis-stress and causal mediation components of the hopelessness theory of depression. *Journal of Personality and Social Psychology, 63*, 667–675.

Monroe, S.M., & Peterman, A.M. (1988). Life stress and psychopathology. In L.H. Cohen (Ed.), *Life events and psychological functioning: Theoretical and methodological issues* (pp. 31–63). Newbury Park, CA: Sage.

Monroe, S.M. & Simons, A.D. (1991). Diathesis-stress theories in the context of life stress research: Implications for the depressive disorders. *Psychological Bulletin, 110*, 406–425.

Nolen-Hoeksema, S., Girgus, J.S., & Seligman, M.E.P. (1986). Learned helplessness in children: A longitudinal study of depression, achievement, and explanatory style. *Journal of Personality and Social Psychology, 51*, 435–442.

O'Hara, M.W., Neunaber, D.J., & Zekoski, E.M. (1984). Prospective study of postpartum depression: Prevalence, course, and predictive factors. *Journal of Abnormal Psychology, 93*, 158–171.

O'Hara, M.W., Rehm, L.P., & Campbell, S.B. (1982). Predicting depressive symptomatology: Cognitive-behavioral models and postpartum depression. *Journal of Abnormal Pscyhology, 91*, 457–461.

O'Hara, M.W., & Zekoski, E.M. (1988). Postpartum depression: A comprehensive review. In R. Kumar & I.F. Brockington (Eds.), *Motherhood and mental illness 2: Causes and consequences* (pp. 17–63). London: Wright.

O'Hara, M.W., Zekoski, E.M., Philipps, L.H., & Wright, E.J. (1990). A controlled prospective study of postpartum mood disorders: Comparison of childbearing and nonchildbearing women. *Journal of Abnormal Psychology, 99*, 3–15.

Pagel, M.D., Becker, J., & Coppel, D.B. (1985). Loss of control, self-blame, and depression: An investigation of spouse caregivers of Alzheimer's disease patients. *Journal of Abnormal Psychology, 94*, 169–182.

Paykel, E.W. (1979). Recent life events in the development of depressive disorders. In R. Depue (Ed.), *The psychobiology of the depressive disorders* (pp. 245–262). New York: Academic Press.

Paykel, E.S. (1982). Life events and early environment. In E.S. Paykel (Ed.), *Handbook of affective disorders* (pp. 146–161). New York: Guildford Press.

Radloff, L. (1977). The CES-D Scale: A self-report depression scale for research in the general population. *Applied Psychological Measurement, 1*, 385–401.

Rehm, L.P. (1977). A self-control model of depression. *Behavior Therapy, 8*, 787–804.

Segal, Z.V., Shaw, B.F., Vella, D.D., Katz, R. (1992). Cognitive and life stress predictors of relapse in remitted unipolar depressed patients: Test of the congruency hypothesis. *Journal of Abnormal Psychology, 101*, 26–36.

Segal, Z.V., Shaw, B.F., & Vella, D.D. (1989). Life stress and depression: A test of the congruency hypothesis for life event content and depressive subtype. *Canadian Journal of Behavioral Science, 21*, 389–400.

Seligman, M.E.P., Peterson, C., Kaslow, N.J., Tanenbaum, R.L., Alloy, L.B., & Abramson, L.Y. (1984). Attributional style and depressive symptoms among children. *Journal of Abnormal Psychology, 93*, 235–238.

Troutman, B.R., & Cutrona, C.E. (1990). Nonpsychotic postpartum depression among adolescent mothers. *Journal of Abnormal Psychology, 99*, 69–78.

Watson, D., & Tellegen, A. (1985). Toward a consensual structure of mood. *Psychological Bulletin, 98*, 219–235.

Weissman, A.R., & Beck, A.T. (November, 1978). *Development and validation of the Dysfunctional Attitude Scale.* Paper presented at the meeting of the Association of Advancement of Behavior Therapy, Chicago.

Whiffen, V.E. (1988). Vulnerability to postpartum depression: A prospective multivariate study. *Journal of Abnormal Psychology, 97*, 467–474.

Williams, J.M.G. (1985). Attributional formulation of depression as a diathesis-stress model: Metalsky et al. reconsidered. *Journal of Personality and Social Psychology, 48*, 1572–1575.

Zuroff, D.C., Igreja, I., & Mongrain, M. (1990). Dysfunctional attitudes, dependency, and self-criticism as predictors of depressive mood states: A 12-month longitudinal study. *Cognitive Therapy and Research, 14*, 315–326.

10
Implications of the Resource Allocation Model for Mood Disorders

CINDY M. YEE

Although classified as a mood disorder in the *Diagnostic and Statistical Manual of Mental Disorders* (American Psychiatric Association, 1987), depression is also known to have important cognitive features. Patients diagnosed with depression often report difficulties in concentration, attention, learning, and memory (Tariot & Weingartner, 1986), and numerous studies document the cognitive impairments that occur with this disorder (e.g., Glass, Uhlenhuth, Hartel, Matuzas, & Fischman, 1981; Silberman, Weingartner, & Post, 1983; Weingartner, Cohen, Murphy, Martello, & Gerdt, 1981; for reviews see McAllister, 1981; W.R. Miller, 1975; Mineka & Sutton, 1992). Tasks requiring sustained effort appear to be particularly likely to produce disruptions during cognitive processing in depression (Cohen, Weingartner, Smallberg, Pickar, & Murphy, 1982; Roy-Byrne, Weingartner, Bierer, Thompson, & Post, 1986). For example, depressed patients have been found to perform relatively poorly on tasks that require effortful or controlled processing (e.g., recalling a long list of items) but have less difficulty on tasks that can be accomplished automatically (e.g., recalling the location of furniture in a room that they just occupied; see Tariot & Weingartner, 1986). In short, considerable evidence indicates that clinical levels of depression are accompanied by cognitive deficits, particularly when depressives are presented with effort-demanding tasks.

The association between depressed mood and cognitive processing deficits has been obtained also in studies of nonclinical subjects. The influence on memory of transient affective states, including sad or depressed moods, has been the focus of a large body of research (for a review see Blaney, 1986). Ellis and colleagues, for example, demonstrated that depressed mood can interfere with the encoding and recall of information (e.g., Ellis, Thomas, McFarland, & Lane, 1985; Ellis, Thomas, & Rodriguez, 1984; Leight & Ellis, 1981). Although Hasher, Rose, Zacks, Sanft, and Doren (1985) failed to replicate this finding, the results of their study have stimulated considerable discussion (see Ellis, 1985; Isen, 1985; Mayer & Bower, 1985; for a response see Hasher, Zacks, Rose, &

Doren, 1985). One possible explanation for the discrepant results is that the range or intensity of naturally occurring moods in nonclinical subjects is not sufficiently different from those of controls (e.g., Hasher, Zacks, et al., 1985). Thus, evidence for cognitive processing difficulties in normal individuals may be less robust than, though consistent with, results obtained from clinical populations.

In view of the prominence accorded to cognitive difficulties associated with depression, an essential next step in this literature is to determine the mechanisms that underlie these abnormalities. As Weingartner and colleagues observed,

Studies of information processing in the depressed state have not systematically explored the cognitive mechanisms that would account for changes in these functions. Defining the structure of depression-related cognitive functions would be important for an understanding of the psychobiology of cognition in the depressed state and its role in the expression of depressive symptoms and would have implications for the development of effective treatment strategies. (1981, p. 42)

One direction that researchers have taken in responding to criticisms such as this involves the *resource allocation model*, which addresses the extent to which depression might alter the amount and availability of attentional resources needed for various cognitive tasks. According to this model, depression increases the information processing load and drains resources that might otherwise be devoted to a task, such that depressed subjects are left with a depleted supply of processing resources that can then interfere with task processing and result in performance decrements (e.g., Ellis & Ashbrook, 1988). The resource allocation model, although not without controversy or complexity, holds particular promise for clarifying the nature of cognitive deficits in depression.

The purpose of this chapter is to outline how the resource allocation model can be used to understand cognitive deficits in mood disorders. The chapter is organized in three sections. In the first, principles of cognitive psychology and human factors research are introduced to provide an overview of resource theory. The predominant paradigm used to study resource theory, known as the *dual-task situation*, is also described, and the value of psychophysiological measures, and particularly event-related potentials (ERPs), for investigating processing resources is addressed. The third section considers the resource allocation model as it applies specifically to depression and includes an illustration of this approach from my own studies of processing resources and early-onset dysthymia. The chapter concludes with a discussion of potential limitations of the resource allocation model in depression and challenges for future research.

An Introduction to Resource Theory, the Dual-Task Paradigm, and Event-Related Potentials

Resource Theory

According to resource theory, a fixed quantity of resources is assumed to be available for processing information at any given moment in time. These resources can be shared during the concurrent performance of two or more tasks until resource limits are exceeded, at which time improved performance on one task is expected to come at the expense of performance on the other tasks (Kahneman, 1973; Navon & Gopher, 1979; Norman & Bobrow, 1975).[1] Patterns of interference can be examined systematically by "overloading" an individual so that, as more and more resources are allocated to one task, performance on another task will decline below the performance levels obtained when the task is performed alone. More recently, researchers have attempted to account for instances when there appears to be a lack of competition for resources between concurrent tasks and proposed a multiple resource model of attention.

The multiple resource model maintains that separate pools of resources exist (e.g., Friedman & Polson, 1981; Sanders, 1983) and, therefore, to the extent that tasks do not compete for the same resources, decrements associated with dual-task performance are minimized (Navon & Gopher, 1979; Wickens, 1980). For example, Wickens (1980, 1984) suggests that different resources are associated with different information processing structures, including stages of processing (perceptual/central and response), modalities of processing (visual and auditory), and codes of processing (spatial and verbal). It follows from this view that a task involving presentation of visual stimuli that may be encoded spatially can be performed easily together with a task involving presentation of auditory stimuli that may be encoded verbally, assuming that there is no overlap in additional processing requirements. Thus, resource theory has been proposed to account for patterns of interference and has been extended to address instances of noninterference in multiple-task performance.

The Dual-Task Paradigm

Much of the empirical basis for resource theory is derived from studies employing dual-task methodology. As the foregoing section indicates, the dual-task paradigm involves the systematic manipulation of resource demands of two concurrent tasks to examine any performance trade-offs that may occur as a function of a limited supply of resources. Two distinct

---

[1] A number of theories have been proposed to account for different aspects of attention. In resource theory, the concept of a limited capacity has been introduced to account for the selective nature of attention.

methods for manipulating resource demands have been employed: (a) shifting task priorities and (b) varying the difficulty level of one or both tasks. To manipulate task priorities, subjects are instructed to maximize their performance on one task (the primary task) while continuing to perform the other task (the secondary task). Resource trade-offs are inferred from shifts in performance relative to single-task conditions, when each of the tasks is performed alone. In addition to varying instructed priorities between tasks, the difficulty level of the primary task can be varied to manipulate resource allocation. An easy primary task will require fewer processing resources, presumably releasing resources for the secondary task, while a difficult primary task will demand greater resources and may leave an inadequate supply for the performance of the secondary task. Applications of both of these procedures are discussed in the section entitled, "An Empirical Illustration."

Event-Related Brain Potentials and Processing Resources

As described thus far, the dual-task methodology appears limited in its capacity for testing aspects of resource theory, because underlying resources must be inferred solely on the basis of changes in performance (Wickens, 1986). That is, simple behavioral outputs, such as reaction time (RT) and accuracy of performance, do not readily reveal or distinguish intermediate processes that may occur between the time a stimulus is presented and a response is made. This difficulty has been addressed by using psychophysiological measures thought to reflect resource allocation, such as the amplitude of the P300 component of the ERP. The ERP is a series of changes in the electrical potential recorded from the brain in response to discrete events. These changes in electrical potential can be measured noninvasively at the scalp. Some components of the ERP are considered to manifest information processing activity in the brain (Donchin, 1981). One such component is the P300, a positive potential that occurs at least 300 ms after an eliciting event.

In prior dual-task studies, investigators have evaluated the effect of task priority on P300 and found that higher priorities are associated with larger P300 amplitudes (Hoffman, Houck, MacMillan, Simons, & Oatman, 1985; Strayer & Kramer, 1990). A separate series of studies has demonstrated that the P300 component elicited by primary task events increases in amplitude with increases in primary task difficulty (Kramer, Wickens, & Donchin, 1985; Wickens, Kramer, Vanasse, & Donchin, 1983; Sirevaag, Kramer, Coles, & Donchin, 1989). In addition, the P300 elicited by secondary task events decreases in amplitude with increases in primary task difficulty (Isreal, Chesney, Wickens, & Donchin, 1980; Kramer, Sirevaag, & Braune, 1987; Kramer, Wickens, & Donchin, 1983; Kramer et al., 1985; Strayer & Kramer, 1990). Thus, the reciprocal effects of primary and secondary task difficulty on P300 amplitude provide converging evidence, along with performance measures, on the underlying resource allocation structure. Furthermore, with regard to multiple resource theory, the P300 appears to be sensitive to the allocation of perceptual and cognitive resources but not to those resources related to response demands (Donchin, Kramer, & Wickens, 1986).[2]

The Resource Allocation Model of Depression

A few suggestions have been made in the literature regarding the effect of depressed mood on resource allocation. Hasher and Zacks (1979), for example, hypothesized that performance deficits on memory tasks that require effortful or controlled processing may result from reductions in attentional capacity or resources among depressed subjects. In an extension of this model, Ellis postulated that a sad or depressed mood can limit the resources available for task performance and thereby interfere with encoding and retrieval of information. Accordingly, recall and memory performance are more likely to be impaired on difficult than easy tasks because such tasks demand greater processing resources (Ellis, 1985; Ellis & Ashbrook, 1988).

Theoretical formulations such as these offer potential insights into our understanding of cognitive deficits and depression, but they also raise some important issues. For instance, what specific factors might account for the depletion of resources in the depressed state? Another consideration is the mechanism by which processing resources are reduced. Is the pool of available resources smaller when people are in a depressed state? Or do performance deficits stem from a failure to respond appropriately to task demands despite a normal pool of processing resources? Researchers have begun to explore these issues to varying degrees.

Several proposals have been offered to explain declines in processing resources during depression. Ellis and Ashbrook (1988), for example, suggest that depressed mood may increase processing of (a) irrelevant or distracting features of a task and (b) events that are unrelated to the task, or extra-task processing. They also propose that a portion of the pool of cognitive resources is absorbed by contemplation of one's sad or depressed mood. This position can be extended to include clinical levels of depression and ruminations that are directed toward associated symptoms, such as feelings of worthlessness, excessive guilt, and indecision. Indirect support for this view is offered by Ingram (1990), who observed that the level of

---

[2] The N200 and frontal negative O-wave components of the ERP also have been found to be sensitive to aspects of the dual-task situation (e.g., Kramer & Strayer, 1988; Yee, Miller, & Kramer, 1987). The P300 component, however, has received by far the most extensive examination in studies of resource allocation and, therefore, is the focus of this report.

self-focused attention appears to be increased in clinical as well as subclinical levels of depression. These proposals, which are not mutually exclusive, represent important conceptual advances and await empirical confirmation.

A second issue raised by the resource allocation model concerns whether an apparent reduction in processing capacity stems from a deficiency in the *availability* or in the *allocation* of cognitive resources. The initial assumption has been that depression depletes the amount of available resources (Ellis & Ashbrook, 1988; Hasher & Zacks, 1979). Much of the support for this claim comes from studies demonstrating impaired recall performance in college students, following a depressed-mood-induction procedure, as compared with students who experienced a neutral-mood induction (e.g., Ellis et al., 1984, 1985; Leight & Ellis, 1981). These data are consistent with the hypothesis that subjects in a depressed-mood state possess a lower level of available processing resources, yet the possibility remains that deficits in recall performance are linked to a deficiency in how the resources are allocated or distributed. In recent investigations on the use of strategies in memory tasks, Hertel and colleagues found that deficits in recall performance could be eliminated when depressed subjects were provided with external cues to guide their attention (Hertel & Hardin, 1990; Hertel & Rude, 1991). Hertel and Rude (1991) contend that depression, therefore, may not impair resource capacity but instead hinders an individual's ability to focus attention. Thus, their results can be interpreted as providing support for the hypothesis that depressives are deficient in the allocation, rather than in the amount, of available processing resources (cf. Hertel & Rude, 1991).

Finally, as Ellis and Ashbrook (1988) appropriately advise, results that are based on recall performance rather than any direct measure of resource allocation must be regarded with caution. Recall performance may be very general and as such provides a rough and perhaps inadequate measure of an underlying resource structure. This raises the question of whether any direct test of the resource allocation hypothesis can be developed. It is virtually certain that we will never be able to measure resources directly, because the concept serves as a hypothetical construct rather than as a tangible entity. Like any hypothetical construct, however, the resource concept can be validated by the convergence of independent methods (Campbell & Fiske, 1959).

An Empirical Illustration

Toward this end, I conducted a dual-task study that employed traditional behavioral measures of performance accuracy and RT as well as the ERP methodology described earlier. The study was conducted using a sample of individuals with early-onset dysthymia to evaluate the utility of conceptualizing this disorder in terms of differences in resource allocation. Specifically, the study was designed to examine whether early-onset dysthymia is characterized by a deficit in the amount or allocation of processing resources. Before providing a description of the methods and results of this study, I consider the rationale for studying early-onset dysthymia within the resource allocation framework.

EARLY-ONSET DYSTHYMIA

Early-onset dysthymia is a relatively mild but chronic form of depression that develops during childhood or adolescence and appears to render individuals at increased risk for developing major depression (Akiskal, 1983; Keller, Lavori, Endicott, Coryell, & Klerman, 1983; Klein, 1990; Kovacs et al., 1984). In a series of psychophysiological studies, dysthymic subjects have been shown to have a tendency to underrespond at various stages of the information processing sequence (G.A. Miller, Yee, & Anhalt, 1992; Yee, Deldin, & Miller, 1992; Yee & Miller, 1988). Such a consistent pattern of underresponsiveness could be the result of a depleted supply of available resources, as would be suggested by the resource allocation model of depression. The research presented here attempts to address this issue while also considering the role of early-onset dysthymia as a high-risk condition.

Because early-onset dysthymia constitutes a risk for major depression, it offers a unique opportunity to examine the possible relationship between reductions in available processing resources and vulnerability to major depressive episodes. A depleted supply of resources may place an individual at risk for developing recurrent episodes of major depression. Reductions in processing resources would be consistent with negative reactions to stressful life events, which, in turn, have been shown to be associated with the onset of major depressive episodes (e.g., Hammen, Mayol, deMayo, & Marks, 1986). In fact, Goplerud and Depue (1985) found preliminary evidence of prolonged recovery to stress in a small sample of dysthymic subjects. Although the study I conducted does not address these issues directly, it may contribute to our understanding of early-onset dysthymia as a high-risk condition.

RESOURCE ALLOCATION IN EARLY-ONSET DYSTHYMIA

In this section, I describe briefly the dual-task study that I conducted on a group of early-onset dysthymics; a more detailed version of this study is provided in Yee and Miller (1992). Early-onset dysthymic subjects were identified using the revised General Behavior Inventory (GBI; Depue & Klein, 1988; Depue, Krauss, Spoont, & Arbisi, 1989), a 73-item self-report scale that identifies cases of dysthymia. Recent validation studies have demonstrated considerable agreement between GBI classification and clinical diagnosis of dysthymia (e.g., Depue & Klein, 1988; Depue et al., 1989; Klein, Dickstein, Taylor, & Harding, 1989). We also included two sets of control subjects: normal controls who did not meet any diagnostic criteria and anhedonic subjects who were identified using scales developed by Chapman, Chapman, and Raulin (1976) and who served as the equivalent of a psychiatric control group.

In the laboratory, subjects were asked to perform a visual recognition running memory task by itself, an auditory recognition running memory task by itself, or the two tasks concurrently. To force resource trade-offs, the difficulty level of the visual task was manipulated, as was the emphasis or priority that subjects assigned to each of the two tasks. For the easy visual task, subjects were presented with a series of two-letter character strings (e.g., "FH") and asked to indicate, with one of two button presses, whether the current set of letters was a match or mismatch to the set that was just presented. This condition is labeled "1-back," because subjects were required to hold one set of letters in memory. In the more difficult visual condition, subjects were asked to hold the most recent two sets of letters in memory, at any given time, and indicate whether the current set of letters was a match or mismatch to the set that was presented 2-back. The auditory task also involved running memory; subjects were asked to indicate whether the current tone was a match or mismatch to the tone presented 1-back. The duration of each character string and tone was 15 and 70 ms, respectively. The intertrial interval between trial onsets was 1,300 ms, with visual and auditory trials equally likely and randomly ordered.

Task priority was manipulated by instructing subjects to maximize their performance on one task or the other. A total of four priority conditions was employed: 100/0, 70/30, 30/70, and 0/100, where the first number refers to the priority of the visual task and the second refers to the priority of the auditory task. Under single-task conditions (100/0 and 0/100), subjects performed each task separately. Under dual-task conditions (70/30 and 30/70), the two tasks were performed concurrently. Thus, subjects were required to perform a series of highly structured dual tasks. Before data collection, each subject received extensive practice to ensure that performance exceeded chance levels. During data collection, ERPs were recorded from three midline electrode sites at frontal, central, and parietal areas of the scalp (Fz, Cz, and Pz). Performance accuracy and RT also were monitored. Because trial-to-trial "latency jitter" can be considerable in decision tasks, latency-corrected P300 was computed from single-trial data. For present purposes, only performance accuracy and P300 amplitude at Pz, from correct responses on match trials, are described.

As shown in Figure 10.1, systematic trade-offs were observed in performance accuracy on correct match trials, as a function of processing priority and task difficulty. A comparison of the white bars in Figure 10.1 reveals the expected decline in performance as visual-task priority

10. Resource Allocation and Mood Disorders 279

Accuracy
1-Back

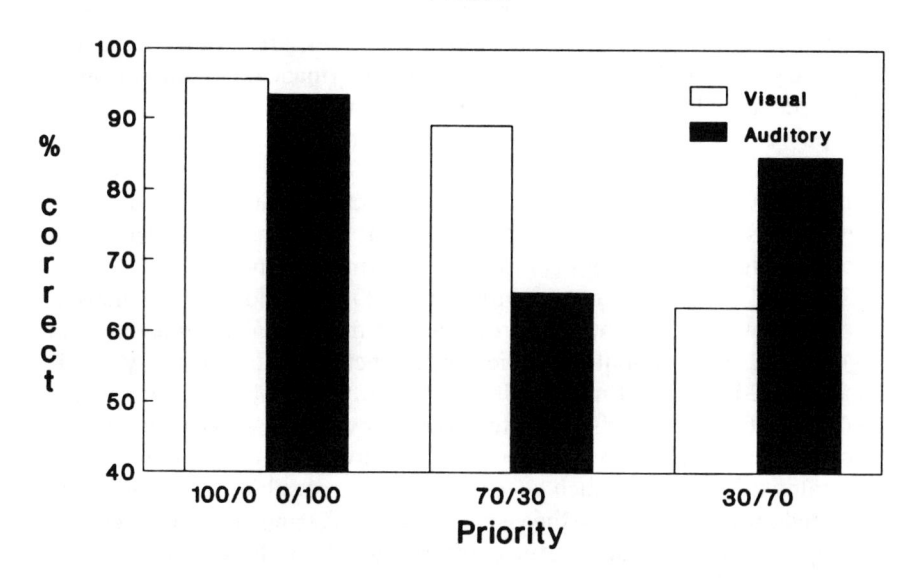

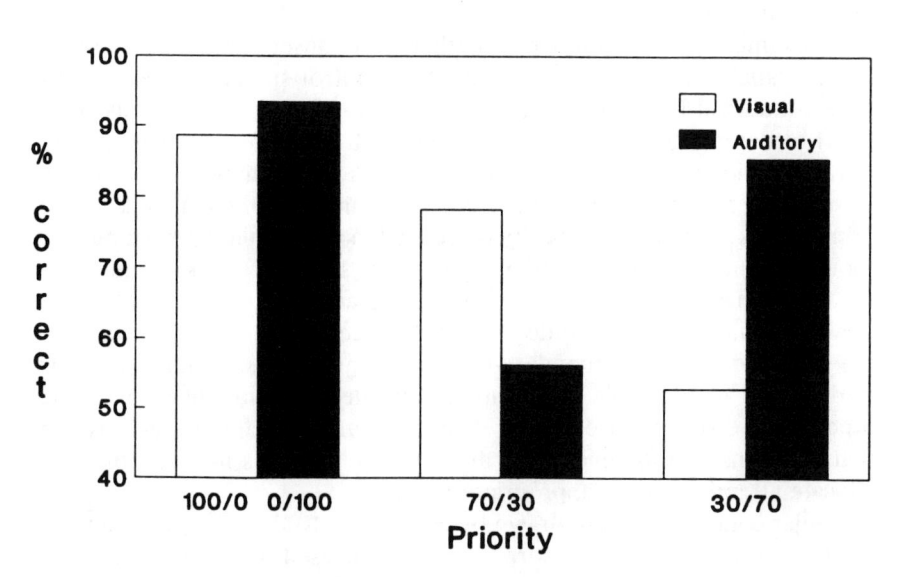

FIGURE 10.1. Visual and auditory performance accuracy on correct match trials, for single-task (100/0 and 0/100) and dual-task (70/30 and 30/70) conditions, in the 1-back and 2-back running memory tasks.

decreased from 100% to 70% to 30% for both 1-back and 2-back tasks. A similar pattern of results was obtained for auditory trials, as shown by the dark bars in Figure 10.1. By comparing each pair of white and dark bars under dual-task conditions, it can be seen that, as subjects devote more resources to the primary task, fewer resources are available to the secondary task, and there is a trade-off in performance. A comparison of the upper and lower panels of Figure 10.1 indicates that the difficulty manipulation also was effective, in that accuracy was lower for the 2-back tasks than for the 1-back tasks. Thus, the performance accuracy data are consistent with predictions of resource allocation models. And, as in previous research (e.g., Yee & Miller, 1988), early-onset dysthymics did not differ significantly from control subjects on performance.

Effects of the priority manipulation on P300 amplitude were somewhat in parallel with those obtained from the accuracy data. As illustrated in Figure 10.2, P300 amplitude varied as a function of task priority for 1-back and 2-back auditory match trials as well as 2-back visual match trials. For the 1-back visual match trials, however, the main effect for priority was qualified by a group × priority interaction. Figure 10.3 illustrates this finding, which suggests that control subjects appropriately allocated fewer resources to the visual task as processing priorities declined. Dysthymics and anhedonics, in contrast, failed to show this effect, as P300 amplitude was not significantly different across conditions. These data suggest that, under certain conditions, early-onset dysthymics (as well as anhedonics) may have difficulty in directing or allocating resources to the appropriate task.

To examine the question of whether early-onset dysthymics possess fewer available processing resources than control subjects, comparisons were conducted between the groups during low-priority conditions (i.e., 30%) to determine if dysthymics are left with fewer resources to devote to the additional demands of a secondary task. Presumably, the majority of available resources should have been consumed by the demands of the primary task. If resource capacity is reduced in dysthymia, P300 amplitude should be attenuated considerably under secondary-task conditions, especially in the presence of a difficult primary task. Results indicated, however, that P300 amplitude was not reduced differentially for any of the groups during the secondary-task conditions. Thus, results from this study found no evidence of attenuation in the total amount of available processing resources in early-onset dysthymia. The data suggest instead that early-onset dysthymics differ from normal controls in their ability to allocate attention to the appropriate tasks.

Similar conclusions were drawn by Hertel and Rude (1991) on the basis of their study of recall performance in depressed subjects. One commonality between the two studies is that the deficits disappeared when the external priority demands or cues became so obvious that they could no longer be neglected or ignored. Interestingly, Krames and MacDonald

P300 Amplitude
1-Back

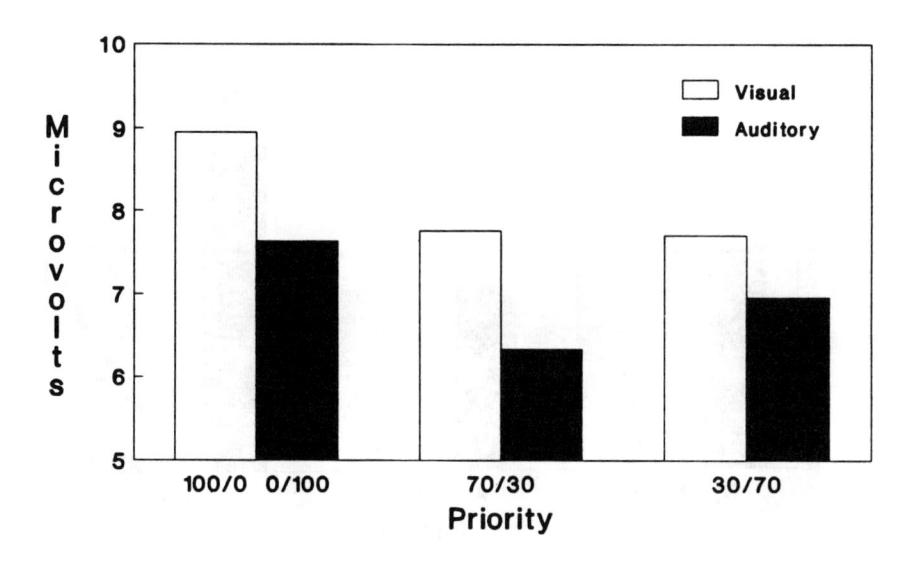

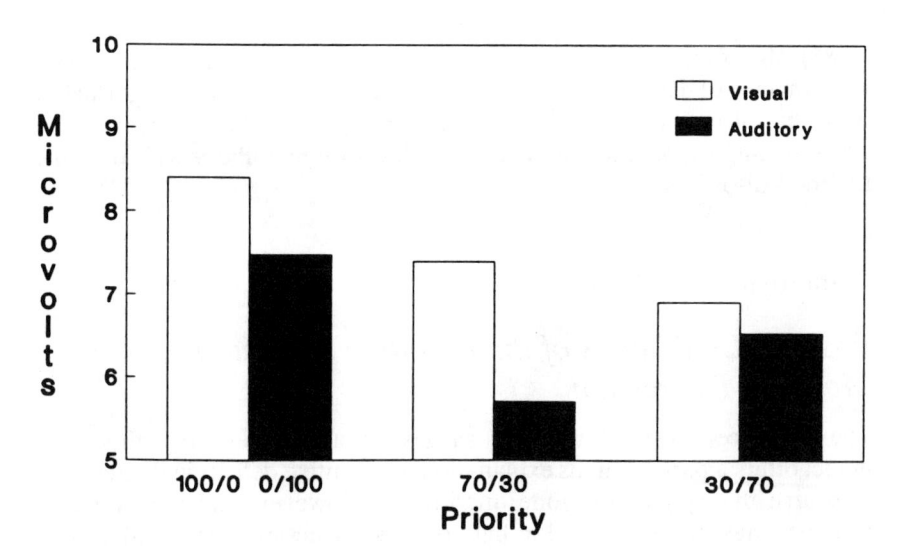

FIGURE 10.2. Visual and auditory P300 amplitude on correct match trials, for single- and dual-task conditions, in the 1-back and 2-back running memory tasks.

P300 Amplitude, Visual Task
1-Back

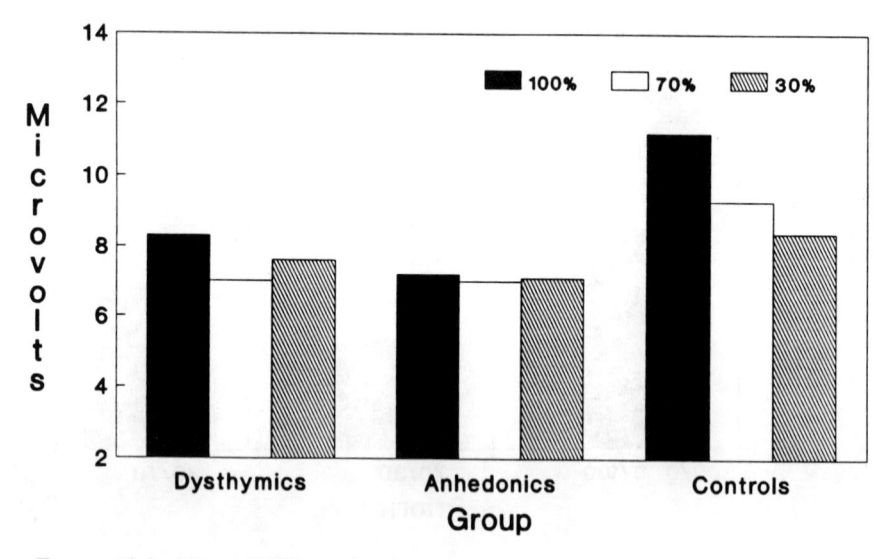

FIGURE 10.3. Visual P300 amplitude on correct match trials for single- and dual-tasks in the 1-back running memory task, recorded from dysthymic, anhedonic, and normal control subjects.

(1985) also found that, as task-relevant demands increased, the recall performance of depressed subjects improved and became indistinguishable from that of nondepressed subjects. Taken together, there appears to be accumulating support for the resource allocation hypothesis as it pertains to mood disorders.

Conclusion

Possible Limitations of the Resource Allocation Model of Depression

The evidence reviewed thus far suggests that the resource allocation model offers a parsimonious explanation of cognitive deficits in depression. Resource theory has not gone uncriticized, however, and the available evidence must be considered in light of these criticisms. One weakness of resource theory is that it lacks conceptual specificity. That is, although the notion of resources offers a compelling theoretical alternative for explaining various phenomena, other theories may be able to account for the same phenomena without making assumptions about resources or limits on resources (Navon, 1984; Navon & Miller, 1987). For example, Navon (1984) has noted that task involvement can be modulated without any change in the supply of mental commodities. Moreover, the theory makes the somewhat arbitrary assumption of a finite supply of resources while also allowing for multiple resource pools. Along similar lines, resource theory has been criticized for lacking precision; a new type of resource pool can be invented to accommodate almost any type of data, so that falsification of the theory becomes unlikely (e.g., Navon, 1984, Neumann, 1987).

These critiques are justified, but the weaknesses inherent in resource theory do not negate the utility of conceptualizing depressive deficits within a resource allocation model. Rather, the concept of resources serves as a convenient framework that allows us to consider the competition between tasks for cognitive processes (Kramer et al., 1985). The perspective taken in my own research is not to use resources to explain cognitive processes but, instead, to employ the notion of resources as a metaphor or context for conceptualizing the processing that occurs under dual-task conditions. Although the shortcomings of this strategy must be borne in mind, such a metaphor is particularly appealing in that it appears to be consistent with the way in which the brain is viewed as processing information within the framework of cognitive psychophysiology and ERPs (e.g., Donchin et al., 1986).

Challenges for Future Research

The emphasis in this chapter has been on how resource theory can be used to help understand prominent symptoms in depression. There also have been suggestions in the literature for extending this theory to account for reductions in attentional capacity or resources in several other populations, including schizophrenic patients (e.g., Gjerde, 1983; Nuechterlein & Dawson, 1984), individuals with anxiety disorders (e.g., Mathews, 1990), and the aged (e.g., Hasher & Zacks, 1979). One goal for future research will be to determine the specificity of resource allocation deficits. In the study described earlier on early-onset dysthymia, group differences were not obtained between the dysthymic and anhedonic subjects. Perhaps similarities in the dysfunctions of attention underlie risk for severe psychopathology. On the other hand, clinical levels of psychiatric disorders may reveal substantially different patterns of resource allocation. It will be important to explore such possibilities.

Another issue that arises is whether additional cognitive processes might be affected by resource limitations or whether reductions in attentional resources impact only on memory operations. In resource allocation models of depression, the assumption has been that memory operations are likely to be impaired (e.g., Ellis & Ashbrook, 1988; Hasher & Zacks, 1979). No empirical evidence, however, suggests that other mental operations might not be similarly influenced. Further research is needed to clarify such a possibility.

In addition, a goal of future research will be to compare and perhaps integrate the resource allocation perspective with other models of depression. For example, the concept of limited resources is entirely compatible with the model proposed by Hammen and colleagues in which chronic and episodic stress are predictive of the severity of depression (Hammen, Davila, Brown, Ellicott, & Gitlin, 1992). The resource allocation model may even suggest a possible underlying mechanism to account for the relationship between stress and severity of depressive reactions.

Finally, just as further research is needed to determine specificity of resource allocation deficits to psychiatric disorders, studies will have to be designed to consider changes in resource allocation that occur as a function of other emotional states (e.g., happiness). Thus far, the preponderance of research on resource theory has focused exclusively on the cognitive mechanisms involved in information processing. Critical exceptions include the work by Hockey, which demonstrates the impact of such stressful factors as exposure to loud noise (e.g., Hockey, 1970b) and loss of sleep (Hockey, 1970a) on dual-task performance. Recognizing the limitations of "dry" information processing models, cognitive psychologists have begun to call for an integration of affect and cognition and have reintroduced the term "energetics" (e.g., Hockey, Coles, & Gaillard, 1986; Sanders, 1983). The concept of energetics was originated by Cannon (1927), Freeman (1931), and Duffy (1934) to incorporate ideas such as the energy-mobilizing function of emotion. The resurrection of this concept in recent years is an attempt to achieve an integration of affective and cognitive processes by encompassing these concepts, with the goal of incorporating this aspect of human behavior into current models of information processing (Hockey et al., 1986). By considering resources in terms of energetics, emotion can, therefore, be assimilated into information processing theories. Various affective states could then be examined, either through the use of mood-induction procedures or affectively laden stimuli, to begin to understand the impact of emotion on resource allocation.

Acknowledgments. Portions of this article were supported in part by National Institute of Mental Health National Research Service Award MH09624 and are based on a doctoral dissertation submitted by the author to the University of Illinois at Urbana-Champaign.

References

Akiskal, H.S. (1983). Dysthymic disorder: Psychopathology of proposed chronic depressive subtypes. *American Journal of Psychiatry, 140*, 11–20.

American Psychiatric Association. (1987). *Diagnostic and statistical manual of mental disorders* (3rd ed., rev.). Washington, DC: Author.

Blaney, P.H. (1986). Affect and memory: A review. *Psychological Bulletin, 99*, 229–246.

Campbell, D.T., & Fiske, D.W. (1959). Convergent and discriminant validation by the multitrait-multimethod matrix. *Psychological Bulletin, 56*, 81–105.

Cannon, W.B. (1927). The James-Lange theory of emotions: A critical examination and an alternative theory. *American Journal of Psychology, 39*, 106–124.

Chapman, L.J., Chapman, J.P., & Raulin, M.L. (1976). Scales for physical and social anhedonia. *Journal of Abnormal Psychology, 85*, 374–382.

Cohen, R.M., Weingartner, H., Smallberg, S.A., Pickar, D., & Murphy, D.L. (1982). Effort and cognition in depression. *Archives of General Psychiatry, 39*, 593–597.

Depue, R.A., & Klein, D.N. (1988). Identification of unipolar and bipolar affective conditions in nonclinical and clinical populations by the General Behavior Inventory. In D.L. Dunner, E.S. Gershon, & J.E. Barrett (Eds.), *Relatives at risk for mental disorders* (pp. 179–202). New York: Raven Press.

Depue, R.A., Krauss, S., Spoont, M.R., & Arbisi, P. (1989). General Behavior Inventory identification of unipolar and bipolar affective conditions in a nonclinical university population. *Journal of Abnormal Psychology, 98*, 117–126.

Donchin, E. (1981). Surprise!... Surprise? *Psychophysiology, 18*, 493–513.

Donchin, E., Kramer, A.F., & Wickens, C. (1986). Applications of brain event-related potentials to problems in engineering psychology. In M.G.H. Coles, E. Donchin, & S.W. Porges (Eds.), *Psychophysiology: Systems, processes, and applications* (pp. 702–718). New York: Guilford Press.

Duffy, E.A. (1934). Emotion: An example of the need for re-orientation in psychology. *Psychological Review, 41*, 184–198.

Ellis, H.C. (1985). On the importance of mood intensity and encoding demands in memory: Commentary on Hasher, Rose, Zacks, Sanft, and Doren. *Journal of Experimental Psychology: General, 114*, 392–395.

Ellis, H.C., & Ashbrook, P.W. (1988). Resource allocation model of the effects of depressed mood states on memory. In K. Fiedler & J. Forgas (Eds.), *Affect, cognition and social behavior* (pp. 25–43). Toronto: C.J. Hogrefe.

Ellis, H.C., Thomas, R.L., McFarland, A.D., & Lane, J.W. (1985). Emotional mood states and retrieval in episodic memory. *Journal of Experimental Psychology: Learning, Memory, and Cognition, 11*, 363–370.

Ellis, H.C., Thomas, R.L., & Rodriguez, I.A. (1984). Emotional mood states and memory: Elaborative encoding, semantic processing, and cognitive effort. *Journal of Experimental Psychology: Learning, Memory, and Cognition, 10*, 470–482.

Freeman, G.L. (1931). Mental activity and the muscular processes. *Psychological Review, 38*, 428–447.

Friedman, A., & Polson, M.C. (1981). Hemispheres as independent resource systems: Limited-capacity processing and cerebral specialization. *Journal of Experimental Psychology: Human Perception and Performance, 7*, 1031–1058.

Gjerde, P.F. (1983). Attentional capacity dysfunction and arousal in schizophrenia. *Psychological Bulletin, 93*, 57–72.
Glass, R.M., Uhlenhuth, E.H., Hartel, F.W., Matuzas, W., & Fischman, M.W. (1981). Cognitive dysfunction and imipramine in outpatient depressives. *Archives of General Psychiatry, 38*, 1048–1051.
Goplerud, E., & Depue, R.A. (1985). Behavioral response to naturally occurring stress in cyclothymia and dysthymia. *Journal of Abnormal Psychology, 94*, 128–139.
Hammen, C., Davila, J., Brown, G., Ellicott, A., & Gitlin, M. (1992). Psychiatric history and stress: Predictors of severity of unipolar depression. *Journal of Abnormal Psychology, 101*, 45–52.
Hammen, C., Mayol, A., deMayo, R., & Marks, T. (1986). Initial symptom levels and the life-event–depression relationship. *Journal of Abnormal Psychology, 95*, 114–122.
Hasher, L., Rose, K.C., Zacks, R.T., Sanft, H., & Doren, B. (1985). Mood, recall, and selectivity effects in normal college students. *Journal of Experimental Psychology: General, 114*, 104–118.
Hasher, L., & Zacks, R.T. (1979). Automatic and effortful processes in memory. *Journal of Experimental Psychology: General, 108*, 356–388.
Hasher, L., Zacks, R.T., Rose, K.C., & Doren, B. (1985). On mood variation and memory: Reply to Isen (1985), Ellis (1985), and Mayer and Bower (1985). *Journal of Experimental Psychology: General, 114*, 404–409.
Hertel, P.T., & Hardin, T.S. (1990). Remembering with and without awareness in a depressed mood: Evidence of deficits in initiative. *Journal of Experimental Psychology: General, 119*, 45–59.
Hertel, P.T., & Rude, S.S. (1991). Depressive deficits in memory: Focusing attention improves subsequent recall. *Journal of Experimental Psychology: General, 120*, 301–309.
Hockey, G.R.J. (1970a). Changes in attention allocation in a multi-component task under loss of sleep. *British Journal of Psychology, 61*, 473–480.
Hockey, G.R.J. (1970b). Effect of loud noise on attentional selectivity. *Quarterly Journal of Experimental Psychology, 22*, 28–36.
Hockey, G.R.J., Coles, M.G.H., & Gaillard, A.W.K. (1986). Energetical issues in research on human information processing. In G.R.J. Hockey, A.W.K. Gaillard, & M.G.H. Coles (Eds.), *Energetics and human information processing* (pp. 3–21). Dordrecht, The Netherlands: Martinus Nijhoff.
Hoffman, J.E., Houck, M.R., MacMillan, III, F.W., Simons, R.F., & Oatman, L.C. (1985). Event-related potentials elicited by automatic targets: A dual-task analysis. *Journal of Experimental Psychology: Human Perception and Performance, 11*, 50–61.
Ingram, R.E. (1990). Self-focused attention in clinical disorders: Review and a conceptual model. *Psychological Bulletin, 107*, 156–176.
Isen, A.M. (1985). Asymmetry of happiness and sadness in effects on memory in normal college students: Comment on Hasher, Rose, Zacks, Sanft, and Doren. *Journal of Experimental Psychology: General, 114*, 388–391.
Isreal, J.B., Chesney, G.L., Wickens, C.D., & Donchin, E. (1980). P300 and tracking difficulty: Evidence for multiple resources in dual-task performance. *Psychophysiology, 17*, 259–273.

Kahneman, D. (1973). *Attention and effort.* Englewood Cliffs, NJ: Prentice-Hall.
Keller, M.B., Lavori, P., Endicott, J., Coryell, W., & Klerman, G.L. (1983). "Double depression": Two-year follow up. *American Journal of Psychiatry, 140,* 689–694.
Klein, D.N. (1990, November). *Dysthymia: A critical examination of the construct.* Invited paper presented at the fifth annual meeting of the Society for Research on Psychopathology, Boulder, CO.
Klein, D.N., Dickstein, S., Taylor, E.T., & Harding, K. (1989). Identifying chronic affective disorders in outpatients: Validation of the General Behavior Inventory. *Journal of Consulting and Clinical Psychology, 57,* 106–111.
Kovacs, M., Feinberg, T.L., Crouse-Novak, M., Paulauskas, S., Pollock, M., & Finkelstein, R. (1984). Depressive disorders in childhood: II. A longitudinal study of the risk for a subsequent major depression. *Archives of General Psychiatry, 41,* 643–649.
Kramer, A.F., Sirevaag, E., & Braune, R. (1987). A psychophysiological assessment of operator workload during simulated flight session. *Human Factors, 29,* 145–160.
Kramer, A.F., & Strayer, D.L. (1988). Assessing the development of automatic processing: An application of dual-task and event-related brain potential methodologies. *Biological Psychology, 26,* 231–267.
Kramer, A.F., Wickens, C.D., & Donchin, E. (1983). An analysis of the processing demands of a complex perceptual-motor task. *Human Factors, 25,* 597–622.
Kramer, A.F., Wickens, C.D., & Donchin, E. (1985). Processing of stimulus properties: Evidence for dual-task integrality. *Journal of Experimental Psychology: Human Perception and Performance, 11,* 393–408.
Krames, L., & MacDonald, M.R. (1985). Distraction and depressive cognitions. *Cognitive Therapy and Research, 9,* 561–573.
Leight, K.A., & Ellis, H.C. (1981). Emotional mood states: Strategies, and state-dependency in memory. *Journal of Verbal Learning and Verbal Behavior, 20,* 251–266.
Mathews, A. (1990). Why worry? The cognitive function of anxiety. *Behavioral Research and Therapy, 28,* 455–468.
Mayer, J.D., & Bower, G.H. (1985). Naturally occurring mood and learning: Comment on Hasher, Rose, Zacks, Sanft, and Doren. *Journal of Experimental Psychology: General, 114,* 396–403.
McAllister, T.W. (1981). Cognitive functioning in the affective disorders. *Comprehensive Psychiatry, 22,* 572–586.
Miller, G.A., Yee, C.M., & Anhalt, J.M. (1992). *ERP incentive effects and potential risk for psychopathology.* Manuscript submitted for publication.
Miller, W.R. (1975). Psychological deficit in depression. *Psychological Bulletin, 82,* 238–260.
Mineka, S., & Sutton, S.K. (1992). Cognitive biases and the emotional disorders. *Psychological Science, 3,* 65–69.
Navon, D. (1984). Resources—A theoretical soup stone? *Psychological Review, 91,* 216–234.
Navon, D., & Gopher, D. (1979). On the economy of the human-processing system. *Psychological Review, 86,* 214–255.

Navon, D., & Miller, J. (1987). Role of outcome conflict in dual-task interference. *Journal of Experimental Psychology: Human Perception and Performance, 13*, 435–448.

Neumann, O. (1987). Beyond capacity: A functional view of attention. In H. Heuer & A.F. Sanders (Eds.), *Perspectives on perception and action* (pp. 361–394). Hillsdale, NJ: Lawrence Erlbaum.

Norman, D.A., & Bobrow, D.G. (1975). On data-limited and resource-limited processes. *Cognitive Psychology, 7*, 44–64.

Nuechterlein, K.H., & Dawson, M.E. (1984). Information processing and attentional functioning in the developmental course of schizophrenic disorders. *Schizophrenia Bulletin, 10*, 160–203.

Roy-Byrne, P.P., Weingartner, H., Bierer, L.M., Thompson, K., & Post, R.M. (1986). Effortful and automatic cognitive processes in depression. *Archives of General Psychiatry, 43*, 265–267.

Sanders, A.F. (1983). Towards a model of stress and human performance. *Acta Psychologica, 53*, 61–97.

Silberman, E.K., Weingartner, H., & Post, R.M. (1983). Thinking disorder in depression. *Archives of General Psychiatry, 40*, 775–780.

Sirevaag, E.J., Kramer, A.F., Coles, M.G.H., & Donchin, E. (1989). Resource reciprocity: An event-related brain potentials analysis. *Acta Psychologica, 70*, 77–97.

Strayer, D.L., & Kramer, A.F. (1990). Attentional requirements of automatic and controlled processing. *Journal of Experimental Psychology: Learning, Memory, and Cognition, 16*, 67–82.

Tariot, P.N., & Weingartner, H. (1986). A psychobiologic analysis of cognitive failures. *Archives of General Psychiatry, 43*, 1183–1188.

Weingartner, H., Cohen, R.M., Murphy, D.L., Martello, J., & Gerdt, C. (1981). Cognitive processes in depression. *Archives of General Psychiatry, 38*, 42–47.

Wickens, C.D. (1980). The structure of attentional resources. In R. Nickerson (Ed.), *Attention and performance VIII* (pp. 239–258). Hillsdale, NJ: Lawrence Erlbaum.

Wickens, C.D. (1984, April–May). *The multiple resources model of human performance: Implications for display design.* In AGARD/NATO Proceedings, Williamsburg, VA.

Wickens, C.D. (1986). Gain and energetics in information processing. In G.R.J. Hockey, A.W.K. Gaillard, & M.G.H. Coles (Eds.), *Energetics and human information processing* (pp. 373–389). Dordrecht, The Netherlands: Martinus Nijhoff.

Wickens, C.D., Kramer, A., Vanasse, L., & Donchin, E. (1983). The performance of concurrent tasks: A psychophysiological analysis of the reciprocity of information processing resource. *Science, 221*, 1080–1082.

Yee, C.M., Deldin, P.J., & Miller, G.A. (1992). Early stimulus processing in dysthymia and anhedonia. *Journal of Abnormal Psychology, 101*, 230–233.

Yee, C.M., & Miller, G.A. (1988). Emotional information processing: Modulation of fear in normals and dysthymics. *Journal of Abnormal Psychology, 97*, 54–63.

Yee, C.M., & Miller, G.A. (1994). A dual-task analysis of resource allocation in dysthymia and anhedonia. *Journal of Abnormal Psychology, 103*, 625–636.

Yee, C.M., Miller, G.A., & Kramer, A.F. (1987). Resource demands of a recognition running-memory task. (Abstract) *Psychophysiology, 24*, 622.

11
Depression and the Behavioral High-Risk Paradigm

STEVEN D. HOLLON

Without doubt, some people are at greater risk for depression than others. Depression tends to be both self-limiting and recurrent; that is, any given episode tends to go away, even in the absence of treatment, but the vast majority of people who get depressed will experience multiple episodes (Consensus Development Panel, 1985). Given that most people will never have even a single clinical episode, this means that risk for the disorder is bimodally distributed; a minority of people in the general population will suffer the majority of the episodes of clinical depression.

Exactly why that is so remains a matter of considerable debate. A long-standing controversy exists as to whether depression is biological or psychological. In all likelihood, it is both (Shelton, Hollon, Purdon, & Loosen, 1991). Depression clearly runs in families, and good evidence indicates that at least some component of that familial transmittal is genetic. At the same time, attitudes, values, and coping styles can be learned, and early life experiences likely shape the nature of the developing personality and confer risk (Garber, 1992). Stressful life events clearly play a role in triggering the experession of depression, and there is reason to think that vulnerability can be acquired, both biological and psychological, as a consequence of experience with the disorder or its causes (Nolen-Hoeksema, Girgus, & Seligman, 1992; Post, 1992).

A behavioral high-risk paradigm seeks to identify markers of vulnerability in individuals not currently in distress. Such indices can be used to predict risk and to provide insight into the causal process. Even if they are not themselves involved in the causal chain, they must at least covary with those causal processes and can be used to select samples for further study. Covariation can have only three explanations; if two phenomena are related, then either the first must cause the second, the second must cause the first, or the two must share some common causal process. To the extent that vulnerability markers can be identified before the onset of the disorder, either they or their correlates must be involved in the causal process, and it can be quite informative to select samples on the basis of their presence or absence.

The four chapters in this section represent a diverse array of efforts to deal with and identify markers of vulnerability. In Chapter 7, Klein and Anderson focus on what they call exophenotypic (as distinct from genetic or biological) indices—subsyndromal behaviors similar in kind but not intensity to the symptom clusters that comprise the more fully developed syndromes of depression and mania. In essence, people who characteristically behave in ways similar to persons with depression or mania can be shown to be at greater risk for the full clinical syndrome. In Chapter 8, Sponheim, Allen, and Iacono review the psychophysiological high-risk studies; electrodermal hypoactivation and increased left frontal electroencephalographic (EEG) alpha activity both appear to be stable traits found in individuals at elevated risk for depression even when euthymic. These markers can be used to select samples at risk and suggest the operation of specific biological abnormalities in the etiology of depression. In Chapter 9, O'Hara reviews the evidence for cognitive diatheses; although mixed, it appears to suggest that the propensity to attribute negative life events to some stable defect in the self or related patterns of information processing may contribute to risk. As for the psychophysiological processes, some indices are stable across time, whereas others appear to be more dependent on current affective state. In Chapter 10, Yee does not so much describe a behavioral high-risk paradigm as present a model to account for the effects of depression on information processing; her resource allocation model suggests that depression alters the amount and availability of attentional resources available for cognitive tasks and predicts that depression is more likely to disrupt processing that requires effort than that which is more nearly automatic. Whether this distinction is itself evident in individuals at risk remains to be seen; if it is, then the dual-task paradigm she describes could be used to identify markers of risk in euthymic individuals.

These four chapters give a picture of the range that is possible in potential markers of risk, running as they do from overt behaviors to psychophysiological indices to measures of information processing and beliefs. There is, of course, no logical restriction on what could be a marker of risk; any phenomenon that differentiates persons at risk from those who are not (or, more accurately, that covaries with level of risk) can qualify as a marker. Markers can be detected on purely atheoretical grounds, and that has sometimes been the case, but, ultimately, they have their greatest utility when embedded in a theory that specifies their relation to underlying causal processes. The relation between theory and observation is always one of iteration. Some phenomenon commands attention and leads to the formulation of an initial hypothesis. This hypothesis is then subjected to potential disconfirmation through the systematic collection of additional observations chosen so as to test its validity. These additional observations either strengthen the initial impression or lead to its revision. This process is then repeated, alternating between theory and observation in a fashion that converges on an increasingly precise understanding of the causal structure of the phenomenon under study. In essence, science always moves back and forth between constructs (conjectures) and operations (observations), hopefully "bootstrapping" its way to greater knowledge (Meehl, 1978). With that process in mind, several observations need to be made about the logical pitfalls inherent in the process of applying high-risk paradigms in the pursuit of underlying causal structures.

Sample Selection Constrains What Can Be Detected

First, it is important to keep in mind that how one defines the high-risk sample to be studied constrains the nature of the markers that can be detected and the nature of the underlying causal processes that they can reveal. For example, defining a sample on the basis of biological parentage (risk appears to be elevated in the offspring of depressed parents) is particularly likely to facilitate the detection of processes linked to genetically transmitted biological parameters. Such a strategy does little to quantify the extent to which such factors are involved in the etiology of depression (beyond saying that their role is nonnegligible), but it is a perfectly reasonable way of isolating those processes that do play a role in genetic transmission. Similarly, identifying a sample on the basis of some type of negative life event (e.g., divorce or loss of a parent during childhood) may facilitate identifying psychosocial markers of risk or their related causal referents, but it will provide what is at best a distorted picture of the relative contributions of various causal processes.

The issue in both cases is that depression is probably an etiologically heterogeneous collection of disorders; it is likely that causal process for most individuals represent some combination of heredity and environment, but that different subtypes of the disorder, or different individuals within the subtypes probably inherit different levels of vulnerability and are subsequently exposed to differing levels of environmental risk. How one defines the sample at risk reflects the underlying model of interest; defining the sample on any basis other than "caseness" is likely to lead to a bias in the representation of inherited versus acquired causal processes. That is a problem only if statements are made about the generality of the markers or causal process detected, but it is a high-probability error that is difficult to avoid.

Existing studies strongly suggest variability in the heritability of different subtypes of depression. Genetic factors appear to play a major role in the etiology of the bipolar disorders, whereas their role in the unipolar disorders is less clear (Depue & Monroe, 1978; Nurnberger & Gershon, 1982). More severe unipolar depressions appear to be highly heritable; less severe unipolar depressions appear to be less so, but their apparent lower heritability may be an artifact of the greater difficulty in detecting the occurrence of less severe episodes. The picture is further complicated by indications that bipolar disorder may be underdiagnosed. Both diagnostic studies involving multiple informants in the Amish (Egeland, 1983) and studies of cyclothymia in nonclinical populations (Depue et al., 1981) suggest that mild hypomania may go undetected in most clinical settings, particularly when partients serve as their own sole informants. This suggests that bipolarity may be more common than is typically recognized and that many patients who are considered unipolar are, in fact, unrecognized bipolars. If true, then much of our current thinking regarding differences between the two subtypes may have to be revised, and particular caution will have to be exercised not to confound the two disorders when trying to apply a behavioral high-risk paradigm.

Stable Traits or Latent Predispositions?

One source of potential variability that can confound the theory-testing process concerns the nature of the purported markers of vulnerability. Given that people can be shown to differ in degree of risk (most people will never become depressed, but most people who become depressed will have multiple episodes), some processes must distinguish people at differing levels of risk; that is, some type of stable diathesis must exist. That does not mean, however, that this diathesis can always be detected, at least using currently existing methodologies. It is quite possible for an underlying diathesis to be latent predisposition; a propensity that becomes activated only under certain conditions. This is clearly the case for at least some forms of biological dysregulation (Depue, Kleiman, Davis, Hutchinson, & Krauss, 1985; Ehlers, Frank, & Kupfer, 1988; Siever & Davis, 1985), and it also appears to be the case for social dysfunction (Suomi, 1991). Information processing may work in a similar fashion; cognitive theories of depression have long posited the existence of negative schemata that are relatively quiescent under normal conditions but that become activated under stress (Hollon, 1992; Riskind & Rholes, 1984). Recent studies have suggested that subjects at risk for depression are more likely to endorse negative beliefs under conditions of challenge or stress than normal controls; few differences are noted in the absence of any such challenge (Miranda, 1992; Miranda & Persons, 1988; Miranda, Persons, & Byers, 1990; Teasdale & Dent, 1987).

This suggests that it is not only a matter of what is assessed, but the conditions under which assessment takes place. Of course, any potential process that contributes to risk must exist in physical reality; hypothetical constructs do not actually cause anything, they must be real, and they must have an underlying physical structure. That is not to say, however, that we necessarily have the technological capacity to measure this underlying structure. Measurement technologies are most likely to advance if there is reason for them to be pursued; that is most likely to be the case if potential latent predispositions are first specified on the basis of theory.

Moreover, individuals may exhibit important differences in terms of precisely what constitutes an adequate challenge. For example, individuals with greater interpersonal needs appear to be more prone to becoming depressed in response to losses, while persons with greater achievement needs may be somewhat more vulnerable to frustrations in that domain (Hammen, Marks, Mayol, deMayo, 1985; Zuroff & Mongrain, 1987). This notion of "content specificity" suggests that what constitutes a challenge may differ across indivduals and necessitate efforts at idiographic assessment.

It is of interest that attention in this regard has largely centered around interpersonal and achievement domains. Since the time of Freud, theorists in the area of depression have focused on threats to subjective well-being in the domains of love and work, including Arieti's distinction between claiming (dependent) versus self-blaming subtypes and Beck's recent distinction between sociotropic versus autonomous personality styles as risk factors for depression (Karasu, 1990).

This distinction takes on particular importance when one notes that in addition to the pharmacological induction via reserpine (Goodwin & Bunney, 1971), the two major psychological paradigms for inducing "depression" in infrahuman species involve separation from care givers (McKinney, Suomi, & Harlow, 1971) and exposure to uncontrollable stress (Seligman, 1975). With respect to the latter, learned helplessness, it is additionally of interest that a phenomenon supposedly based purely on associative learning appears to be easiest to induce in those species that are most socially dependent during infancy (e.g., rats appear to be more susceptible to helplessness induction than cats, although the latter are more intelligent, and dogs are more susceptible still). It may be that depression represents a biopsychosocial phenomenon in which a biological propensity prewired into the species to facilitate attachment during infancy may come to be triggered under conditions of overwhelming stress later in life. In this regard, it is worthy of note that the three major approaches to the treatment of depression that have fared best in the treatment outcome literature are pharmacotherapy, cognitive therapy, and interpersonal psychotherapy, as each represents one of the components of that larger biopsychosocial model (Munoz, Hollon, McGrath, Rehm, & Vandenbos, 1994). This correspondence may reflect an underlying unity in the causal structure of the disorder.

Multiple Causal Processes and Chains

Depression likely is the consequence of multiple causal processes. This multiplicity can take at least two forms. First, depression probably is a heterogenous disorder that encompasses a number of different subtypes with distinct causal structures (e.g., Abramson, Metalsky, & Alloy, 1989; Klein, 1974). Second, even within a given subtype, the underlying causal structure probably can best be described as an interrelated series of events, each influenced by processes more distal ("upstream") in the chain and each, in turn, exerting an influence on more proximal ("downstream") events.

This suggests that it is very unlikely that any given process or factor will either be necessary or sufficient in the etiology of depression. Heterogeneity with respect to causality rules out the former, and the notion of sequential causal chains makes unlikely the latter. This means that efforts to uncover the causal structure of the disorder must almost certainly be probabilistic in nature and open to the possibility that different subtypes may exist and that they may or may not share a particular causal component.

Moreover, it is important to keep in mind that science never tests discrete variables; rather, it always tests causal models. For example, specificity design logic suggests that for a variable to be said to be causal of a disorder, it must not only covary with and be temporally antecedent to that disorder, it must also be specific in the sense of not covarying with some other disorder different from the one in question (Garber & Hollon, 1992). This logic is based on the notion that the same process cannot cause two different disorders, if those disorders are seen as being distinct from one another. However, if etiology is based on the joint co-occurrence of two or more events (i.e., if two or more processes interact to cause a disorder), then it is quite possible for any one of those events to be part of the causal structure of multiple disorders.

For example, Kendler and colleagues have suggested that depression and anxiety share an underlying genetic vulnerability, with subsequent environmental events determining which of the disorders is likely to be expressed (Kendler, Heath, Martin, & Eaves, 1987). Nonspecificity for the genetic component in that model hardly implies noncausality. Rather, it is the contribution of the environmental processes with which it interacts that determines the specific nature of the disorder expressed; specificity design logic holds, but only at the level of the causal model tested, not with respect to any single variable or construct within the model.

Acquisition of Risk

Vulnerability to depression apparently can either be inherited or acquired. In some instances, the same processes may be introduced by either route. For example, recent studies have suggested that having a negative explanatory style (also known as attributional style) may increase risk for depression under conditions of stress (Peterson & Seligman, 1984). Because this phenomenon was first hypothesized in the context of social-cognitive theory, it is generally presumed that it is acquired as a consequence of early learning experiences (Abramson, Seligman, & Teasdale, 1978). However, in a recent study contrasting identical and fraternal twins, explanatory style was found to be quite heritable, with genetic factors accounting for about half of the variability in its expression (Schulman, Keith, & Seligman, 1993). Conversely, prepubescent children who became depressed (typically as a consequence of parental conflict) were found to become more negative in their explanatory style, a change that remained stable over time even after their distress began to resolve (Nolen-Hoeksema, Girgus, & Seligman, 1992).

Biological processes may well work in a similar fashion. People not only differ with respect to the genes they inherit, but their propensity for biological dysregulation may be further affected by subsequent events. For example, Post (1992) has described a phenomenon called "kindling" in which exposure to noxious life events alters the way in which genes are transcribed at the neuronal level, resulting in an increased sensitivity to subsequent events and the possible onset of spontaneous episodes. Post has proposed that this neuronal sensitization may account for the classic observation that external precipitants can be more easily specified for earlier episodes than for later ones. Permanent changes in sensitization within the postsynaptic neuron may be one mechanism by which biological vulnerability can be acquired as a consequence of life experience.

This suggests that risk is hardly static over time and that either psychological or biological processes can be acquired as a consequence of life experiences that confer risk. Application of a behavioral high-risk paradigm may have to take into account where in the history of the disorder risk is assessed and what impact subsequent events can have on both the acquisition of such risk and its expression.

Temporality of Risk

A growing consensus in the field holds that a distinction needs to be made between the return of symptoms associated with a recent episode, or *relapse*, versus the onset of a wholly new episode, or *recurrence* (Frank et al., 1991). Most episodes of depression last for 6 to 9 months if left untreated, and there is reason to think that patients treated to remission pharmacologically are at elevated risk for symptom return if medication ceases before the end of this period (Prien & Kupfer, 1986). This suggests that the episode may have an underlying course that is distinct from its symptomatic expression and that drugs may suppress symptoms but not alter the course of the underlying episode (Hollon, Evans, & DeRubeis, 1991).

If true, this could have real implications for the application of high-risk paradigms, because the processes maintaining the life of an existing episode may not be the same as those that initiated it or that confer risk for subsequent episodes. Studies that seek to contrast subjects with a history of prior depression versus those without may need to exercise caution not to mix patients who are merely remitted (i.e., currently asymptomatic, but not yet "out of episode") with those who have truly recovered (i.e., those patients for whom the prior episode has truly run its course). The reasons are twofold: (a) Patients who are only remitted are likely to be abnormal with respect to two sets of causal process (or their correlates), those vulnerability factors that confer risk for new episodes and those mechanisms that maintain the current episode. (b) Efforts to follow the sample over time are likely to be confounded by the inability to determine whether symptom return reflects the reemergence of the existing episode or the onset of a new one.

At this time, there is no good way to distinguish relapse from recurrence on the basis of manifest symptomatology or to determine just who has fully recovered as opposed to merely remitted other than on the basis of clinical course. At the very least, subjects with a history of clinical depression should have been symptom free for a minimum of 4 to 6 months before they are considered fully recovered from the previous episode, and even that may not be sufficient. Considerable variability is seen in the course of recovery, and the current practice of providing maintenance medication on an ongoing basis for euthymic patients only further serves to obscure the underlying clinical course. Few studies using a known-groups behavioral high-risk design to contrast subjects who were formerly depressed versus those without such a history have shown sufficient care in subject selection to support such a distinction.

As described elsewhere, this temporality in underlying course could be turned to good advantage methodologically (Hollon et al., 1990). Because drugs appear to be largely symptom suppressive, it should be possible to distinguish among the consequences of being depressed (symptoms), the causal processes maintaining the episode (mechanisms), and the vulnerability factors that confer risk for onset (predispositions) by following a sample of pharmacologically treated depressed patients across time and comparing them to subjects who have never been depressed. At the point at which they begin treatment, depressed patients should be elevated with respect to symptoms, mechanisms, and predispositions. Shortly after initial remission, they should no longer be symptomatic, but should still be elevated with respect to both mechanisms and predispositions, because they are still "in episode" and at elevated risk for both relapse and recurrence. After several months of continued remission, the underlying episode should have run its course and the formerly depressed patients should show elevations only with respect to stable predispositions. This logic can be checked by discontinuing medication for some of the sample shortly after remission and for another subset of the sample shortly after recovery; rates of symptom return should be about three times as high for the patients who have merely remitted compared with those who have fully recovered, and up to five times as high for the latter compared with patients maintained on medications or subjects with no history of depression.

The key point for the current discussion is that it is not sufficient to define risk on the basis of prior history; it is important to try to ascertain where subjects are with respect to the course of the underlying disorder. Certainly, studies that rely on patients recently remitted as their "at-risk" group should be viewed with some suspicion, particularly if those patients are still being continued on medication. Such a sample can be used to detect stable predispositions that confer risk, but it may have difficulty distinguishing such factors from the more transient causal processes that define the underlying episode.

Similarly, it is important to distinguish between samples consisting of subjects at risk with prior histories of depression versus those without. Because risk can be acquired, it is quite possible that vulnerability factors that predict to recurrence may differ from those that predict initial risk. Moreover, because experience with depression can leave residual consequences that are not themselves causal, markers of vulnerability in a previously depressed sample are particularly likely to be epiphenomenal correlates of true causal processes. In either case, temporality appears to be an important feature of the underlying causal structure of depression.

Integration Across Domains

Up to this point, biological and psychological factors have been treated as if they were mutually exclusive. This is hardly the case. All behavior rests on a biological substrate, but one that is responsive to and facilitates interaction with the external environment. Exposure to various life events and experiences is likely to influence the underlying structure and function of the brain, and changes at the biochemical level are likely to influence psychological functioning (Shelton et al., 1991).

In some respects, distinctions between different regions of the brain and their related psychological functions may be more important than those between psychological versus biological functions. Basic affective processes are largely mediated by lower visceral-emotive neural systems (i.e., the limbic system and the brain stem), but these systems are subject to influence from, and provide input to, higher somatic-cognitive (cortical) systems (Panksepp, 1986). Both process information from the environment, and both play a role in coordinating behavioral responses to that input. Although clearly central, the limbic system is not alone in regulating affect. For example, depressed patients show reduced left anterior cortical activation relative to normal controls, a pattern that persists into remission (Henriques & Davidson, 1990) and predicts response to parental separation in infants (Davidson & Fox, 1989). Strokes in this region of the cortex are particularly likely to produce a secondary depression (Robinson, Kubos, Starr, Rao, & Price, 1984). Davidson and Tomarken (1989) have speculated that this pattern of cortical activation may serve as a diathesis that lowers the threshold for negative emotions and deficits in approach behavior.

Most critical for the present discussion, subjects evidencing such a pattern of cortical activation appear to be particularly likely to exhibit a depressotypic explanatory style (Davidson, Abramson, Tomarken, Wheeler, 1990). Correspondence between a psychophysiological measure of cortical activation and self-reports of information processing proclivities suggests a fundamental coherence between structure and function, because the region of the cortex involved appears to play an important role in planing and judgment, particularly as it relates to the regulation of affective processes. That this same psychological proclivity appears to be both inherited (Schulman et al., 1993) and acquired (Nolen-Hoeksema et al., 1992), predictive of risk in both normal and remitted clinical samples (Hollon, DeRubeis, & Seligman, 1991), and particularly susceptible to change following psychosocial (as opposed to pharmacological) interventions (Hollon et al., 1990) suggests that distinctions between regions of neural activation and the functions they subserve may be more important than any simple distinction between psychological and biological functions.

Moreover, evidence suggests that the basic defect in depression may not lie so much in individual differences in responsivity to stress as in an inability to return to baseline following initial perturbation. For example, Goplerud and Depue (1985) had cyclothyic, dysthymic, and normal control subjects monitor their responses to stressful life events over a period of several days. Subjects with a history of affective disorder did not become more distressed than normal controls, but they did take longer to return to baseline. Metalsky and colleagues have reported a similar finding with respect to response to examination stress; subjects who did poorly on the examination tended to become distressed, but that distress was most likely to persist among those subjects who also had a negative attributional style (Metalsky, Halberstadt, & Abramson, 1987). Finally, Weiss and Simson (1985) have found near universal but transient depletion of norepinephrine in neurons projecting from the brain stem to the limbic system in mice subjected to inescapable shock; only those animals that lack the homeostatic capacity to recover from that temporary depletion go on to show behavioral effects closer to the time course exhibited by depression in humans. These converging lines of research suggest that depression may be more a consequence of a failure of homeostatic regulation to dampen down the stress response following exposure to noxious life events than of any undue responsivity in that system in the first place.

Conclusions

The application of behavioral high-risk paradigms to the study of the affective disorders clearly has great merit. Given what is already known about the distribution of the disorder, some individuals must be at greater risk than others. Detecting factors that predict subsequent risk can only contribute to the process of theory building and the mobilization of clinical resources to prevent the disorder or its recurrence.

It is likely that the affective disorders are heterogenous in their causal structures, that causality with respect to any given subtype will best be described in terms of a chain of sequential events, and that some vulnerability factors may be latent in nature and require activation in order to be adequately assessed. Moreover, evidence indicates that risk can either be inherited or acquired, that the nature of risk may change over time, and that biological and psychological processes are highly interrelated. Efforts to adapt behavioral high-risk paradigms to take these likely attributes of the affective disorders into account can only facilitate the detection of vulnerability factors and their underlying causal structures.

References

Abramson, L.Y., Metalsky, G., & Alloy, L.B. (1989). Hopelessness depression: A theory-based subtype of depression. *Psychological Review, 96*, 358–372.

Abramson, L.Y., Seligman, M.E.P., & Teasdale, J.D. (1978). Learned helplessness in humans: Critique and reformulation. *Journal of Abnormal Psychology, 87*, 49–74.

Consensus Development Panel (1985). NIMH/NIH consensus development conference statement on mood disorders: Pharmacological prevention of recurrences. *American Journal of Psychiatry, 142*, 469–476.

Davidson, R.J., Abramson, L.Y., Tomarken, A.J., & Wheeler, R.E. (1990). *Asymmetrical anterior temporal brain activity predicts beliefs about the causes of negative life events.* Unpublished manuscript.

Davidson, R.J., & Fox, N.A. (1989). Frontal brain asymmetry predicts infants' response to maternal separation. *Journal of Abnormal Psychology, 98*, 127–131.

Davidson, R.J., & Tomarken, A.J. (1989). Laterality and emotion: An electrophysiological approach. In F. Boller & J. Grafman (Eds.), *Handbook of neuropsychology* (Vol. 3, pp. 419-441). Amsterdam, The Netherlands: Elsevier Science.

Depue, R.A., Kleiman, R.M., Davis, P., Hutchinson, M., & Krauss, S. (1985). The behavioral high risk paradigm and bipolar affective disorder: VIII. Serum free cortisol in nonpatient cyclothymic subjects selected by the General Behavior Inventory. *American Journal of Psychiatry, 142*, 175-181.

Depue, R.A., & Monroe, S.M. (1978). The unipolar-bipolar distinction in the depressive disorders. *Psychological Bulletin, 85*, 1001-1029.

Depue, R.A., Slater, J.F., Wolfstetter-Kausch, H., Klein, D., Goplerud, E., & Farr, D. (1981). A behavioral paradigm for identifying persons at risk for bipolar depressive disorders: A conceptual framework and five validation studies. (Monograph) *Journal of Abnormal Psychology, 90*, 381-437.

Egeland, J.A. (1983). Bipolarity: The iceberg of affective disorders? *Comprehensive Psychiatry, 24*, 337-344.

Ehlers, C.L., Frank, E., & Kupfer, D.J. (1988). Social *zeitgebers* and biological rhythms: A unified approach to understanding the etiology of depression. *Archives of General Psychiatry, 45*, 948-952.

Frank, E., Prien, R.F., Jarrett, R.B., Keller, M.B., Kupfer, D.J., Lavori, P.W., Rush, A.J., & Weissman, M.M. (1991). Conceptualization and rationale for consensus definitions of terms in major depressive disorder: Remission, recovery, relapse, and recurrence. *Archives of General Psychiatry, 48*, 851-855.

Garber, J. (1992). Cognitive models of depression: A developmental perspective. *Psychological Inquiry, 3*, 235-240.

Garber, J., & Hollon, S.D. (1992). What can specificity designs say about causality in psychopathology research? *Psychological Bulletin, 110*, 129-136.

Goodwin, F.K., & Bunney, W.E., Jr. (1971). Depression following reserpine: A reevaluation. *Seminars in Psychiatry, 3*, 435.

Goplerud, E., & Depue, R.A. (1985). Behavioral response to naturally occurring stress in cyclothymia and dysthymia. *Journal of Abnormal Psychology, 94*, 128-139.

Hammen, C., Marks, T., Mayol, A., & deMayo, R. (1985). Depressive self-schemas, life stress, and vulnerability to depression. *Journal of Abnormal Psychology, 94*, 308-319.

Henriques, J.B., & Davidson, R.J. (1990). Regional brain electrical asymmetries discriminate between previously depressed and healthy control subjects. *Journal of Abnormal Psychology, 99*, 22-31.

Hollon, S.D. (1992). Cognitive models of depression from a psychobiological perspective. *Psychological Inquiry, 3*, 250-253.

Hollon, S.D., DeRubeis, R.J., & Seligman, M.E.P. (1991). Cognitive therapy and the prevention of depression.*Applied and Preventive Psychology, 1*, 89-95.

Hollon, S.D., Evans, M.D., & DeRubeis, R.J. (1990). Cognitive mediation of relapse prevention following treatment for depression: Implications of differential risk. In R.E. Ingram (Ed.), *Psychological aspects of depression* (pp. 117-136). New York: Plenum Press.

Karasu, T.B. (1990). Toward a clinical model of psychotherapy for depression: I. Systematic comparison of three psychotherapies. *American Journal of Psychiatry, 147*, 133-147.

Kendler, K.S., Heath, A.C., Martin, N.G., & Eaves, L.J. (1987). Symptoms of anxiety and symptoms of depression. *Archives of General Psychiatry, 144*, 451-457.

Klein, D.F. (1974). Endogenomorphic depression: A conceptual and terminological revision. *Archives of General Psychiatry, 31*, 447–454.

McKinney, W.T., Suomi, S.J., & Harlow, H.F. (1971). Depression in primates. *American Journal of Psychiatry, 127*, 1313–1320.

Meehl, P.E. (1978). Theoretical risks and tabular asterisks: Sir Karl, Sir Ronald, and the slow progress of soft psychology. *Journal of Consulting and Clinical Psychology, 46*, 806–834.

Metalsky, G.I., Halberstadt, L.J., & Abramson, L.Y. (1987). Vulnerability to depressive mood reactions: Toward a more powerful test of the diathesis-stress and causal mediation components of the reformulated theory of depression. *Journal of Personality and Social Psychology, 52*, 386–393.

Miranda, J. (1992). Dysfunctional thinking is activated by stressful life events. *Cognitive Therapy and Research, 16*, 473–483.

Miranda, J., & Persons, J. B. (1988). Dysfunctional attitudes are mood-state dependent. *Journal of Abnormal Psychology, 97*, 76–79.

Miranda, J., Persons, J.B., & Byers, C.N. (1990). Endorsement of dysfunctional beliefs depends on current mood state. *Journal of Abnormal Psychology, 99*, 237–241.

Munoz, R.F., Hollon, S.D., McGrath, E., Rehm, L.P., & Vandenbos, G.R. (1994). On the AHCPR *Depression in Primary Care* guidelines: Further considerations for practitioners. *American Psychologist, 49*, 42–61.

Nolen-Hoeksema, S., Girgus, J.S., & Seligman, M.E.P. (1992). Predictors and consequences of childhood depressive symptoms: A 5-year longitudinal study. *Journal of Abnormal Psychology, 101*, 405–422.

Nurnberger, J.I., & Gershon, E.S. (1982). Genetics. In E.S. Paykel (Ed.), *Handbook of affective disorders* (pp. 126–145). New York: Guilford Press.

Panksepp, J. (1986). The anatomy of emotions. In R. Plutchik & H. Kellerman (Eds.), *Emotion: Theory, research, and experience* (pp. 91–124). New York: Academic Press.

Peterson, C., & Seligman, M.E.P. (1984). Causal explanations as a risk factor for depression: Theory and evidence. *Psychological Reviews, 91*, 347–374.

Post, R. (1992). Transduction of psychosocial stress into the neurobiology of recurrent affective disorder. *American Journal of Psychiatry, 149*, 999–1010.

Prien, R.F., & Kupfer, D.J. (1986). Continuation drug therapy for major depressive episodes: How long should it be maintained? *American Journal of Psychiatry, 143*, 18–23.

Riskind, J.H., & Rholes, W.S. (1984). Cognitive accessibility and the capacity of cognitions to predict future depression: A theoretical note. *Cognitive Therapy and Research, 8*, 1–12.

Robinson, R.G., Kubos, K.L., Starr, L.B., Rao, K., & Price, T.R. (1984). Mood disorders in stroke patients: Importance of location of lesion. *Brain, 107*, 81–93.

Schulman, P., Keith, D., & Seligman, M.E.P. (1993). Is optimism heritable? A study of twins. *Behaviour Research and Therapy, 31*, 569–574.

Seligman, M.E.P. (1975). *Helplessness: On depression, development, and death.* San Francisco: Freeman.

Shelton, R.C., Hollon, S.D., Purdon, S.E., & Loosen, P.T. (1991). Biological and psychological aspects of depression. *Behavior Therapy, 22*, 201–228.

Siever, L.J., & Davis, K.L. (1985). Overview: Toward a dysregulation hypothesis of depression. *American Journal of Psychiatry, 142*, 1017–1031.

Suomi, S. (1991). Uptight and laid-back monkeys: Individual differences in the response to social challenges. In S.E. Brauth, W.S. Hall, & R.J. Dooling (Eds.), *Plasticity of development* (pp. 27–56). Cambridge, MA: MIT Press.

Teasdale, J.D., & Dent, J. (1987). Cognitive vulnerability to depression: An investigation of two hypotheses. *British Journal of Clinical Psychiatry, 26*, 113–126.

Weiss, J.M., & Simson, P.G. (1985). Neurochemical basis of stress-induced depression. *Psychopharmacology Bulletin, 21*, 447–457.

Zuroff, D.C., & Mongrain, M. (1987). Dependency and self-criticism: Vulnerability factors for depressive affective states. *Journal of Abnormal Psychology, 96*, 14–22.

Index

Anhedonia, 4–15, 50–52, 55–71, 184

Brain-wave biofeedback, 57, 234

Camberwell Family Interview, 89 ff
Cerebral asymmetry, 66–67, 226–229, 236, 239, 240, 241, 298
Cognitive neuroscience, vi ff
Cognitive resources, 68–69, 77, 272–273, 275–276, 277 ff
Cross-cultural issues, 37–38, 76, 110, 111, 187

Depressive personality, 207–208

Expressed emotion, 88 ff, 139

Facial expression, 52, 58, 128, 134

General Behavior Inventory, 202, 210–213

Hypomanic Personality Scale, 4, 32–33, 213–214

Impulsive Nonconformity Scale, 4, 28–30
Intense Ambivalence Scale, 4, 31–32

Learned helplessness, 251–252, 254–260

Magical Ideation Scale, 4, 21–27, 38–39, 52–55
Meehl, vii, 3–4, 181, 191, 192, 291
Movement abnormality, 128–132, 143–145, 188, 189

Orienting, 61–63, 75, 158, 169–171, 223–226, 236, 237, 240, 241

Perceptual Aberration Scale, 4, 15–21, 24–27, 38–39, 52–55, 72–74
Physical Anhedonia Scale, 4–11, 50–52, 55–71

Relapse, 88, 93 ff, 186, 295–296
Risk factors, 3 ff, 47 ff, 92, 122–127, 159–160, 181, 199–204, 250 ff, 254–264, 289

Schizotaxia, 3
Sensory gating, 73–74, 163–165
Social Anhedonia Scale, 4, 12–15
Somatic Symptoms Scale, 4, 30–31